Fortschritte der Chemie organischer Naturstoffe

Progress in the Chemistry of Organic Natural Products

40

Founded by L. Zechmeister
Edited by W. Herz, H. Grisebach, G. W. Kirby

Authors:
P. A. Cadby, R. G. Cooke, J. M. Edwards,
W. Heller, C. W. Jefford, E. Lederer,
P. Lefrancier, Sukh Dev, Ch. Tamm

Springer-Verlag
Wien New York 1981

Dr. W. Herz, Professor of Chemistry, Department of Chemistry,
The Florida State University, Tallahassee, Florida, U.S.A.

Prof. Dr. H. Grisebach, Biologisches Institut II, Lehrstuhl für Biochemie der Pflanzen,
Albert-Ludwigs-Universität, Freiburg i. Br., Federal Republic of Germany

G. W. Kirby, Sc. D., Regius Professor of Chemistry, Chemistry Department,
The University, Glasgow, Scotland

With 21 Figures

© 1981 by Springer-Verlag/Wien

Softcover reprint of the hardcover 1st edition 1981

Library of Congress Catalog Card Number AC 39-1015

ISSN 0071-7886

ISBN-13: 978-3-7091-8613-8   e-ISBN-13: 978-3-7091-8611-4
DOI: 10.1007/978-3-7091-8611-4

Contents

List of Contributors

Cadby, Dr. P. A., Département de Chimie Organique, Section de Chimie, Université de Genève, 30, quai Ernest-Ansermet, CH-1211 Genève 4, Suisse.

Cooke, R. G., D. Sc., Chemistry Department, University of Melbourne, Parkville 3052, Australia.

Edwards, J. M., Ph. D., Pharmacy School, The University of Connecticut, Storrs, CT 06268, U.S.A.

Heller, Dr. W., Biologisches Institut II, Lehrstuhl für Biochemie der Pflanzen, Albert-Ludwigs-Universität, Schänzlestrasse 1, D-7800 Freiburg i. Br., Bundesrepublik Deutschland.

Jefford, Prof. Dr. C. W., Département de Chimie Organique, Section de Chimie, Université de Genève, 30, quai Ernest-Ansermet, CH-1211 Genève 4, Suisse.

Lederer, E., Ph. D., D. ès Sc., Professor Emeritus, Université Paris Sud; Director Emeritus, Institut de Chimie des Substances Naturelles, C.N.R.S., F-91190 Gif-sur-Yvette, France.

Lefrancier, P., D. ès Sc., Institut Choay, 10 Passage Morel, F-92120 Montrouge, France.

Sukh Dev, Dr., Malti-Chem Research Centre, Nandesari, Vadodara 391 340, India.

Tamm, Prof. Dr. Ch., Institut für Organische Chemie, Universität Basel, St. Johanns Ring 19, CH-4056 Basel, Schweiz.

Chemistry of Synthetic Immunomodulant Muramyl Peptides

By P. LEFRANCIER, Institut Choay, Montrouge, France, and E. LEDERER, Laboratoire de Biochimie, C.N.R.S., Gif-sur-Yvette, and Institut de Biochimie, Orsay, France

With 15 Figures

Contents

Abbreviations. The following abbreviations will be used throughout this chapter:

Mur	muramic acid
DAP	diaminopimelic acid
Ac	acetyl
Boc	Tert-butyloxycarbonyl
Z	Benzyloxycarbonyl
For	Formyl

Bzl   benzyl
Me   methyl
Ts   tosyl
Ms   mesyl
NBzl   nitrobenzyl
OSu   succinimide ester
DMF   dimethylformamide
AcOH   acetic acid

MDP   muramyl dipeptide (N-acetyl-muramyl-L-alanyl-D-isoglutamine)
WSA   Water Soluble Adjuvant
AL   multi-poly-D,L-alanyl-poly-L-lysine

I. From Freund's Adjuvant to MDP

Freund's adjuvant (mycobacterial cells in a water-in-oil emulsion containing the antigen in the water phase) (*18*) is used for stimulating the production of antibodies against the antigen used. A five to ten fold titer of antibodies is obtained with complete Freund's adjuvant in comparison with incomplete adjuvant lacking Mycobacteria. It also induces delayed hypersensitivity to the antigen (*9, 72*).

The active compounds in Mycobacteria, which are responsible for their immunostimulant activity are part of the cell wall which is a macro-molecular insoluble net-work having two distinctly different components: the peptidoglycan (Fig. 1) and a mycolate of an arabinogalactan. These are linked by phosphodiester bridges.

Fig. 1 shows the structure of the monomer of the peptidoglycan of Mycobacteria; N-acetyl D-glucosamine is linked to N-glycolyl muramic acid (*2*) which carries a diamidated tetrapeptide L-Ala-D-isoGln-*meso*-DAP-D-Ala.

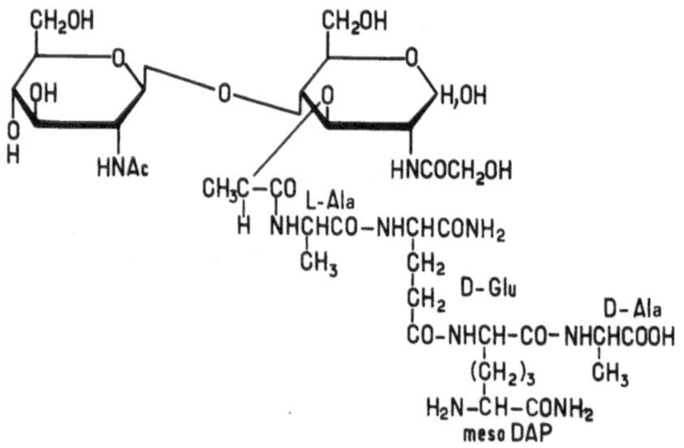

Fig. 1. Peptidoglycan of Mycobacteria (*40*)

References, pp. 38—47

This same structure is shown in a simplified manner in the lower left part of Fig. 2, which illustrates a hypothetical structure of the mycobacterial cell wall, in which mycolic acids [Myc] are esterified to an arabinogalactan. The latter is itself linked by a phosphodiester bridge to the peptidoglycan (*40, 41*).

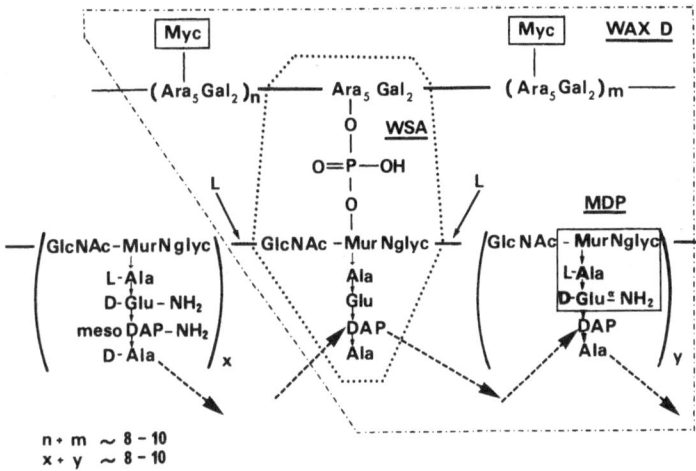

Fig. 2*. Simplified scheme of the mycobacterial cell wall and its adjuvant active derivatives: Wax D, WSA, MDP [quoted from (*40*)]

It has been known since 1958 that the chloroform soluble "Wax D" (enclosed by frame – – – – – – in Fig. 2) is adjuvant active (*71*). It was believed for a long time that biological activity was due essentially to the mycolic acids as typical constituents of mycobacteria.

In 1972, however, ADAM et al. (*3*) showed that hydrosoluble products, obtained by lysozyme treatment of purified cell walls, or delipidated cells were more active than Wax D and cell walls. Among these, the best defined, called WSA *(Water Soluble Adjuvant)* (enclosed by frame … in Fig. 2) has a molecular weight of approximately 20.000 daltons and consists of an arabinogalactan linked to a peptidoglycan. It can also be prepared from *Nocardia* strains (*4*); similar products from Mycobacteria where obtained by MIGLIORE and JOLLES by autolysis of human strains (*52*) and by HIU from hydrogenolysis products of BCG (*21*).

---

* Fig. 2 is a modified version of Fig. 6 of a 1971 review (*40*); the production of hydrosoluble WSA preparations by lysozyme digestion of cell walls can only be explained by the absence of mycolic acids on some of the arabinogalactan moieties. WSA, MW $\sim$20,000 daltons contains about 15 Ara_5Gal_2 units, one phosphodiester bridge and 5 peptidoglycan units. The WSA unit shown here has a mol. weight of about 2.000 daltons.

Further degradation of low molecular weight fractions obtained from lysozyme digestion of cell walls then led to the conclusion that an N-acyl muramyl dipeptide should be the minimal adjuvant active structure (small square frame of Fig. 2). And indeed, N-acetyl-muramyl-L-alanyl D-isoglutamine (1) first synthesized by MERSER *et al.* (*49, 50*) was found to be fully active (*16*). KOTANI *et al.* (*29, 30*) independently came to the same conclusion.

II. Synthesis of N-acetyl-muramyl-L-alanyl-D-isoglutamine (MDP)

This section deals exclusively with the synthesis of MDP (1) itself (*20, 30, 33, 35, 44, 50*). Details of earlier preparations of N-acetyl-muramyl-peptides (*5, 8, 39*), although dealing with analogues of MDP, are mentioned occasionally.

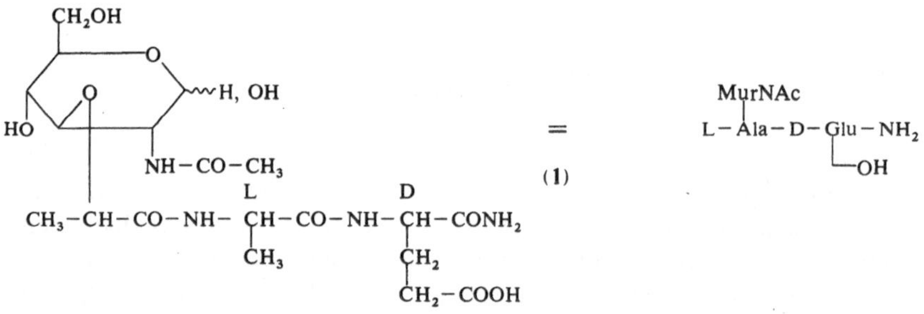

The general synthetic scheme can be summarized as follows: suitably protected derivatives of both carbohydrate and peptide moieties are prepared first, then coupled and subsequently all the protecting groups are eliminated to give the free glycopeptide. Each step will be reviewed below.

1. Protected Dipeptide Derivatives

The fully protected dipeptide derivative, BOC-L-alanyl-D-isogluta-mine benzyl ester, was synthesized by conventional methods (Fig. 3).

References, pp. 38—47

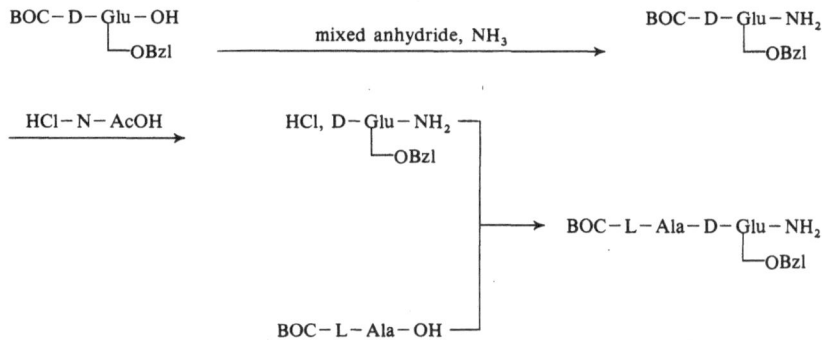

Fig. 3. Synthesis of *t*-butyloxycarbonyl-L-alanyl-D-isoglutamine benzyl ester (*43*)

The α-amino and γ-carboxyl functions are blocked by t-butyloxy-carbonyl and benzyl ester groups respectively. The coupling is performed using the N,N′-dicyclohexylcarbodiimide with N-hydroxysuccinimide (*33*), mixed anhydride (*33, 43*), or activated ester (*43*) methods. The key intermediate, BOC-D-isoglutamine benzyl ester, is prepared by a process which allows the non-equivocal synthesis of this isoglutamine derivative (Fig. 3) (*43*).

The preparation of another protected dipeptide derivative, BOC-L-alanyl-D-isoglutamine *t*-butyl ester, was reported without details, the coupling being performed by means of the succinimide ester method (*35*).

2. Protected N-acetyl-muramyl Derivatives

In the synthesis of N-acetyl-muramyl-peptides, and obviously of MDP, the most frequently used protected muramyl derivative is 1-O-benzyl-4,6-O-benzylidene-N-acetyl-muramic acid. This allows the mild and possibly simultaneous elimination of the benzylidene group and the benzyl group. Starting from D-glucosamine, either α- or β-benzyl-4,6-O-benzylidene-N-acetyl-muramide can be obtained (Fig. 4) (*17, 25, 53*). In spite of its slower hydrogenolysis rate (*44*), the α-isomer which is easier to prepare is generally preferred (*33, 44*) to the β-isomer (*50*).

The preparation of 4,6-O-anisylidene-N-acetyl-muramic acid, without protection at C-1, but bearing an easily removable blocking group at C-4 and C-6, was reported (*35*). Recently, 1-α-benzyl-4,6-O-isopropylidene-N-acetyl-muramide has been prepared (*20*).

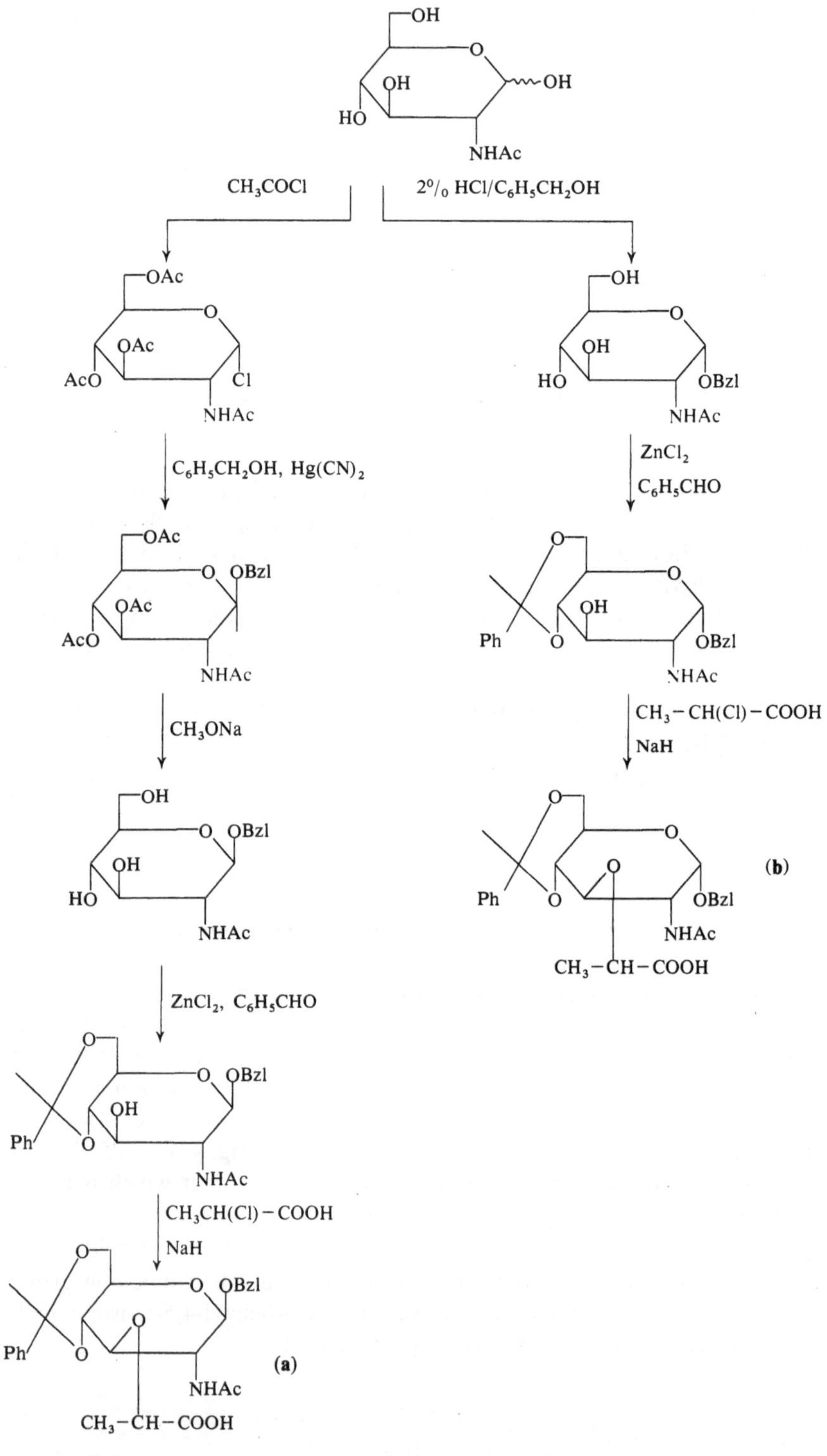

Fig. 4. Synthesis of benzyl-2-acetamido-4,6-benzylidene-3-O-(D-1-carboxyethyl)-2-deoxy-β-D-glucopyranoside [β-benzyl-4,6-benzylidene-N-acetyl-muramide] (a) and of its α-isomer [α-benzyl-4,6-benzylidene-N-acetyl-muramide] (b). (17, 25, 53)

3. Coupling of Protected Muramic Acid and Dipeptide Derivatives

Several different methods were used to couple muramic acid with the peptide moiety. In the first synthesis of MDP, N-ethyl-5-phenyl-is-oxazolium-3'-sulfonate (WOODWARD's reagent K) was used (50). This condensing reagent had been applied in earlier preparations of various N-acetyl-muramyl-peptides (8, 39). A mixture of acetonitrile-dimethyl-formamide (2:1), is used as solvent but the protected muramic acid derivative is poorly soluble and the reaction does not always proceed well.

The mixed anhydride method gives more reliable results (33, 44) and is simple, rapid and cheap, especially if a modified process (6) is applied (44).

Activated ester coupling methods have also been used. The 1-suc-cinimidyl ester of 1-α-O-benzyl-4,6-O-benzylidene-N-acetyl-muramide, prepared using N,N-'-dicyclohexylcarbodiimide, was allowed to react with the dipeptide derivative (33). The 5-norbornene-2,3-dicarboxyimidyl ester of 4,6-O-anisylidene-N-acetyl-muramide can be readily coupled with the triethyl-ammonium salt of L-alanyl-D-isoglutamine (35).

4. Deprotecting Procedures

Various ways have been proposed to eliminate the commonly used benzyl ester, benzyl glycosidic and benzylidene groups.

a) *Stepwise removal.* It had been reported (8), that the hydrogeno-lysis of 1-α-O-benzyl-4,6-O-benzylidene-N-acetyl-muramide does not pro-ceed easily and that previous removal of the benzylidene groups by acid cleavage is advantageous. In the first synthesis of MDP (50), the fully protected glycopeptide was treated with 60% acetic acid for 1 h at 100°, then purified through a silica gel column using chloroform-methanol (6:1) as eluent, and finally hydrogenolyzed for 3 h in the presence of 5% palladium on charcoal.

This two step procedure was also applied to the isopropylidene group which is more labile than the benzylidene group (20).

b) *One step removal.* Due to the conditions, under which acid hydro-lysis of benzylidene groups is carried out, the previous procedure is not devoid of drawbacks, especially because of the chemical instability of the peptide chain. Thus, in spite of its inherent difficulties, a single step hydrogenolysis is generally preferred (5, 30, 33, 39, 44). The time of hydrogenation (in acetic acid in presence of 5% palladium on charcoal), depends on the α- or β-substitution at the anomeric carbon, being respectively 48 h and 10 h. MDP is then purified through an AGIX2 column (acetate form) eluted with 0,1 N acetic acid solution (*44*) or through a silica gel column, using *n*-butanol-acetic acid-water (65 : 10 : 25) as eluent (*47*).

Reduction by sodium in liquid ammonia has also been used to split the benzyl ester, benzyl glycosidic and benzylidene groups (*73*). The reaction takes place rapidly and smoothly, but some artefacts may be formed in such an alkaline system. MDP is then purified through a silica gel column using *n*-butanol-acetic acid-pyridine-water (30 : 20 : 6 : 24) as eluent.

Recently, a modified procedure was employed giving 4,6-O-anisylidene-N-acetyl-muramyl-L-alanyl-D-isoglutamine. The single blocking group was easily removed with 60% acetic acid at room temperature (*35*).

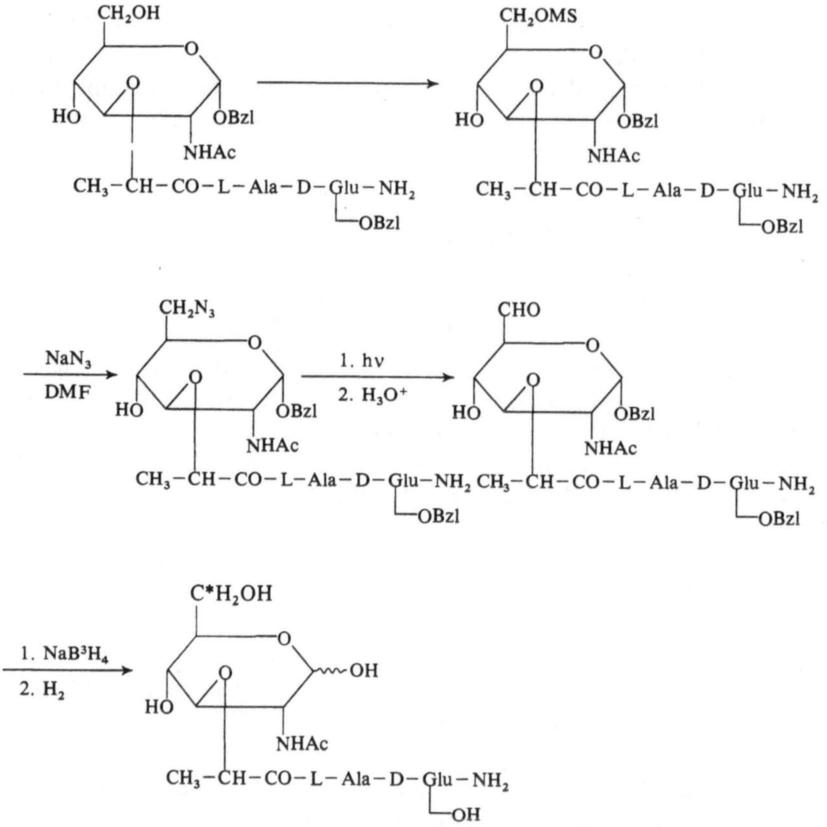

Fig. 5. Synthesis of [6-^{3}H]-N-acetyl-muramyl-L-alanyl-D-isoglutamine (*15*)

5. Synthesis of Labelled MDP

The mechanism of action of MDP and its metabolism can be studied using radioactive derivatives.

MDP has been labelled with ^{14}C at the C-1 of the lactyl moiety. The synthesis of the labelled compound follows the procedure of MERSER *et al.* (*50*) using D-L-bromo-2-propionic-^{14}C-1 acid (*57*).

For preparing (^{14}C-alanine)-MDP, a modified synthesis has been described using 4,6-O-anisylidene-N-acetyl-muramyl-^{14}C-L-alanyl-D-iso-glutamine (*35*) which can be deprotected by mild acid treatment.

More recently, the synthesis of (6-^{3}H)-N-acetyl-muramyl-L-alanyl-D-isoglutamine was reported (*15*). The specific tritiation at the C-6 of the muramyl moiety of MDP is shown in Fig. 5.

III. Synthesis of Other N-acetyl-muramyl-dipeptides, Analogues and Derivatives of MDP

1. Modifications of the Peptide Moiety

To study the relationship between the structure of the dipeptide moiety and the biological activities of MDP, numerous analogues and derivatives have been synthesized.

The preparation of analogues containing other amino acids to replace the L-alanyl or D-isoglutaminyl residues is achieved according to the previously discussed methods used for MDP and will not be commented upon unless necessary.

a) *Analogs with a diastereoisomeric peptide chain.* The replacement of L-alanine and D-isoglutamine by their enantiomeric forms gives the three diastereoisomeric analogues (**2**) (*32, 44*), (**3**) (*44*), (**4**) (*32, 33, 44*).

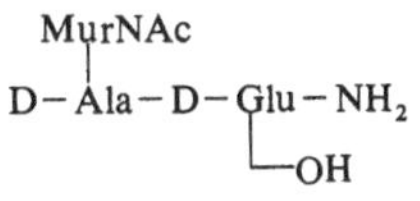

MurNAc
D−Ala−D−Glu−NH$_2$
        └−OH

(**2**)

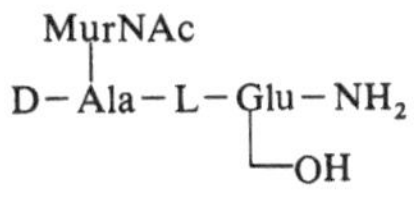

MurNAc
D−Ala−L−Glu−NH$_2$
        └−OH

(**3**)

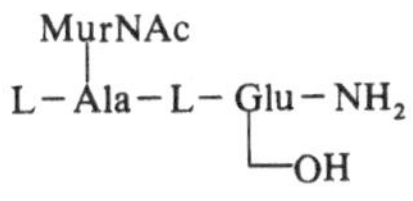

MurNAc
L−Ala−L−Glu−NH$_2$
        └−OH

(**4**)

b) *Replacement of L-alanine by another amino acid.* In some peptido-glycans, L-serine or glycine are linked to N-acetyl-muramic acid, in place of the more frequently occurring L-alanine (*63*). The two corresponding N-acetyl-muramyl-dipeptides (**5**) and (**6**) have been synthesized (*31, 44*).

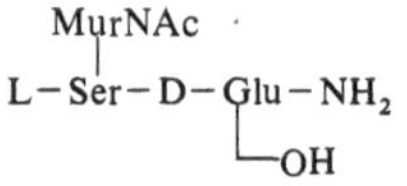

MurNAc ·   MurNAc
L−Ser−D−Glu−NH₂   Gly−D−Glu−NH₂
          L−OH   L−OH

(**5**) 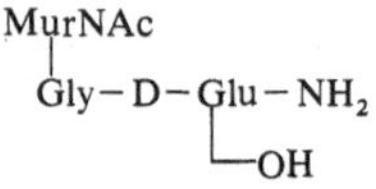   (**6**)

Other similar analogues were prepared (*47*). The substituting amino-acyl residue is either a natural (**7** and **8**) or an unatural one (**9**).

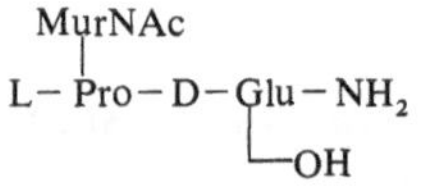

MurNAc   MurNAc
L−Pro−D−Glu−NH₂   L−Leu−D−Glu−NH₂
          L−OH   L−OH

(**7**) 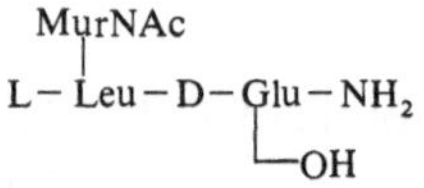   (**8**)

MurNAc
β−Ala−D−Glu−NH₂
          L−OH

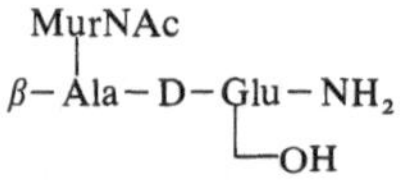

(**9**)

Various analogues (**10, 11, 12**), in which L-alanine is replaced by a N-methyl-amino acid, were prepared with the aim of protecting the compounds against possible enzymatic degradation (*47*).

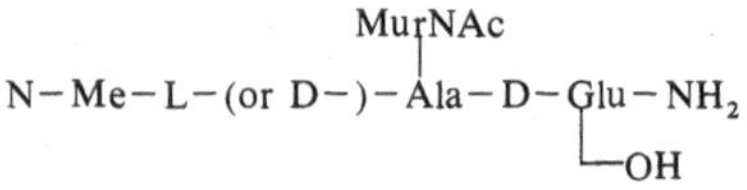

MurNAc   MurNAc
N−Me−L−(or D−)−Ala−D−Glu−NH₂   N−Me−L−Leu−D−Glu−NH₂
                L−OH   L−OH

(**10a**), (**10b**) 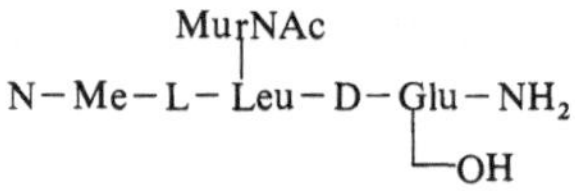   (**11**)

MurNAc
N−Me−L−Val−D−Glu−NH₂
          L−OH

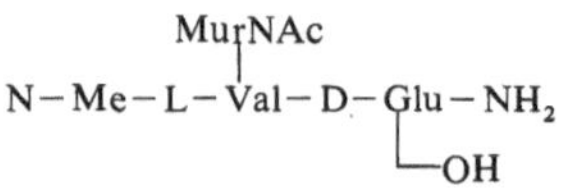

(**12**)

In these syntheses, the corresponding N-methylamino acids are prepared as described by CHEUNG and BENOITON (9).

c) *Replacement of D-isoglutamine by another amino acid.* The prominent role played by the D-isoglutaminyl residue was deduced from the biological activities shown by a few analogues in which it is replaced by a more or less structurally related amino acid, namely (13) (33), (14, 15) (33, 44), (16) (44), (17 and 18) (47).

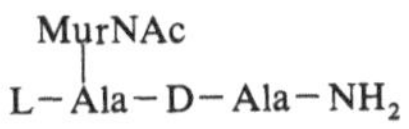

MurNAc
|
L−Ala−D−Ala−NH₂

(13)

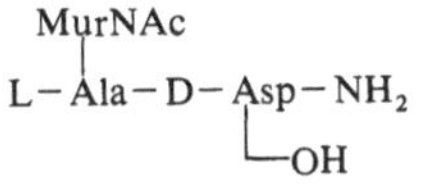

MurNAc
|
L−Ala−D−Asp−NH₂
            └−OH

(14)

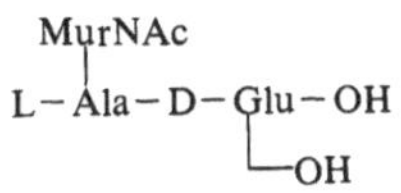

MurNAc
|
L−Ala−D−Glu−OH
          └−OH

(15)

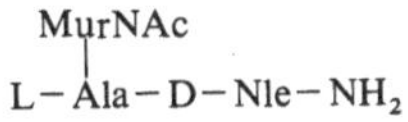

MurNAc
|
L−Ala−D−Nle−NH₂

(16)

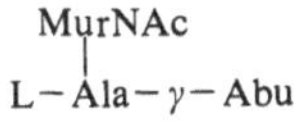

MurNAc
|
L−Ala−γ−Abu

(17)

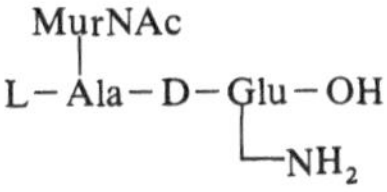

MurNAc
|
L−Ala−D−Glu−OH
          └−NH₂

(18)

d) *N-acetyl-muramyl-L-alanyl-D-glutamic acid derivatives.* In compound (15), the α- or/and the γ-carboxyl functions can be substituted by various groups, thus giving ester or amide functions.

e) *Esters of N-acetyl-muramyl-L-alanyl-D-glutamic acid.* The γ-methyl ester of MDP (19) and the di-methyl ester of N-acetyl-muramyl-L-alanyl-D-glutamic acid (20) are directly obtained from the corresponding free glycopeptides by means of diazomethane (44).

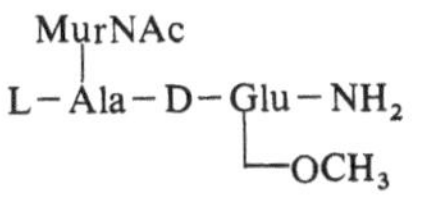

MurNAc
|
L−Ala−D−Glu−NH₂
          └−OCH₃

(19)

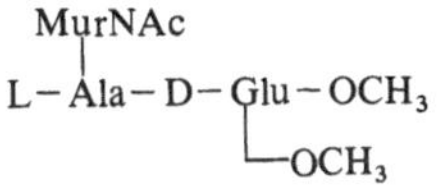

MurNAc
|
L−Ala−D−Glu−OCH₃
          └−OCH₃

(20)

All the other compounds listed below are synthesized according to the previously described stepwise method used for MDP.

The synthesis of the α-methyl ester of N-acetyl-muramyl-L-alanyl-D-glutamic acid (**21**) (*45*) and the α-methyl ester of N-acetylmuramyl-L-alanyl-D-glutamine (**22**) (*47*), starts with the preparation of BOC-L-Ala-D-Glu(OBzl) and Z-L-Ala-D-Gln respectively which were esterified by means of diazomethane.

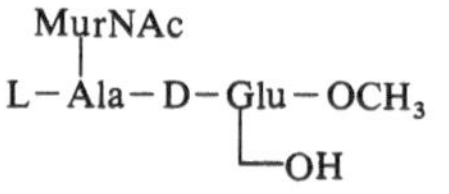

MurNAc
L−Ala−D−Glu−OCH₃
           └OH
(21)

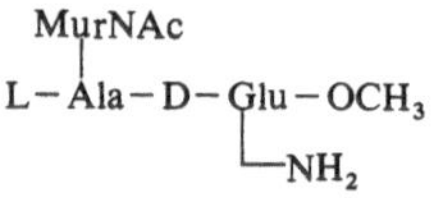

MurNAc
L−Ala−D−Glu−OCH₃
           └NH₂
(22)

Various derivatives in which the α- or γ-carboxyl function of the D-glutamic acid residue is esterified by an alkyl chain (**23** to **27**), have been synthesized (*46*).

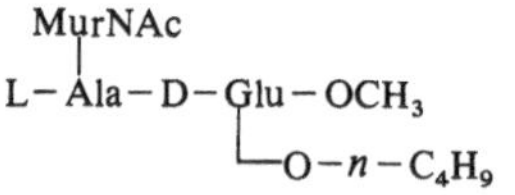

MurNAc
L−Ala−D−Glu−OCH₃
           └O−*n*−C₄H₉
(23)

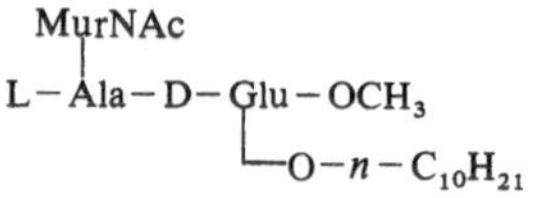

MurNAc
L−Ala−D−Glu−OCH₃
           └O−*n*−C₁₀H₂₁
(24)

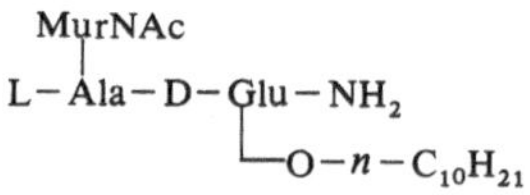

MurNAc
L−Ala−D−Glu−NH₂
           └O−*n*−C₁₀H₂₁
(25)

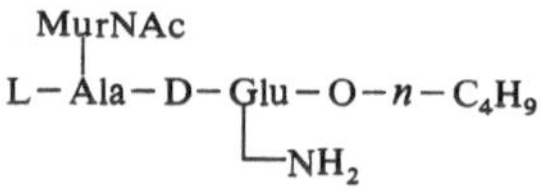

MurNAc
L−Ala−D−Glu−O−*n*−C₄H₉
           └NH₂
(26)

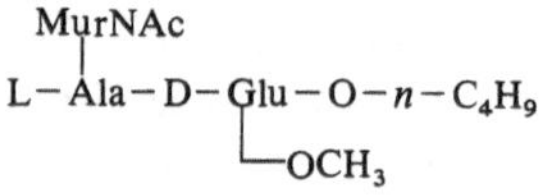

MurNAc
L−Ala−D−Glu−O−*n*−C₄H₉
           └OCH₃
(27)

All but one (**27**) of these compounds are synthesized through the key reaction of the caesium salt of BOC-D-Glu-α-methyl ester or BOC-D-Gln or of BOC-L-Ala-D-isoGln with either 1-bromo-*n*-butane or 1-bromo-*n*-decane, according to the esterification method described by WANG *et al.* (*70*).

The same method of esterification is used for the preparation of BOC-D-Glu-α-*n*-butyl-γ-methyl diester in the synthesis of (**27**):

$$BOC-D-Glu(OBzl) \xrightarrow[\text{2. } Br-n-C_4H_9]{\text{1. } Cs_2CO_3} BOC-D-Glu(OBzl)-O-n-C_4H_9$$

$$\xrightarrow{H_2/Pd} BOC-D-Glu-O-n-C_4H_9 \xrightarrow[\text{2. } ICH_3]{\text{1. } Cs_2CO_3}$$

$$BOC-D-Glu(OCH_3)-O-n-C_4H_9 \longrightarrow (27)$$

f) *Amides of N-acetyl-muramyl-L-alanyl-D-glutamic acid.* N-acetyl-muramyl-L-alanyl-D-glutamic-α,γ-di-amide (**28**), -α-amide-γ-methyl-amide (**29**), α,γ-di-methylamide (**30**), or α-methylamide (**31**) where synthesized by ammonolysis or methyl amminolysis of BOC-L-Ala-D-Glu(OBzl)-NH₂, of BOC-L-Ala-D-Glu(OBzl)₂ or of BOC-D-Glu(OBzl)-OSu (*44*).

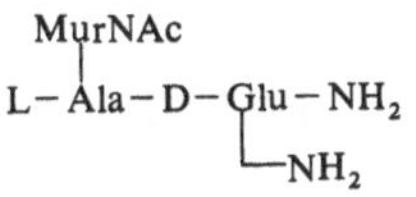

```
    MurNAc                          MurNAc
      |                               |
L-Ala-D-Glu-NH₂                L-Ala-D-Glu-NH₂
          └NH₂                            └NH-CH₃

     (28)                            (29)
```

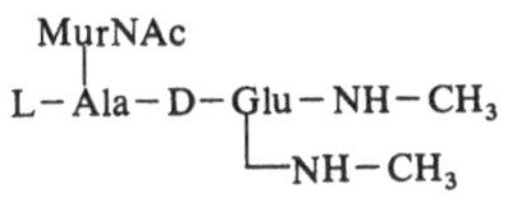

```
    MurNAc                          MurNAc
      |                               |
L-Ala-D-Glu-NH-CH₃             L-Ala-D-Glu-NH-CH₃
          └NH-CH₃                         └OH

     (30)                            (31)
```

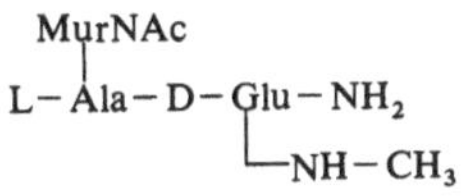

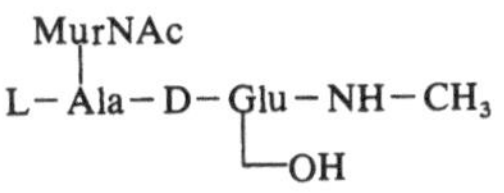

The γ-n-butylamide of N-acetyl-muramyl-L-alanyl-D-isoglutamine (**32**) is reported without details of its preparation (*32*).

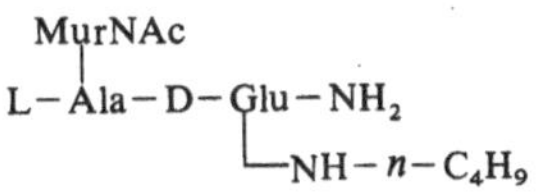

```
    MurNAc
      |
L-Ala-D-Glu-NH₂
          └NH-n-C₄H₉

     (32)
```

2. Modifications of the Carbohydrate Moiety

In some analogues of MDP, only the N-acetyl-muramyl portion has been modified.

The muramic acid moiety is important for the expression of the biological activity of MDP, since NaBH₄ reduction leads to the inactive muraminitol derivative (**33**) and β-elimination gives the lactyl-dipeptide (**34**) which is inactive or even seems to inhibit adjuvant activity (*1*).

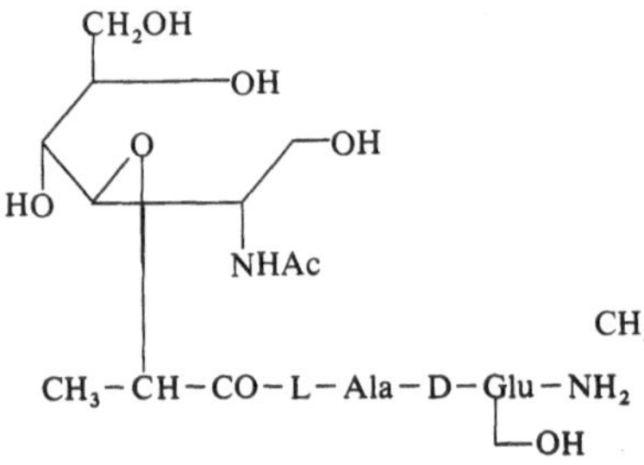

(33) 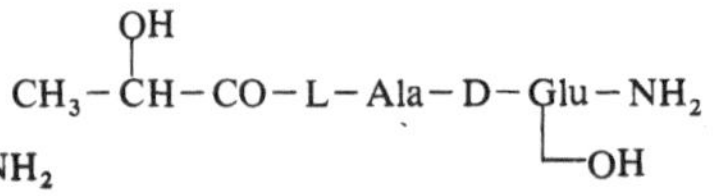   (34)

Some analogues in which N-acetyl-muramic acid is replaced by another carbohydrate have been reported.

The 2-deacetamido-MDP (35) is obtained by the procedure described for MDP *via* benzyl-4,6-O-benzylidene-3-O-(D-1-carboxyethyl)-2-deoxy-α-D-glucopyranoside (47). More recently, the D-manno-analogue (36) has been prepared *via* benzyl-2-acetamido-5,6-O-isopropylidene-3-O-(D-1-carboxyethyl)-2-deoxy-α-D-manno-furanoside (20).

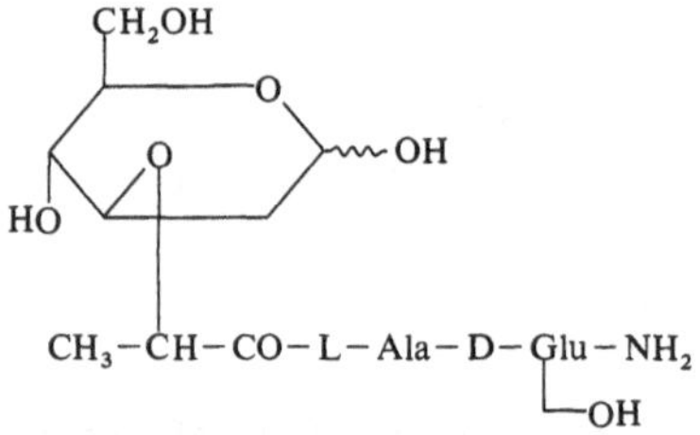

(35) 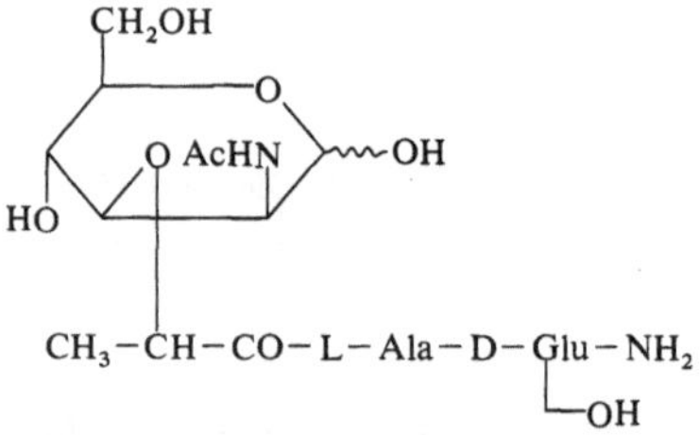   (36)

The preparation of benzyl-2,3,6-O-benzyl-4-O-(D-1-carboxyethyl)-β-D-glucopyranoside (67) allowed the synthesis of 4-O-(D-2-propionyl-L-alanyl-D-isoglutamine)-D-glucopyranose (37) (47).

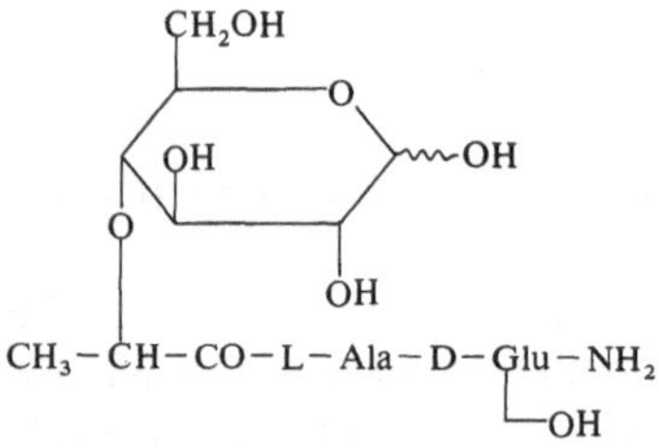

(37) 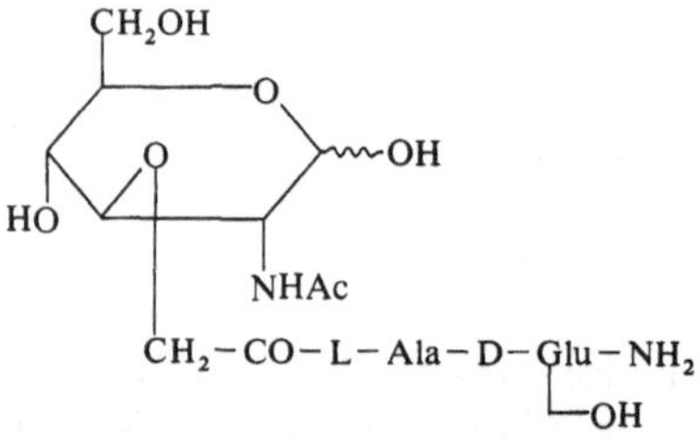   (38)

The condensation of chloroacetic acid with benzyl-2-acetamido-4,6-benzylidene-2-deoxy-β-D-glucopyranoside in the usual manner afforded benzyl-2-acetamido-4,6-O-benzylidene-3-O-acetyl-β-D-glucopyranoside, thus leading to the synthesis of N-acetyl-normuramyl-L-alanyl-D-isoglutamine (**38**) (norMDP) (*45*).

Many derivatives of MDP have been described in which the hydroxyl groups of muramic acid, especially the primary hydroxyl, are substituted; in the following only details of the preparation of suitably protected N-acetyl-muramic acid derivatives will be discussed.

Starting with α- and β-methyl-4,6-O-benzylidene-N-acetyl-muramide (*25*), the α-methyl glycoside of MDP (**39**) and its β-anomer (**40**) were synthesized (*45*).

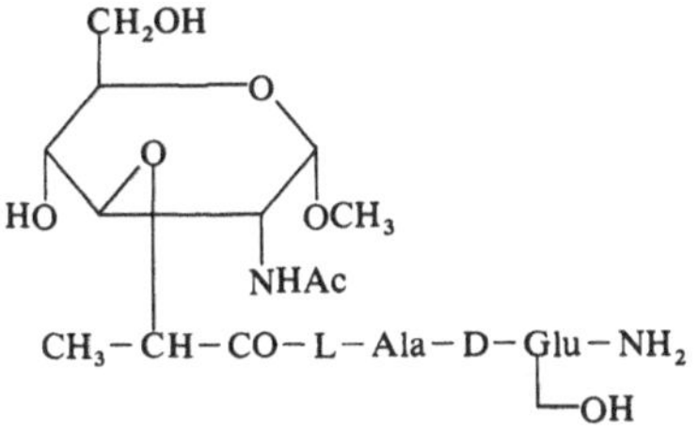

(39) 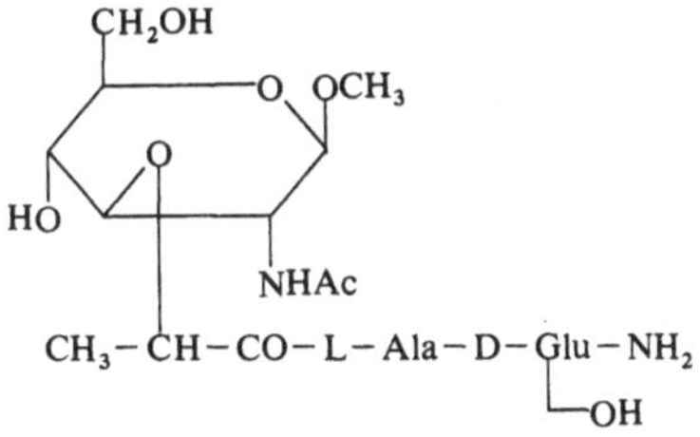   (40)

In the same way, β-*p*-nitro-phenyl-4,6-O-benzylidene-N-acetyl-muramide (*25*) permitted the preparation of the fully protected β-*p*-nitrophenyl-glycoside of MDP. One step catalytic hydrogenation of the latter eliminated all the protecting groups and reduced the nitro to an amino group to give the *p*-aminophenylglycoside of MDP (**41**) (*45*). This could be cross-linked by glutaraldehyde as mentioned later (*54*).

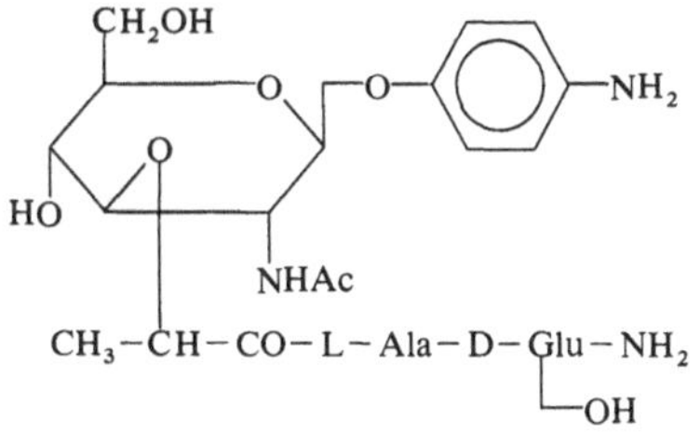

(41)

After debenzylidenation of benzyl-4,6-O-benzylidene-N-acetyl-muramyl-L-alanyl-D-isoglutamine benzyl ester in the usual manner by 60% acetic acid at 100°, the ensuing 4,6-diol derivative can be acylated specifically on the primary hydroxyl function by means of a large excess of an acyl chloride to give 6-O-acyl derivatives of MDP after the usual hydrogenolytic deprotection. Several such compounds have been described, (**42**—**46**) (*36*).

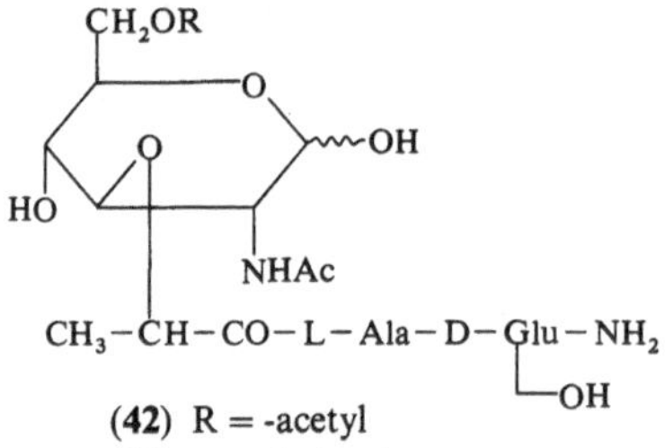

CH_2OR

OH

HO

NHAc

$CH_3-CH-CO-L-Ala-D-Glu-NH_2$
　　　　　　　　　　　　　　└OH

(**42**) R = -acetyl
(**43**)　　-butyryl
(**44**)　　-octanoyl
(**45**)　　-lauroyl
(**46**)　　-stearoyl

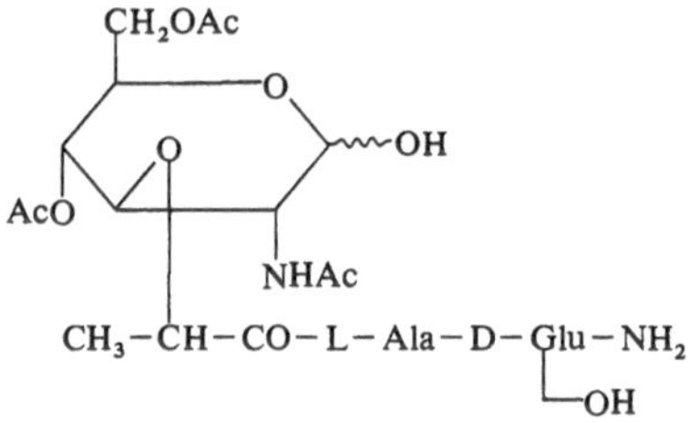

CH_2OAc

OH

AcO

NHAc

$CH_3-CH-CO-L-Ala-D-Glu-NH_2$
　　　　　　　　　　　　　　└OH

(**47**)

If the acylation is achieved by means of acetic anhydride in pyridine, the 4,6-di-O-acetyl derivative of MDP (**47**) is obtained (*36, 45*).

An interesting family of related compounds, (**48**—**55**), namely long chain fatty esters of MDP, has been synthesized by a Japanese group (*34, 37, 66*).

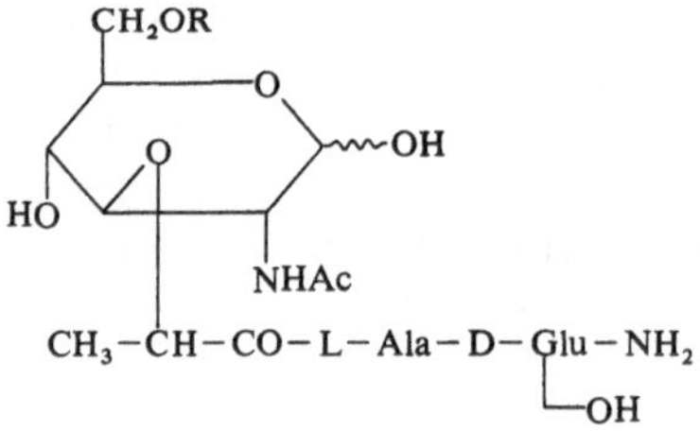

CH_2OR

OH

HO

NHAc

$CH_3-CH-CO-L-Ala-D-Glu-NH_2$
　　　　　　　　　　　　　└OH

(**48**) R = -mycoloyl
(**49**)　　-nocardomycoloyl
(**50**)　　-corynomycoloyl
(**51**)　　-triacontanoyl
(**52**)　　-2-tetradecylhexadecanoyl
(**53**)　　-3-hydroxy-2-tetradecylhexadecanoyl
(**54**)　　-2-docosyltetracosanoyl
(**55**)　　-3-hydroxy-2-docosylhexacosanoyl

Compound (**48**), 6-O-mycoloyl-MDP results from acylation of the primary hydroxyl group with natural mycolic acid isolated from mycobacterial cells. Its synthesis is illustrated in Fig. 6 (*34, 66*). The key intermediate (**a**), a partially protected N-acetyl-muramic acid derivative, can be substituted by a tosyl group at C-6, which can be exchanged with the potassium salt of the mycolic acid in the presence of 18-crown-6 following the method of Polonsky *et al.* (*58*). This mild acylation procedure avoids the use of the acid chloride or trifluoracetic anhydride method, which cannot be applied in the present instance because of the presence of a β-hydroxyl group in mycolic acid. The pentachlorophenyl ester of the muramic acid moiety prepared by exchange ester reaction (19) is used to afford the fully protected glycopeptide (e), thus minimizing the formation of the internal ester (f) under the influence of a dehydrating agent such as N,N'-dicyclohexylcarbodiimide.

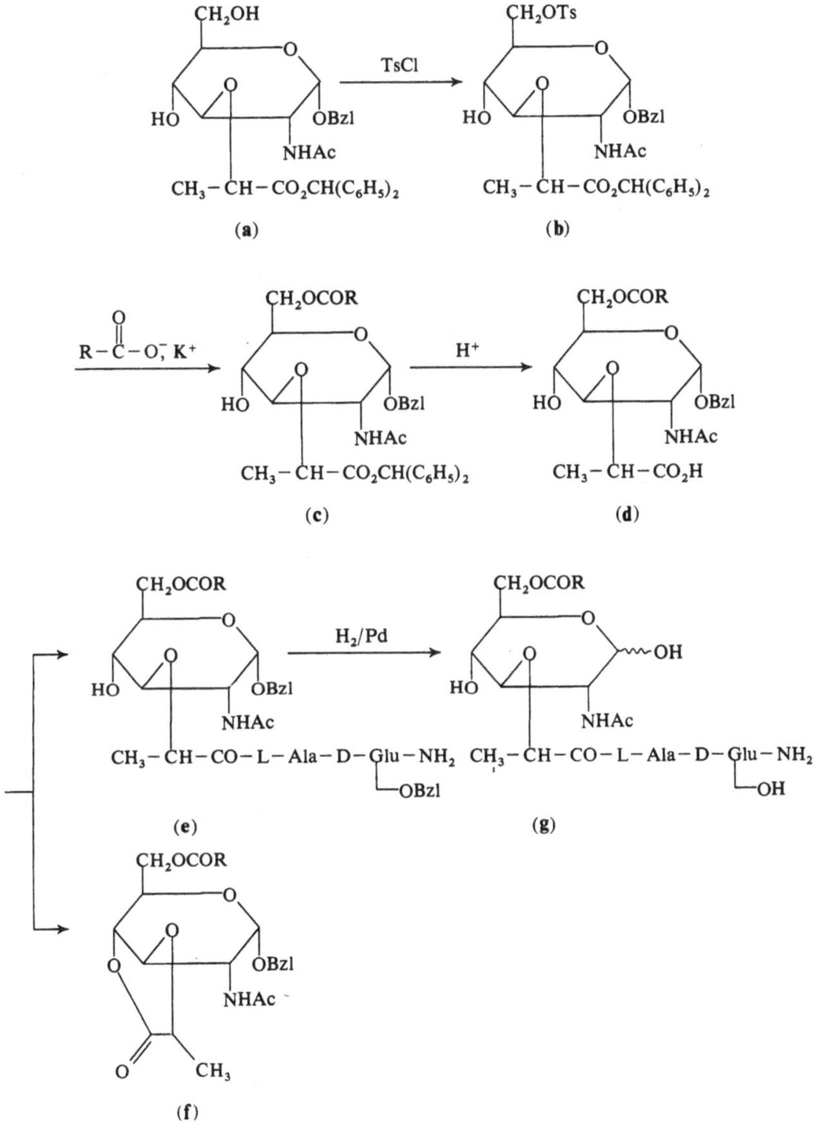

Fig. 6. Synthesis of 6-O-Mycoloyl-N-acetyl-muramyl-L-alanyl-D-isoglutamine (48) (34, 66)
R = mycoloyl

This procedure was also applied to the synthesis of the closely related 6-O-nocardomycoloyl-MDP (**49**) (*66*) and 6-O-corynomycoloyl-MDP (**50**) (*47, 66*). In these instances the internal ester (**f**) was not formed when N,N'-dicyclo-hexylcarbodiimide in presence of N-hydroxysuccinimide was used as coupling reagent (*37*). Similar 6-O-acyl-derivatives of MDP (**51—55**) have been synthesized in the same manner, the synthetic fatty acyl group being linear, α-branched and α-branched-β-hydroxylated (*37*).

IV. Synthesis of N-acetyl-muramyl-tri-, tetra-, and -pentapeptides, and of Some Analogs Bearing a Lipophilic Group at the C-terminal End

The monomeric unit of many peptidoglycans has the structure N-acetyl-D-glucosaminyl-(1-4)-N-acetyl-muramyl-L-alanyl-D-isoglutaminyl-L-lysyl-D-alanine shown below.

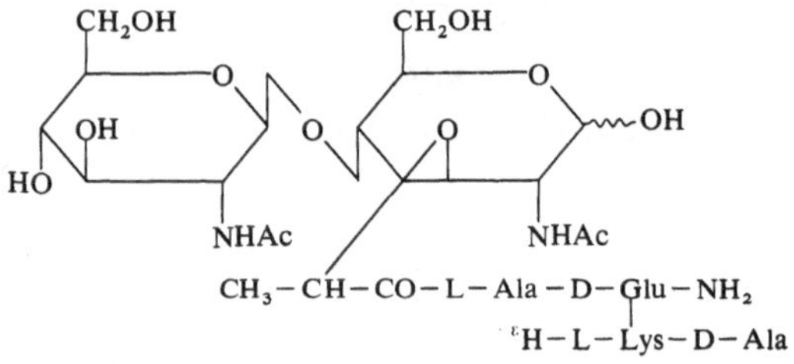

Some slight modifications of this unit also occur in nature (*63*). Here, only two will be mentioned, namely those in which the α-amide function of the D-glutamyl residue has been replaced by an amino acid residue, and those in which the L-lysyl residue has been replaced by an α,ε-diaminopimelyl residue. Synthetic glycopeptides related to these structures have been reported.

1. Substitution of the α-amide Group of MDP by a Free or Amidated Amino Acid

In some peptidoglycans, the α-carboxyl function of the D-glutamyl residue is substituted with glycine (*M. lysodeikticus*), glycine amide (*A. arthrocyaneus*), or D-alanine amide (*Arthrobacter* sp.). The corresponding N-acetyl-muramyl-tripeptides (**56—58**), have been synthesized (*32, 44, 45*).

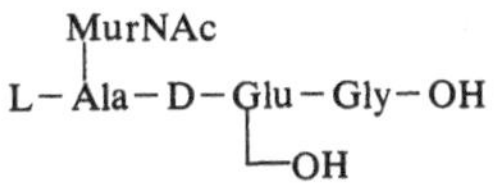

MurNAc
|
L−Ala−D−Glu−Gly−OH
        └−OH

(56)

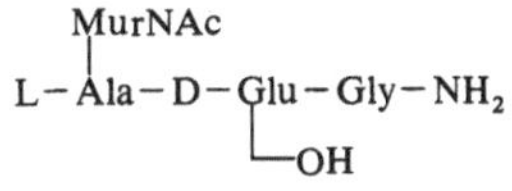

MurNAc
|
L−Ala−D−Glu−Gly−NH$_2$
        └−OH

(57)

MurNAc
|
L−Ala−D−Glu−D−Ala−NH$_2$
        └−OH

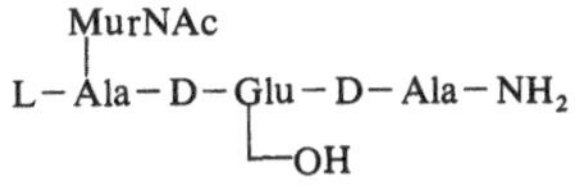

(58)

BOC-D-Glu(OBzl)-Gly-OBzl is elongated at the N-terminal with BOC-L-Ala and then with α-benzyl-4,6-O-benzylidene-2-acetamido-muramide leading to (56), after final hydrogenolysis of all the protecting groups (32, 44). The same general procedure is applied to the synthesis of (57), but it starts with the preparation of BOC-D-Glu(OBzl)-Gly-NH$_2$ by aminolysis of the mixed anhydride intermediate of BOC-D-Glu(OBzl)-Gly-OH (45). Due to its solubility in water such a glycine amide peptide is not easy to handle. Another way of preparation was proposed (32): BOC-L-Ala-D-Glu(OBzl)-Gly-OEt is hydrogenated and the resulting ethyl ester amminolyzed to give BOC-L-Ala-D-Glu-Gly-NH$_2$ which is coupled with the 5-norbornene-2,3-dicarboxyimidyl ester of the protected muramic acid.

Compound (58) was reported without details of its preparation (32).

2. Lengthening of the Peptide Chain at the Carboxyl Function of MDP

a) *Lysine containing N-acetyl-muramyl-peptides*. The synthesis of the N-acetyl-muramyl-pentapeptides (59 and 60) was reported (39). N-acetyl-muramyl-L-alanyl-γ-D-glutamyl-L-lysyl-D-alanyl-D-alanine (59), was identified with the glycopeptide of a biosynthetic precursor of the *S. aureus* peptidoglycan. The synthesis of the peptide moiety was achieved by step-wise elongation from the C-terminal using *p*-nitrophenyl ester as coupling method, the α-amino functions being protected with the *t*-butyloxycarbonyl group, the C-terminal carboxyl function and the ε-amino function of the lysyl residue with nitrobenzyl ester and benzyloxycarbonyl groups respectively. The resulting peptide derivatives were coupled with α-benzyl-4,6-O-benzylidene-N-acetyl-muramide by means of N-ethyl-5-phenyl-isoxazo-lium-3'-sulfonate (WOODWARD's Reagent K). The free glycopentapeptide was obtained by catalytic hydrogenation.

```
       MurNAc                              MurNAc
        |                                   |
L − Ala − D − Glu − OH            L − Ala − D − Glu − L − Lys − D − Ala − D − Ala
              └─ L − Lys − D − Ala − D − Ala            └─ OH

              (59)                                      (60)
```

The synthesis of the N-acetyl-muramyl tetrapeptide (61) (33, 44), which has the peptide sequence of the monomeric glycopeptide unit of many peptidoglycans, and that of its diastereoisomer (62) (45) has been reported. The fully protected tetrapeptide was prepared by coupling the two fragments BOC-L-Ala-D-isoGln and Lys(Z)-D-(or L)-Ala-OBzl by means of the mixed anhydride method.

```
       MurNAc                              MurNAc
        |                                   |
L − Ala − D − Glu − NH₂           L − Ala − D − Glu − NH₂
              └─ L − Lys − D − Ala               └─ L − Lys − L − Ala

              (61)                                      (62)
```

Three N-acetyl-muramyl-tripeptides were synthesized which bear on the carboxyl function of MDP either a L-lysyl, (63) (33, 44) a L-alanyl, (64), or a D-alanyl, (65) (45) residue. All of these compounds were obtained by the general stepwise method of synthesis. The hydrogenolysis step leading to (61 and 63) was complete only in presence of 0.5 equivalents of hydrochloric acid (32).

```
       MurNAc                              MurNAc
        |                                   |
L − Ala − D − Glu − NH₂           L − Ala − D − Glu − NH₂
              └─ L − Lys                          └─ L − Ala

              (63)                                      (64)
```

```
       MurNAc
        |
L − Ala − D − Glu − NH₂
              └─ D − Ala

              (65)
```

The glycopeptide (66), substituted on both the α and γ-carboxyl function of the D-glutamic residue by an amino-acyl and a dipeptidyl residue, was prepared via the key intermediate BOC-D-Glu-Gly-OBzl. The latter was synthesized as its dicyclohexylamine salt by treatment of the anhydride of BOC-D-glutamic acid with glycine benzyl ester in presence of dicyclohexylamine. This tripeptide derivative was subsequently coupled with L-Lys(Z)-D-Ala-OBzl (*32*).

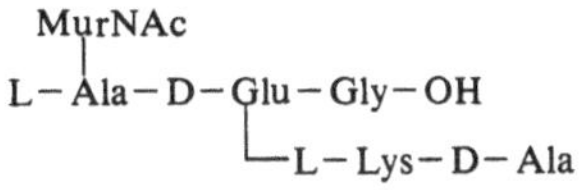

MurNAc
|
L—Ala—D—Glu—Gly—OH
└—L—Lys—D—Ala

(**66**)

b) α,ε-*diaminopimelyl (DAP) containing N-acetyl-muramyl-peptides.* α,ε-diaminopimelic acid (2,6-diamino-heptane-1,7-dioic) replaces L-lysine in many peptidoglycans and is a branching point of the peptide chain.

The glycopeptide (**67**) which is a fragment of the peptidoglycan of *P. peterssonii* and *S. ventriculi*, contains the less frequently occuring L-L-isomer. It was synthesized (*5*) by condensation of the fragments and subsequent one-step hydrogenolysis. Thus, by means of N-ethoxy-

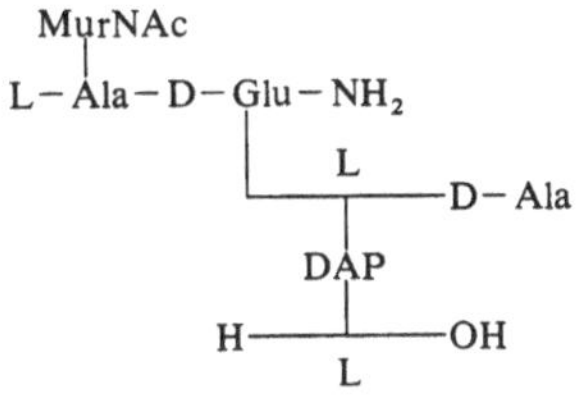

MurNAc
|
L—Ala—D—Glu—NH$_2$
    |
    L
    └————D—Ala
    |
    DAP
    |
H————OH
    L

(**67**)

carbonyl-2-ethoxy-1,2-di-hydroquinoline, 1-α-benzyl-4,6-O-benzylidene-N-acetyl-muramyl-L-alanine and N′-carbobenzoxy-(L$_2$)-L,L-α,ε-diaminopimelyl-(L$_1$)-D-alanine-nitrobenzyl ester were coupled. The preparation of the latter derivative is illustrated in Fig. 7. Formyl-(L)-bromonorvalyl-D-alanine-nitrobenzyl ester is condensed with dimethyl carbobenzoxy aminomalonate. The ensuing derivative is first saponified, then decarboxylated to give the unsymetrically acylated N$_1$-carbobenzoxy-N$_2$-formyl-L-L (and D,D)-α,ε-diaminopimelyl-D-alanine. The diastereoisomeric *p*-nitrobenzyl ester can be separated, the L-L isomer being obtained by recrystallization.

N-acetyl-muramyl-tetrapeptides which contain the more frequently occurring α,ε-*meso*-diaminopimelyl residue have apparently not been prepared except by biosynthesis, in spite of the synthesis of non-symetric α,ε-*meso*-diaminopimelyl-tetrapeptides which represents a striking success after a year-long effort (*7*).

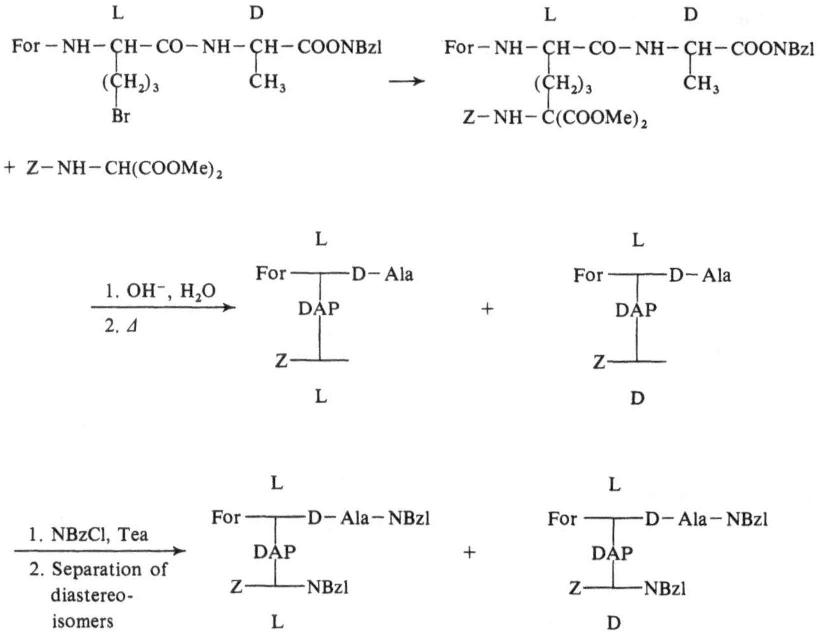
For − NH − CH − CO − NH − CH − COONBzl
(CH2)3
CH3
Br
For − NH − CH − CO − NH − CH − COONBzl
(CH2)3
CH3
Z − NH − C(COOMe)2
+ Z − NH − CH(COOMe)2
1. OH−, H2O
2. Δ
For
D − Ala
DAP
Z
1. NBzCl, Tea
2. Separation of diastereo-isomers
D − Ala − NBzl
NBzl

Fig. 7. Synthesis of N′-Carbobenzoxy-(L_2)-L,L-α_1-diaminopimelyl-(L_1)-D-alanine dinitrobenzyl ester (*5*)

3. Lipophilic Derivatives of N-acetyl-muramyl-L-alanyl-D-isoglutaminyl-L-alanine

Several lipophilic derivatives of MDP in which the carboxyl function of the peptide chain is substituted by a L-alanyl residue esterified by a medium or long chain alcohol (**68**—**72**) have been reported (*46*).

The synthesis of these compounds is performed by a simplified procedure which consists in coupling the various L-alanine alkyl esters with MDP, the carboxyl of which is previously activated by means of hydrosoluble carbodiimide in the presence of N-hydroxybenzotriazole. Almost pure products are obtained, and can be further purified either by ion-exchange chromatography for the less lipophilic or by silica gel chromatography for the more lipophilic derivatives.

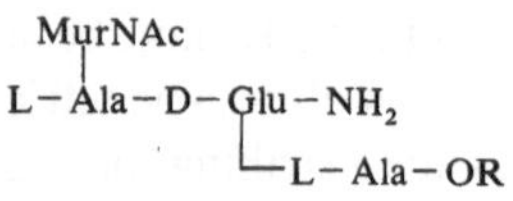
MurNAc
L − Ala − D − Glu − NH2
L − Ala − OR

(**68**) R = *n*-butyl
(**69**) benzyl
(**70**) *n*-decyl
(**71**) *n*-pentadecyl
(**72**) *n*-eicosyl

MurNAc
|
L − Ala − D − Glu − NH$_2$
└─ L − Ala − OCH$_2$ − CHOH − CH$_2$OCOC$_{86}$H$_{172}$O

(**73**)

A strongly lipophilic compound (**73**) was also prepared (**46**) from glycerol-1-mycolate (**12**) isolated from the strain AN5 of *Mycobacterium bovis**. It is synthesized according to the previously described procedure, the starting material being 1-(BOC-L-alanyl)-glycerol-3-mycolate. This compound is obtained by heating the potassium salt of BOC-L-alanine with 1-tosyl-glycerol-3-mycolate in the presence of 18-crown-6 in boiling benzene.

V. Synthesis of N-acetyl-β-D-glucosaminyl-(1-4)-N-acetyl-muramyl-peptides

To study the influence on the immunostimulant activity of MDP of a longer carbohydrate moiety, which mimics more closely the glycan strand of the peptidoglycan, N-acetyl-β-D-glucosaminyl-(1-4)-N-acetyl-muramyl-L-alanine and -L-alanyl-D-isoglutamine, (**74**) and (**75**) respectively, were synthesized.

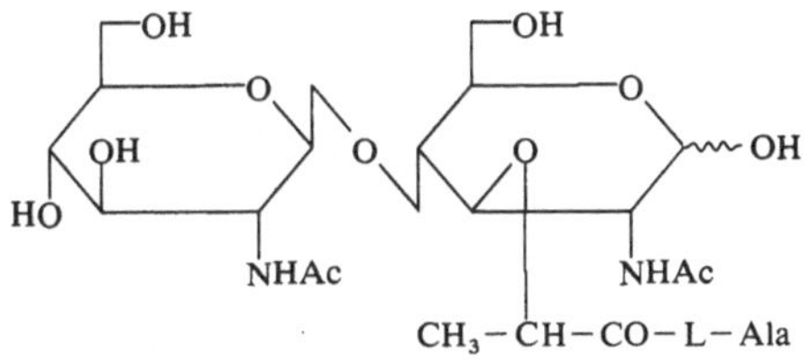

(**74**)

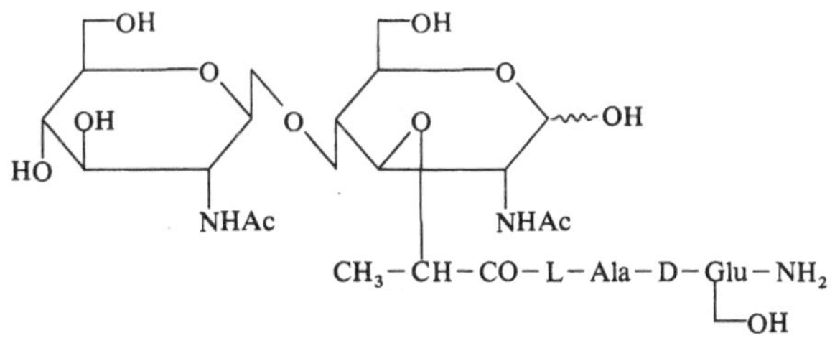

(**75**)

---

* The glycerol-1-mycolate is a mixture of glycerol esters with several analogous and homologous mycolic acids (**41, 59**). The formula given here is an average and approximate one.

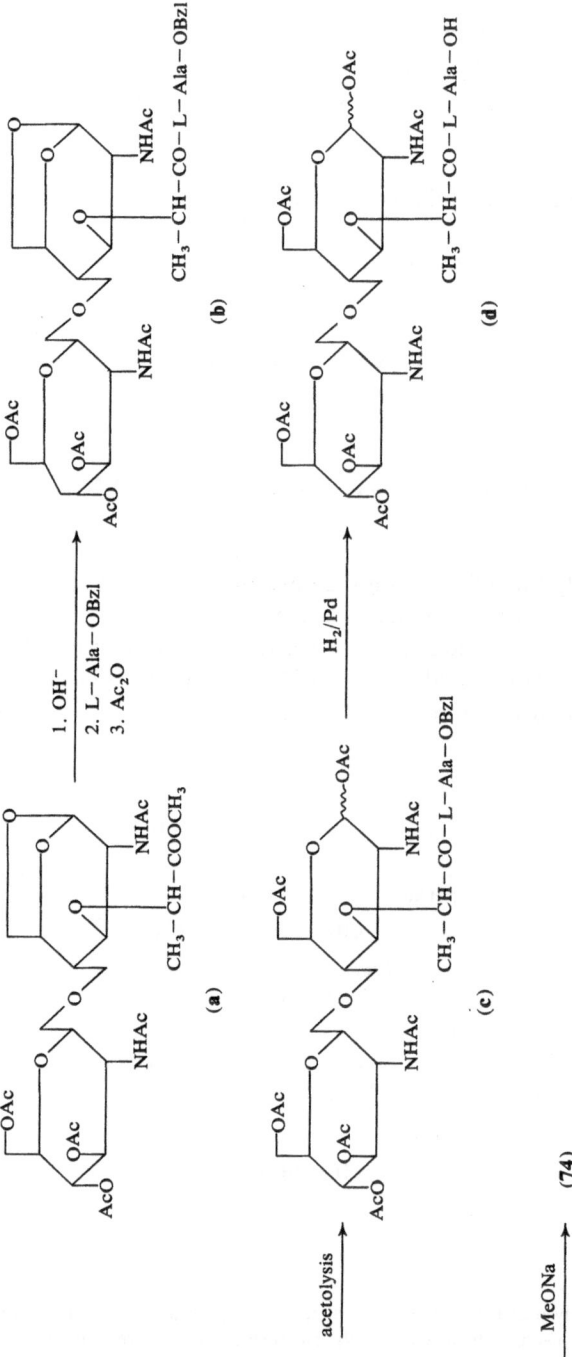

Fig. 8. Synthesis of N-acetyl-β-D-glucosaminyl-(1-4)-N-acetyl-muramyl-L-alanine (**74**) (*51*)

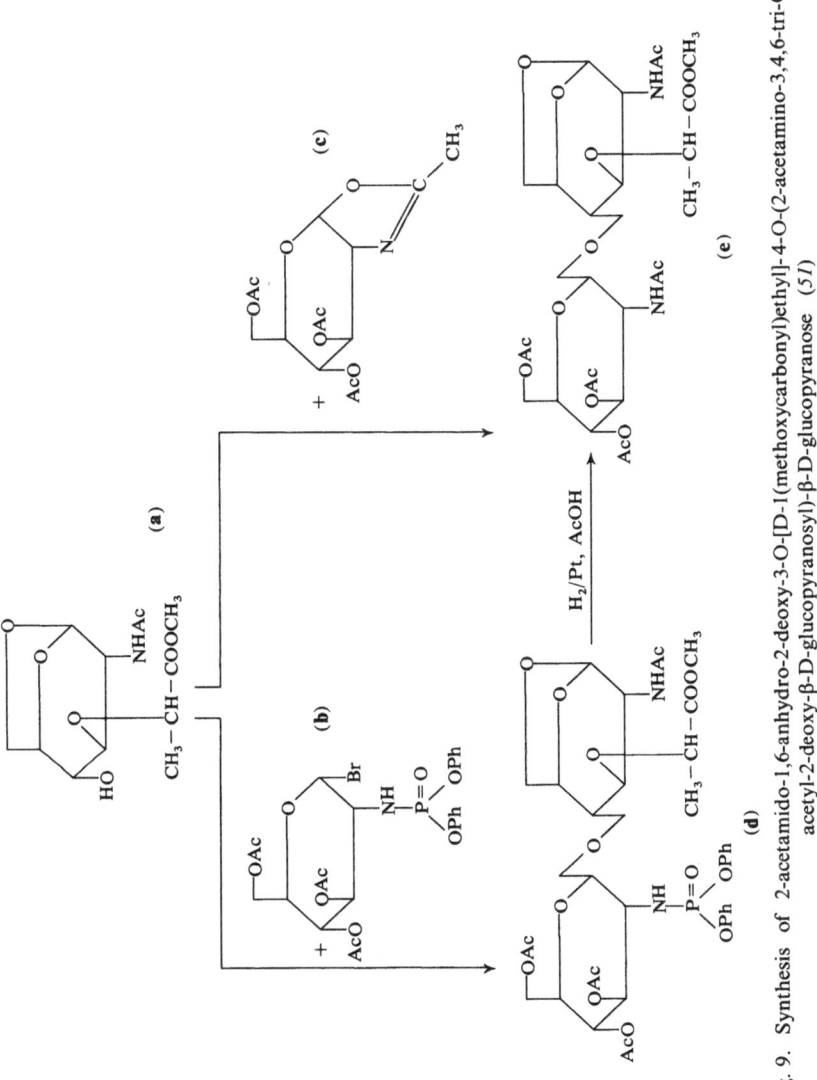

Fig. 9. Synthesis of 2-acetamido-1,6-anhydro-2-deoxy-3-O-[D-1(methoxycarbonyl)ethyl]-4-O-(2-acetamino-3,4,6-tri-O-acetyl-2-deoxy-β-D-glucopyranosyl)-β-D-glucopyranose (51)

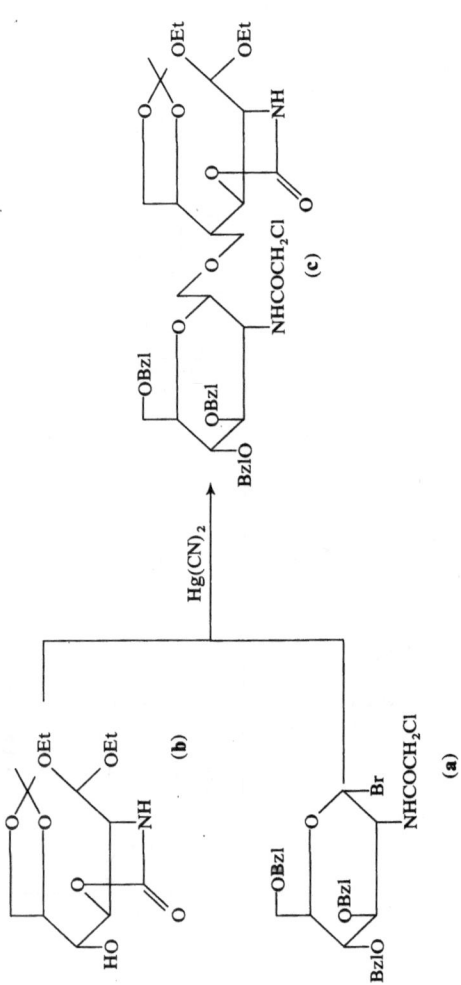

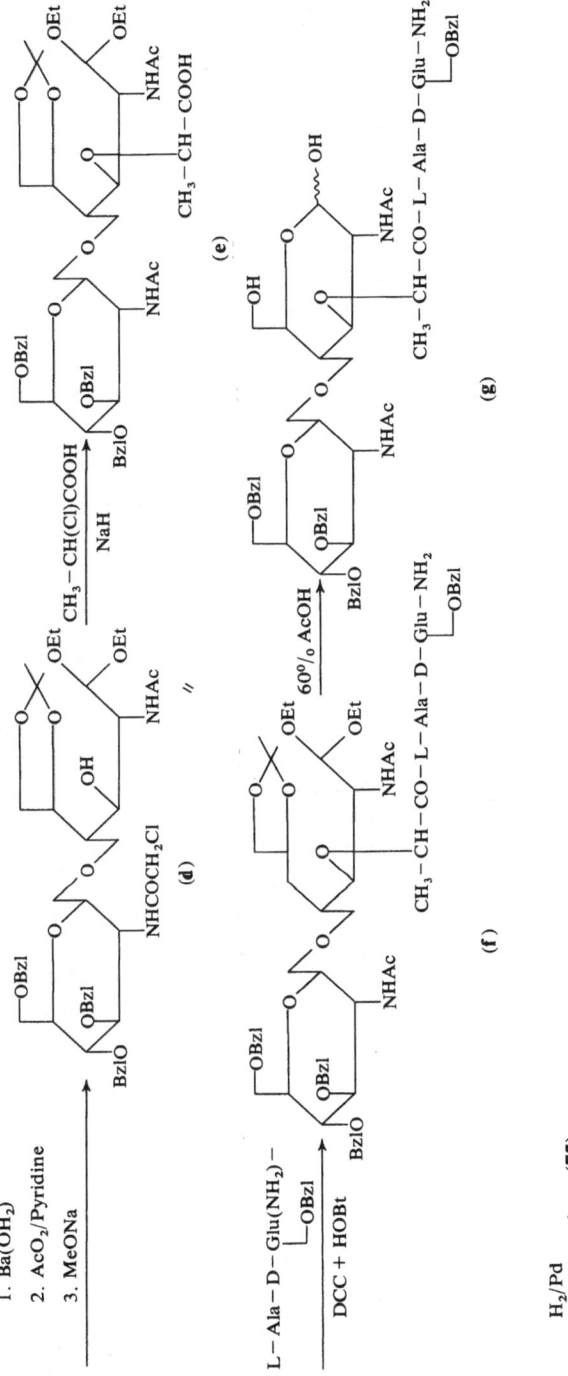

Fig. 10. Synthesis of N-acetyl-β-D-glucosaminyl-(1-4)-N-acetyl-muramyl-L-alanyl-D-isoglutamine (75) (38)

The synthesis of (74) is shown in Fig. 8 (51). It starts with the preparation (described below) of 2-acetamido-1-6-anhydro-2-deoxy-3-O-[D-1(methoxycarbonyl)ethyl]-4-O-(2-acetamido-3,4,6-tri-O-acetyl-2-deoxy-β-D-glucopyranosyl)-β-D-glucopyranose. After saponification of this fully protected disaccharide derivative (a), the ensuing intermediate is first coupled with L-alanine benzyl ester by means of WOODWARD's reagent K, then acetylated to give (b), which is easier to handle than the non acylated derivative. Compound (74) is obtained after three successive treatments: mild acetolysis by a mixture of acetic anhydride, acetic acid, and sulfuric acid, then catalytic hydrogenolysis and carefully controlled deacetylation.

The fully protected disaccharide derivative is obtained in two different ways from the starting material, 2-acetamido-1,6-anhydro-2-deoxy-3-O-(D-1-(methoxycarbonyl)ethyl)-β-D-glucopyranose, as illustrated in Fig. 9. The derivative (a) is coupled with 3,4,6-tri-O-acetyl-2-deoxy-2-diphenyloxyphosphorylamino-α-D-glucopyranoside bromide (b) (74) according to the Königs-Knorr reaction, but the yield of the coupling is better if it is condensed in the presence of p-toluene sulfonic acid with the methyl oxazolide derivative of peracetylated D-glucosamine (c) (75).

The synthesis of (75) follows a different route, shown in Fig. 10 (38). Starting from 3,4,6-tri-O-benzyl-D-glucosamine (24), 3,4,6-tri-O-benzyl-N-dichloroacetyl-D-glucosaminyl bromide (a) is prepared which is then condensed by means of a Königs-Knorr reaction with an open chain form of protected glucosamine, namely 2-N,3-O-carbonyl-5,6-O-carbonyl-5,6-O-isopropylidene-D-glucosamine diethyl acetal (b). The disaccharide derivative is hydrolysed under alkaline conditions to remove the dichloroacetyl group and to open the oxazolidone ring, and the resulting compound is N-acetylated to yield (d), N-acetyl-4-O-(N-acetyl-3,4,6-tri-O-benzyl-β-D-glucosaminyl)-5,6-O-isopropylidene-D-glucosamine diethyl acetal. The free hydroxyl at C-3 in the open chain moiety of this compound is replaced in the usual manner by a D-lactic ether group, the carboxyl function of which allows the coupling with the dipeptide derivative, using N,N'-dicyclohexylcarbodiimide and N-hydroxysuccinimide to give N-acetyl-4-O-(N-acetyl-3,4,6-tri-O-benzyl-β-D-glucosaminyl-5,6-O-isopropylidene-muramyl-L-alanyl-D-isoglutamine benzyl ester acetal (g). The final deprotection is performed in two steps, first by treatment in 60% aqueous acetic acid (60°, 70 min.), the ketal and acetal groups being removed simultaneously, and then by catalytic hydrogenation to give (75).

The synthesis of the disaccharide dipeptide (76), an analogue of (75), has been described recently (20). The tetra-O-benzoyl-α-D-glucopyranosyl bromide was condensed by a Königs-Knorr reaction with benzyl-2-acetamido-6-O-benzoyl-3-O-(D-1-carboxyethyl)-2-deoxy-α-D-glucopyranoside to give the protected disaccharide in high yield. After saponification, coupling with the suitable peptide intermediate and the final hydrogenolysis were performed as usual to give (76).

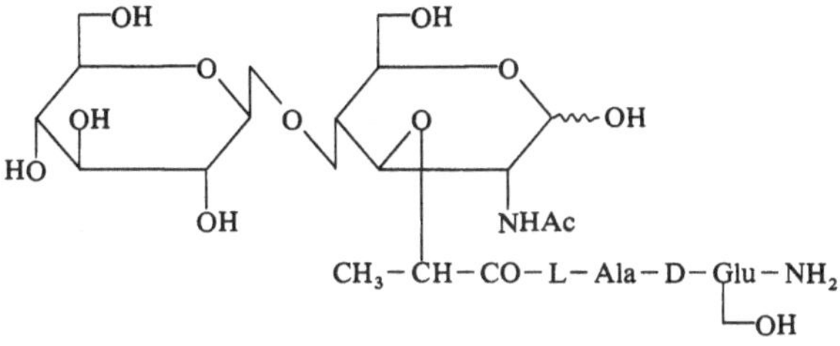

CH$_3$-CH-CO-L-Ala-D-Glu-NH$_2$
                              └-OH

(76)

VI. Synthesis of Oligomers and Conjugates of MDP

1. Synthesis of Oligomers of MDP

As MDP is excreted very rapidly (55), several oligomers have been synthesized, in order to obtain compounds which might have more long lasting biological effects.

Thus (77) (47) is a dimeric derivative of N-acetyl-muramyl-L-alanyl-D-isoglutamine-L-lysine [MDP-Lys, (63)], in which the ε-amine functions of each monomeric glycotripeptide unit are linked by an adipimidyl residue, the coupling being achieved by means of dimethyl adipimidate (23, 48).

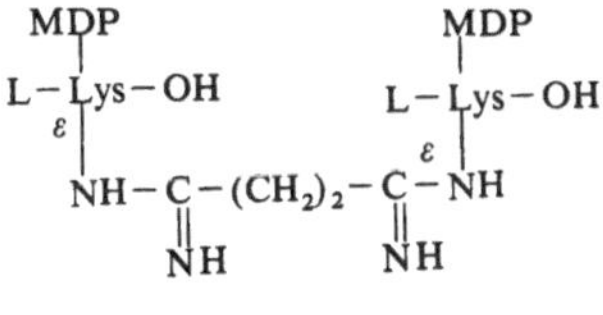

MDP   MDP
 | |
L-Lys-OH   L-Lys-OH
  ε | ε |
NH-C-(CH$_2$)$_2$-C-NH
    ‖   ‖
    NH   NH

(77)

The trimer, (78), of MDP-L-Lys, has been synthesized (47) via N^α-BOC-N^ε[N^α-BOC-N^ε-(N^α-BOC-N^ε-Z-L-Lys)-L-Lys]-L-Lys-NH$_2$. After debutyloxycarbonylation of this protected tri-lysyl derivative, MDP was coupled by means of N,N'-dicyclohexylcarbodiimide in presence of N-hydroxybenzotriazole.

$$\begin{array}{ccc} \text{MDP} & \text{MDP} & \text{MDP} \\ | & | & | \\ \text{H}-\text{L}-\text{Lys}\!-\!\!-\!\!-\!\text{L}-\text{Lys}\!-\!\!-\!\!-\!\text{L}-\text{Lys}-\text{NH}_2 \end{array}$$

(78)

The β-D-p-aminophenyl glycoside (41) of MDP, has been cross-linked with glutaraldehyde to give an oligomer with a molecular weight of approximately 6.000 daltons, indicating that eight to ten monomeric molecules have undergone cross-linking (54).

2. Synthesis of Conjugates of MDP

Interesting immunostimulant activities might be expected by conjugating MDP to macromolecules. Such compounds have been prepared with several proteins as carriers *via* carbodiimide or phenylisothiocyanate intermediates (61). Moreover, MDP and some analogues have been attached covalently to a synthetic carrier, multi-poly-D-L-alanyl-poly-L-lysine (hereafter denoted *AL*) (64) by means of N,N'-dicyclohexylcarbodiimide in presence of N-hydroxybenzotriazole to give (79—81) (47)

[N-acetyl-muramyl-L-alanyl-D-isoglutamine]-*AL*   (79)

[N-acetyl-muramyl-D-alanyl-D-isoglutamine]-*AL*   (80)

[N-acetyl-muramyl-β-alanyl-D-isoglutamine]-*AL*   (81)

VII. Mass Spectrometry of MDP and Analogues

Das and Nebelin (11, 52a) have checked the structures of several MDP analogues; permethylated peptides (10, 68) were analyzed by electron impact mass spectrometry.

Fig. 11 shows the structure of permethylated MDP with the principal fragmentations (dotted lines); Fig. 12 shows the corresponding mass spectrum. The molecular ion peak is at m/e 618 and a "sequence peak" at m/e 417, accompanied by m/e 389 (loss of CO) and 357 (loss of MeOH). Loss of the entire pyranose ring gives the ion at m/e 342. The complementary C-terminal fragment to m/e 417 is at m/e 203. Peaks at m/e 260 and 228 (260-MeOH) contain the intact pyranose ring and are known to occur in the spectra of permethylated-N-acetyl hexosamines. In these fragments the entire lactyl peptide chain has been eliminated.

References, pp. 38—47

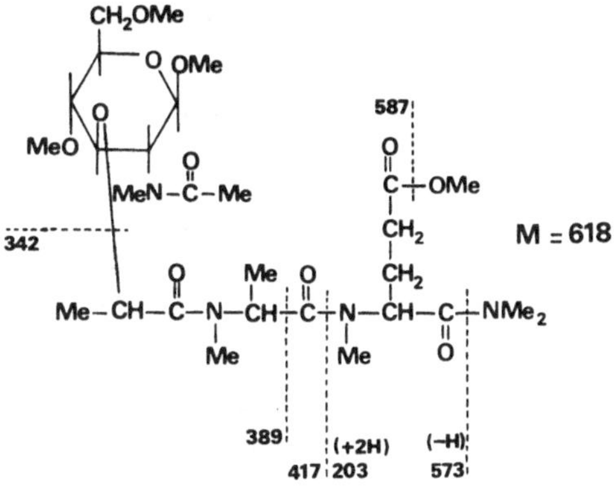

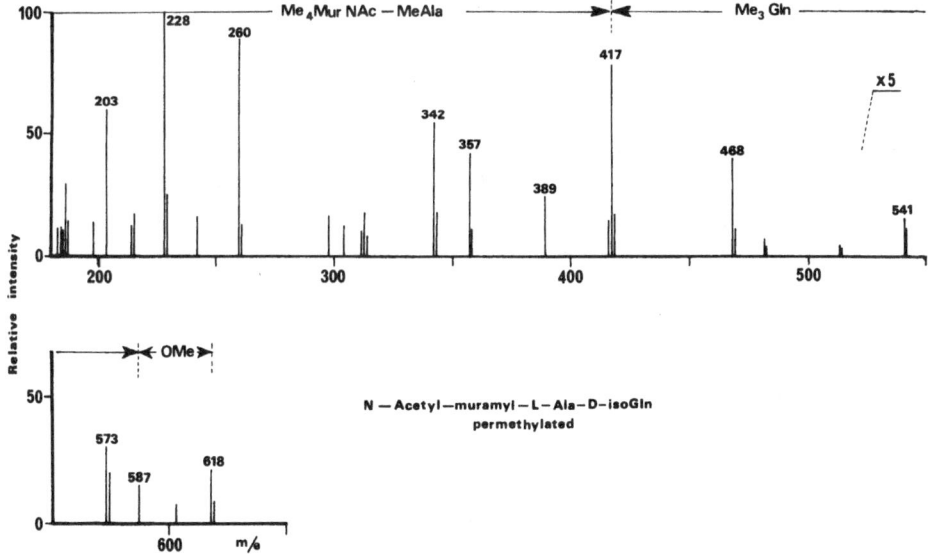

Fig. 11

Fig. 13 shows the formula of permethylated MurNAc-L-Ala-D-isoGln-L-Lys-D-Ala with the principal fragmentations. The molecular ion is at m/e 901.

Fig. 14 shows some fragments derived from the muramyl moiety, which are most characteristic and abundant in all spectra of muramyl peptides. The fragment on the left side is the result of elimination of the

Fig. 12

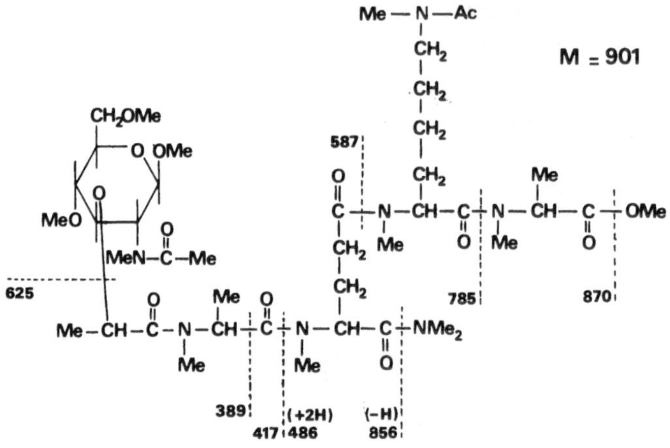

Fig. 13

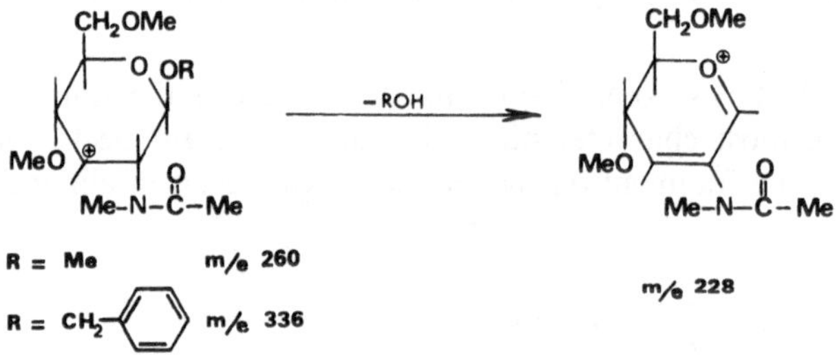

Fig. 14

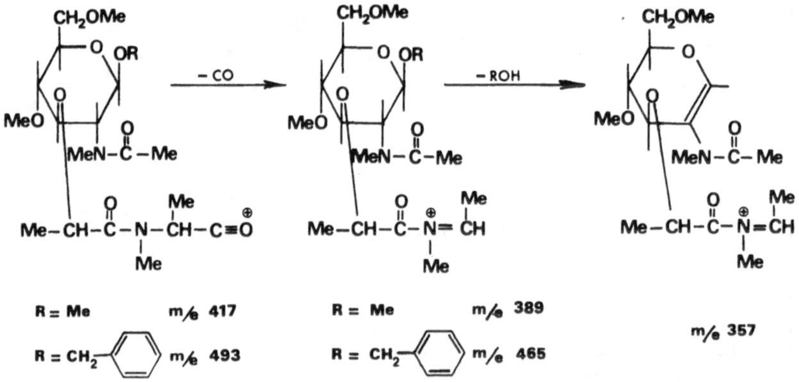

Fig. 15

lactyl peptide from the 3-position of the pyranose ring of N-acetyl glucosamine. The benzyl glycoside of MDP, gives the corresponding fragment at m/e 336. Specific elimination of the glycosidic rest OR from C-1 yields fragment 228, which is the base peak in most of the spectra.

Fig. 15 shows three fragments which are also present in all spectra of muramylpeptides; m/e 417 for the methylglycoside, or m/e 493 for the benzylglycoside, which after loss of CO may yield m/e 389 or 465 respectively, both of which then give the oxonium ion at m/e 357 (52a).

VIII. ^{13}C-NMR Spectrometry of MDP and Derivatives

The ^{13}C-NMR spectra of N-acetylmuramyl-L-alanyl-D-glutamic acid (82), methyl N-acetylmuramyl-L-alanyl-D-glutamate (83), N-acetylmura-myl-L-alanyl-D-isoglutamine (84) and N-acetyl-D-glucosamine (85) were recorded (60) and the carbon shifts assigned, as listed in Table 1, by comparison of the signals of the MDP derivatives (82—84) with those of model (85) and the use of literature data for monosaccharides (69) and peptides (13, 22). The spectral behavior of the N-acetylmuramyl moiety reveals that replacement of the 2-hydroxy group of glucose by an acetamido unit results in shielding of C(2) and etherification of the 3-hydroxy group leads to deshielding of C(3). Due to the presence of two anomers in

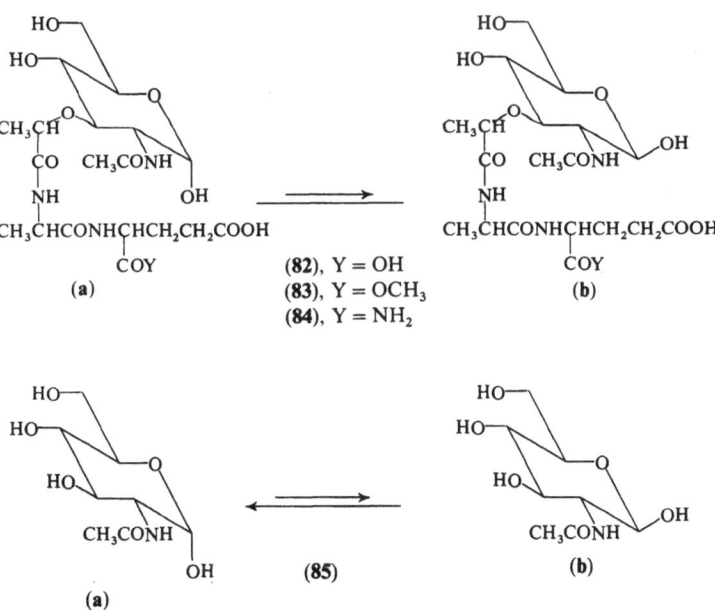

aqueous solution the N-acetylmuramyl residue shows signal twinning. Comparison of the intensities of the non-overlapping signals of like carbons of each anomer with each other after the establishment of anomer equilibrium reveals a $\alpha : \beta$ anomer ratio of 2.1 for the MDP derivatives (**82—84**) and 1.6 for the model (**85**). Finally, differentiation of the keto-carbons is based on the unique, low-field position of the terminal carboxy group, the N-acetyl carbonyl shift of (**85**) and the recorded alanine carbonyl shift (*60*).

Table 1. *Carbon Shifts of N-Acetylmuramyl-L-alanyl-D-glutamic Acid Derivatives and N-Acetyl-D-glucosamine*[a]

	84a	84b	82a	82b	83a	83b	85a	85b
C(1)	91.0	95.0	91.0	95.0	91.0	95.0	91.0	95.3
C(2)	53.8	56.3	53.8	56.3	53.8	56.3	54.6	57.2
C(3)	79.5	82.4	79.5	82.5	79.7	82.6	71.1	74.3
C(4)	69.2	68.9	69.1	68.9	69.1	68.9	70.5	70.3
C(5)	71.6	75.7	71.6	75.8	71.6	75.8	71.8	76.2
C(6)	60.7	60.9	60.7	60.8	60.7	60.7	61.0	61.0
Ac Me	22.2	22.4	22.2	22.4	22.2	22.4	22.4	22.7
CO	174.6	175.2	174.7	175.3	174.9	175.6	174.5	174.8
Lac C(α)	77.7	78.0	77.8	78.0	77.7	78.0		
C(β)	18.7	18.7	18.7	18.7	18.8	18.8		
CO	173.8	174.1	173.8	174.0	173.8	174.1		
Ala C(α)	49.7	49.7	49.7	49.7	49.9	49.9		
C(β)	17.0	17.0	16.9	16.9	16.8	16.8		
CO	175.5	175.5	175.5	175.5	175.8	175.8		
Glu C(α)	52.1	52.1	52.2	52.2	52.8	52.8		
C(β)	26.0	26.0	26.4	26.4	26.3	26.3		
C(γ)	30.1	30.1	31.2	31.2	30.3	30.3		
C(δ)	176.8	176.8	177.5	177.5	176.8	176.8		
CO	174.6	174.6	173.1	173.1	174.9	174.9		

[a] δ values in parts per million downfield from TMS; δ(TMS) $=\delta$ (dioxane) $+ 66.6$ ppm.

IX. Analysis of MDP

Nuclear magnetic resonance spectroscopy and mass spectrometry have been used for the analysis of MDP and its analogues, as previously discussed, and, sometimes also infra-red spectroscopy (*20, 73*). In this section, other analytical procedures useful in analysis and determination of physico-chemical specification of commercially available MDP will be reviewed briefly.

The homogeneity of MDP can be checked by descending paper chromatography using *n*-butanol-acetic acid-water (5 : 1 : 2), ethyl acetate-pyridine-acetic acid-water (5 : 5 : 1 : 3) or ethyl acetate-pyridine-water (8 : 2 : 1) (50, 56) or by high voltage electrophoresis on paper at pH 3.6 (56), the detection being effected according to SHARON (65). More frequently, silica gel TLC is used with n-butanol-pyridine-acetic acid-water (30 : 20 : 6 : 24 and 30 : 24 : 6 : 20), n-butanol-acetic acid-water (4 : 1 : 5, upper phase, and 4 : 1 : 1), ethyl acetate-pyridine-acetic acid-water (5 : 5 : 1 : 3), the detection being effected by chlorination (33, 44). The latter method has allowed the study of the chemical stability of MDP which is strongly dependent on the presence of water. Thus, after 10 months at 40°, a weak spot of a by-product appears in lyophilized commercial dosage forms which contain approximately 10% of water (THELY, M., personal commun.).

High performance liquid chromatography of MDP seems to be promising for various analytical purposes, e.g. if it is desired to check its chemical or even stereochemical homogeneity, to study its stability or to quantify its dosage forms. Some preliminary results have been obtained (60). Thus, on a Spherisorb ODS (5 μm) column, eluted with $5 \cdot 10^{-3}$ M ammonium acetate at pH 2.5-acetonitrile (995 : 5), it is possible to separate the α- and β-anomers of MDP and to follow the variation of their ratio with time, as the result of their mutarotation in water which is shown by the decrease of the optical rotation (c = 0.5, water) from $[\alpha]_D^{20} = +37°$ to $[\alpha]_D^{20} = +34°$, after four hours. These results indicate an α/β anomer ratio of 2.0 ± 0.2 in agreement with the ratio found by ^{13}C NMR spectrometry (see above). Using a Lichrosorb 10-NH$_2$ column, eluted with ammonium acetate $5 \cdot 10^{-3}$, pH 3-acetonitrile (10 : 90), the two anomers of MDP can be separated from the two peaks of its diastereoisomer containing a D-alanyl residue. This method can be applied as a routine check of the MDP content and the purity of commercial MDP preparations.

MDP, as all the N-acetyl-muramyl-peptides, can be precipitated by addition of ether from its ethanol or methanol solution, but it occludes a considerable amount of solvent judging from elemental analysis. Repeated lyophilization allows the elimination of all traces of solvent (33).

Addendum

This review was submitted for publication in 1979. Since then a great number of new derivatives of MDP have been synthesized, proving the continuous and ever increasing interest in this field. This addendum contains bibliography published until August 1980; only the most interesting new compounds are mentioned, without details of their synthesis.

a) Modification of the Peptide Moiety

Among many L-amino acids tested, only L-valine and L-α-amino-butyric acid can replace L-alanine in MDP to give enhanced adjuvant active analogues (*76*).

b) Substitution of the D-glutamic Acid Residue

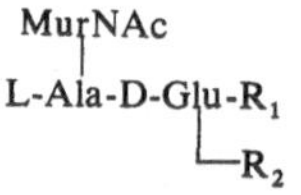

$$
\begin{array}{c}
\text{MurNAc} \\
|\\
\text{L-Ala-D-Glu-R}_1 \\
\llcorner\text{R}_2
\end{array}
$$

(86)

In compound (86) one R is an amide group, the other a medium chain alkyl ester (up to *n*-decyl) or both R are such esters (*77*).

c) Modification of the Carbohydrate Moiety

The 6-hydroxyl group has been replaced by an amino group (87 a—c) (*78—81*) or an acyl amino group (87 d—f) (*78—81*) or the 5-hydroxy-methyl group by a hydrogen atom (88, R=H) or a methyl group (88, R=CH$_3$) (*80*).

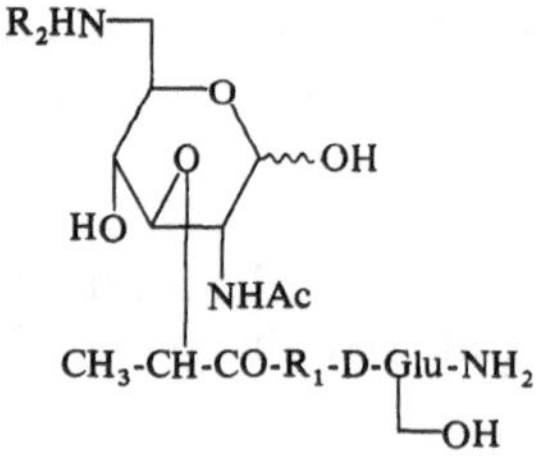

	R_1 =	R_2 =
a	L-Ala-	H
b	L-Val	H
c	L-Ser	H
d	L-Ala	Ac
e	L-Ala	stearoyl
f	L-Ala-	mycoloyl

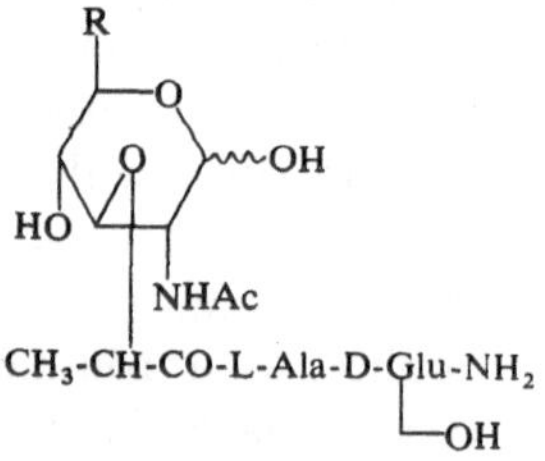

(87)

(88)

The C-2 substituent of the N-acetyl muramyl residue has also been modified to give analogues having a glucosaminyl or glucosyl moiety, respectively (89 a) and (89 b) (*79*).

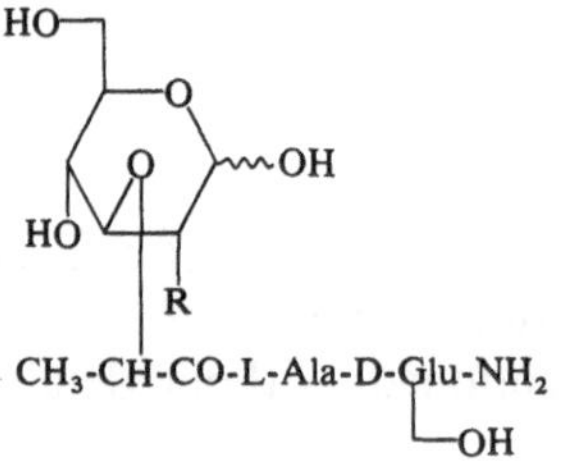

(89) a   R=NH$_2$
       b   =OH

The synthesis of N-acetyl-β-D-glucosaminyl-(1→4)-N-acetyl-muramyl-dipeptide according to various synthetic (*82—84*) and hemi-synthetic (*85*) procedures has been described, and, also, the synthesis of one of its isomers O-(N-acetyl-β-muramyl-L-alanyl-D-isoglutamine)-(1→4)-N-acetyl-D-glucosamine (**90**) (*86*).

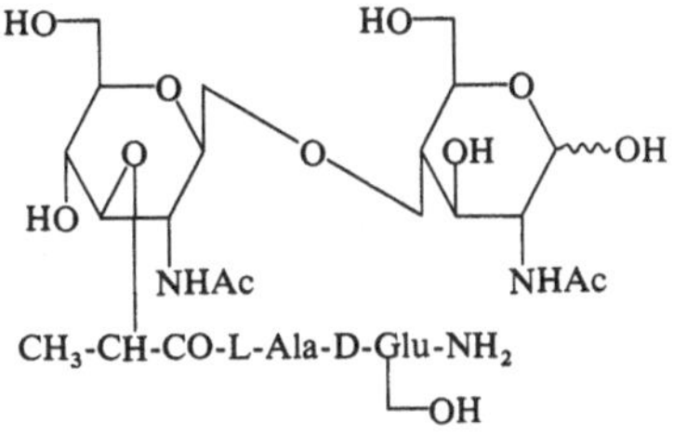
HO
HO
O
O
O
OH
OH
HO
NHAc
NHAc
CH3-CH-CO-L-Ala-D-Glu-NH2
OH

(**90**)

d) Lipophilic Derivatives

6-Quinonyl-MDP (**91 a**) and related compounds (**91 b—d**) have been synthesized to obtain increased antitumor activity (*87, 88*).

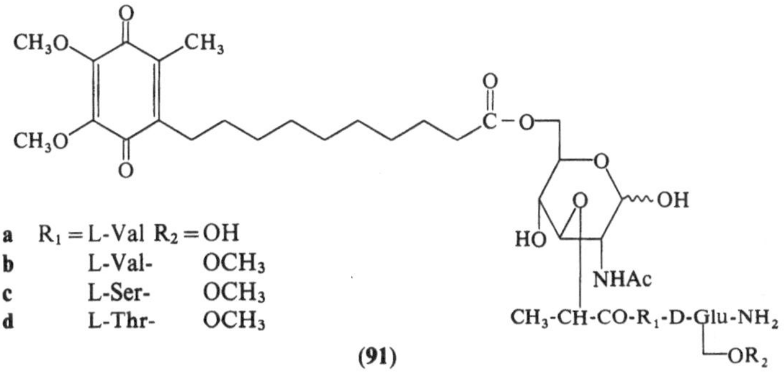
O
CH3O
CH3
CH3O
O
O
C-O
O
O
OH
HO
NHAc
CH3-CH-CO-R1-D-Glu-NH2
OR2

a R_1 = L-Val R_2 = OH
b L-Val- OCH_3
c L-Ser- OCH_3
d L-Thr- OCH_3

(**91**)

More lipophilic compounds such as (**92 a—c**) have been reported (*89*).

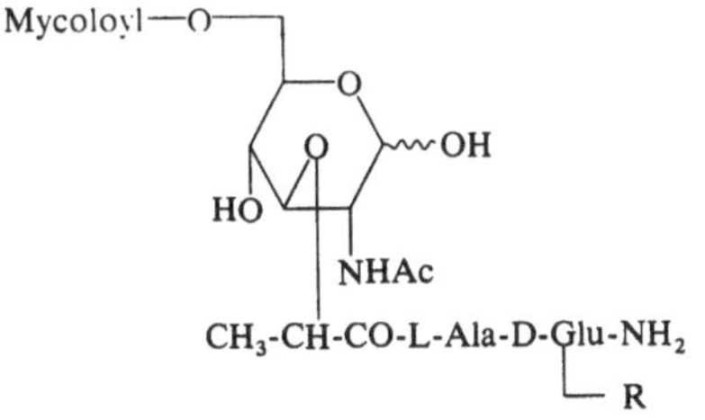
Mycoloyl—O
O
O
OH
HO
NHAc
CH3-CH-CO-L-Ala-D-Glu-NH2
R

(**92**)

a R = L-Lys-(Myc)
b = NH-$(CH_2)_2$-OMyc
c = N^{ε}(6-O-mycoloyl-MDP)-Lys

e) New Types of Peptide Adjuvants

A nona-peptide (93) (*90*) and a lipopeptide (94) (*91, 92*), structurally related to the peptide moiety of bacterial peptidoglycans, have been reported to have adjuvant activities.

H-L-Ala-D-Glu-NH$_2$
        └─L -Lys(Ac)-D-Ala-(Gly)$_5$-OMe

(93)

Lauroyl-L-Ala-D-Glu
        └──────────OH
              │
        D D, L L Dap
      H Gly────────┴────────NH$_2$

(94)

f) Biological Activities of MDP and Its Derivatives

General reviews on the biological aspect of these glycopeptides have been published recently (*76, 93, 94*).

Acknowledgements. The authors are grateful to Dr. B. C. Das and Professor E. Wenkert for their help in writing the respective chapters on mass and NMR spectrometry.

The synthetic work of one of us (P. L.) would not have been possible without the most competent and efficient help of M. Derrien, I. Lederman, M. Level, M. Petitou, J. L. Amiot and M. Zuber.

The work of the authors has been pursued over the last years under the auspices of G.I.R.P.I. (Groupement d'Intérêt Economique — Institut pour la Recherche et la Production d'Immunostimulants) with essential personal contributions of professor L. Chedid, Dr. J. Choay and professor F. Gros which are gratefully acknowledged. Many syntheses described in this review are the result of stimulating discussions with them.

The research of E. L. and his colleagues at the Institut de Chimie des Substances Naturelles (Gif sur Yvette) and the Institut de Biochimie (Université Paris-Sud, Centre d'Orsay) was supported, in part by CNRS, DGRST, Fondation pour la Recherche Médicale, Ligue Nationale Française contre le cancer and the Cancer Research Institute, N. Y.

References

1. Adam, A., M. Devys, V. Souvannavong, P. Lefrancier, and E. Lederer: Correlation of structure and adjuvant activity of N-acetyl-muramyl-L-alanyl-D-isoglutamine (MDP), its derivatives and analogues. Anti-adjuvant and competition properties of stereoisomers. Biochem. Biophys. Res. Comm. **72,** 339 (1976).

2. Adam, A., J.-F. Petit, J. Wietzerbin-Falzspan, P. Sinaÿ, D. W. Thomas, and E. Lederer: L'acide N-glycolyl-muramique, constituant des parois de *Mycobacterium smegmatis:* identification par spectrométrie de masse. FEBS Letters **4,** 87 (1969).

3. ADAM, A., R. CIORBARU, J.-F. PETIT, and E. LEDERER: Isolation and properties of a macromolecular water-soluble immunoadjuvant fraction from the cell wall of *Mycobacterium smegmatis*. Proc. Natl. Acad. Sci. **69**, 851 (1972).

4. ADAM, A., R. CIORBARU, J.-F. PETIT, E. LEDERER, L. CHEDID, A. LAMENSANS, F. PARANT, M. PARANT, J. P. ROSSELET, and F. M. BERGER: Preparation and biological properties of watersoluble adjuvant fractions from delipidated cells of *Mycobacterium smegmatis* and *Nocardia opaca*. Infection and Immunity **7**, 855 (1973).

5. ARENDT, A., A. KOLODZIEJEZYK, and T. SOKOLOWSKA: Synthesis of mucopeptide fragments of bacterial cell walls containing diaminopimelic acid. Part IX. Synthesis of pentapeptide related to bacterial mureine fragment. Roczniki Chem. Ann. Soc. Chim. Polonorum. **48**, 1921 (1974).

6. BEYERMAN, H. C., E. W. B. DE LEER, and J. FLOOR: On the repetitive excess mixed anhydride method for the synthesis of peptides. Synthesis of the sequence 1—10 of human growth hormone. Rec. Trav. Chim. Pays. Bas. **92**, 481 (1973).

7. BRICAS, E.: Survey of the Synthetic Work in the field of the bacterial Cell Wall Peptides. *In* The Chemistry of Polypeptides (P. G. KATSOYANNIS, Ed.), p. 205. New York: Plenum Press. 1973.

8. CHATURVEDI, N. C., M. C. KHOSLA, and N. ANAND: Synthesis of analogs of bacterial cell wall glycopeptides. J. Med. Chem. **9**, 971 (1966).

9. CHEUNG, S. T., and N. L. BENOITON: N-methylaminoacids in peptide synthesis. The synthesis of N-tert-butyloxycarbonyl-N-methylamino acids by N-methylation. Can. J. Chem. **55**, 906 (1977).

10. DAS, B. C., and E. LEDERER: Mass spectrometry of peptide derivatives: Scope and Limitations. *In* Peptides 1971 (H. NESVABDA, Ed.), p. 253. Amsterdam: North Holland Publishing Co. 1973.

11. DAS, B. C., and N. NEBELIN: Structural Characterization of N-acetyl-muramyl-peptides by mass spectrometry. *In* Peptides 1978 (I. Z. SIEMON and G. KUPRYSZEWSKI, Eds.).

12. DEFAYE, J., and E. LEDERER: Synthèse d'un α-monomycolate de glycerol. Bull. Soc. Chim. Biol. **38**, 1301 (1956).

13. DESLAURIERS, R., and I. C. P. SMITH: Conformation and Structure of Peptides, Chapter 1 in: Topics in Carbon-13 N.M.R. Spectroscopy (G. C. LEVY, Ed.), Vol. 2. New York, N. Y.: Wiley-Interscience. 1976.

14. DESLAURIERS, R., I. C. P. SMITH, and R. WALTER: Conformational flexibility of the Neurohypophyseal Hormones oxytocin and Lysine-Vasopressin. A Carbon-13-Spin-Lattice Relaxation Study of back-bone and side chain. J. Amer. Chem. Soc. **96**, 2289 (1974).

15. DURETTE, P. L., A. ROSEGAY, M. A. R. WALSH, and T. Y. SHEN: Synthesis of (6-³H)-N-acetylmuramyl-L-alanyl-D-isoglutamine. Tetrahedron Lett. 291 (1979).

16. ELLOUZ, F., A. ADAM, R. CIORBARU, and E. LEDERER: Minimal structural requirements for adjuvant activity of bacterial peptidoglycan derivatives. Biochem. Biophys. Res. Comm. **59**, 1317 (1974).

17. FLOWERS, H. M., and R. W. JEANLOZ: The synthesis of 2-acetamido-3-O-(D-1-carboxyethyl)-2-deoxy-α-D-glucose (N-acetylmuramic acid) and of benzyl glycoside derivatives of 2-amino-3-O-(D-1-carboxy-ethyl)-2-deoxy-D-glucose (Muramic acid). J. Org. Chem. **28**, 2983 (1963).

18. FREUND, J.: The mode of action of immunologic adjuvants. Advances in Tuberculology **7**, 130 (1956).

19. FUJINO, M., and C. HATANAKA: A new procedure for the pentachlorophenylation of N-protected amino-acids. Chem. Pharm. Bull. **16**, 929 (1968).

20. HASEGAWA, A., Y. KANEDA, M. AMANO, M. KISO, and I. AZUMA: A facile Synthesis of N-acetyl-muramyl-L-alanyl-D-isoglutamine and its Carbohydrate Analogs and their Immunoadjuvant Activities. Agric. Biol. Chem. **42**, 2187 (1978).

21. Hiu, I. J.: Water-soluble and lipid-free fraction from BCG, with adjuvant and anti-tumor activity. Nature New Biology **238**, 241 (1972).
22. Howarth, O. W., and D. M. J. Lilley: Carbon-13-NMR of peptides and proteins. Prog. NMR Spectroscopy **12**, 1 (1978).
23. Hunter, M. J., and M. L. Ludwig: Amidination. Methods in Enzymol. **25**, 585 (1972).
24. Inch, T. D., and H. G. Fletcher, Jr.: Synthesis with partially benzylated sugars. VI. Some solvolytic reactions of 2-acetamido-1-O-acyl-2-deoxy-D-glucopyranose and D-galactopyranose derivatives. J. Org. Chem. **31**, 1810 (1965).
25. Jeanloz, R. W., E. Walker, and P. Sinaÿ: Synthesis of various glycosides of 2-amino-3-O-(D-1-carboxyethyl)-2-deoxy-D-glucopyranose (Muramic acid). Carbohyd. Res. **6**, 184 (1968).
26. Jollès, P., and A. Paraf: Chemical and Biological Basis of Adjuvants. Molecular Biology, Biophysics, Vol. **13**, p. 153. Berlin-Heidelberg-New York: Springer. 1973.
27. Keim, P., R. A. Vigna, R. C. Marshall, and F. R. N. Gurd: Carbon 13-Nuclear Magnetic Resonance of Pentapeptides of glycine containing central residues of aliphatic amino acids. J. Biol. Chem. **248**, 6104 (1973).
28. Keim, P., R. A. Vigna, J. S. Morrow, R. C. Marshall, and F. R. N. Gurd: Carbon 13-Nuclear Magnetic Resonance of pentapeptides of glycine containing central residues of Serine, Threonine, Aspartic and Glutamic Acids, Asparagine and Glutamine. J. Biol. Chem. **248**, 7811 (1973).
29. Kotani, S., Y. Watanabe, T. Shimono, K. Kato, F. Kinoshita, T. Shiba, S. Kusumoto, Y. Tarumi, K. Yokogawa, and S. Kawata: Abstr. Symp. Intern. Immunostimulants Bacteriens. Pasteur Institute, Paris (1974).
30. Kotani, S., Y. Watanabe, F. Kinoshita, T. Shimono, I. Marisaki, T. Shiba, S. Kusumoto, Y. Tarumi, and K. Ikenaka: Immunoadjuvant activities of synthetic N-acetyl-muramyl-peptides or aminoacids. Biken J. **18**, 105 (1975).
31. Kotani, S.: Biological activities of bacterial cell wall peptidoglycans and their sub-units, with special reference to the immuno-adjuvant actions. Seikagaku **48**, 1081 (1976).
32. Kotani, S., F. Kinoshita, Y. Watanabe, J. Morisaki, T. Shimono, K. Kato, T. Shiba, S. Kusumoto, K. Ikenaka, and Y. Tarumi: Effects of Chemical modifications of the glutamic acid residue in N-acetyl-muramyl-peptides on the immunoadjuvancies of the molecules. Biken J. **20**, 125 (1977).
33. Kusumoto, S., Y. Tarumi, K. Ikenaka, and T. Shiba: Chemical synthesis of N-acetylmuramyl-peptides with partial structures of bacterial cell wall and their analogs in relation to immunoadjuvant activities. Bull. Chem. Soc. Japan **49**, 533 (1976).
34. Kusumoto, S., S. Okada, and T. Shiba: Synthesis of 6-O-mycoloyl-N-acetyl-muramyl-L-alanyl-D-isoglutamine with immunoadjuvant activity. Tetrahedron Lett. 4287 (1976).
35. Kusumoto, S., K. Ikenaka, and T. Shiba: Synthesis of N-acetylmuramyl-L-(U-^{14}C-)-alanyl-D-isoglutamine. Tetrahedron Letters 4055 (1977).
36. Kusumoto, S., S. Okada, K. Yamamoto, and T. Shiba: Synthesis of 6-O-acyl-derivatives of immunoadjuvant active N-acetyl-muramyl-L-alanyl-D-isoglutamine. Bull. Chem. Soc. Japan. **51**, 2122 (1978).
37. Kusumoto, S., M. Inage, T. Shiba, I. Azuma, and Y. Yamamura: Synthesis of long chain fatty acid esters of N-acetyl-muramyl-L-alanyl-D-isoglutamine in relation to anti-tumor activity. Tetrahedron Lett. 4895 (1978).
38. Kusumoto, S., K. Yamamoto, and T. Shiba: Synthesis of N-acetyl-β-D-glucosaminyl-(1 4)-N-acetyl-muramyl-L-alanyl-D-isoglutamine. Tetrahedron Lett. 4407 (1978).
39. Lanzilotti, A. E., E. Benz, and L. Goldman: Total Syntheses of N²-1-(2-acetamido-3-O-D-glucosyl)-D-propionyl-L-alanyl-D-α- and γ-glutamyl}-L-lysyl-D-alanyl-D-alanine, and identity of the γ-glutamyl isomer with the glycopeptide of a bacterial cell wall precursor. J. Amer. Chem. Soc. **86**, 1880 (1964).

40. LEDERER, E.: The mycobacterial cell wall. Pure and Applied Chemistry 25, 135 (1971).
41. LEDERER, E., A. ADAM, R. CIORBARU, J.-F. PETIT, and J. WIETZERBIN: Cell walls of myco-bacteria and related organisms; chemistry and immunostimulant properties. Molecular and Cellular Biochemistry 7, 87 (1975).
42. LEDERER, E.: Natural and synthetic immunostimulants related to the mycobacterial cell wall. Medicinal Chemistry V, p. 257. Amsterdam: Elsevier Scientific Publishing Company. 1977.
43. LEFRANCIER, P., and E. BRICAS: Synthèse de la subunité peptidique du peptidoglycane de la paroi de trois bactéries gram-positif et de peptides de structure analogue. Bull. Soc. Chim. Biol. 49, 1257 (1967).
44. LEFRANCIER, P., J. CHOAY, M. DERRIEN, and I. LEDERMAN: Synthesis of N-acetyl-muramyl-L-alanyl-D-isoglutamine, an adjuvant of the immune response and of some N-acetyl-muramyl-peptide analogs. Int. J. Peptide Protein Res. 9, 249 (1977).
45. LEFRANCIER, P., M. DERRIEN, I. LEDERMAN, F. NIEF, J. CHOAY, and E. LEDERER: Synthesis of some new analogs of the immunoadjuvant glycopeptide MDP (N-acetyl-muramyl-L-alanyl-D-isoglutamine). Int. J. Peptide Res. 11, 289 (1978).
46. LEFRANCIER, P., M. PETITOU, M. LEVEL, M. DERRIEN, J. CHOAY, and E. LEDERER: Synthesis of N-acetyl-muramyl-L-alanyl-D-glutamic α-amide (MDP) or α-methyl ester derivatives, bearing a lipophilic group at the C-terminal peptide end. Int. J. Peptide Protein Res. 14, 437 (1979).
47. LEFRANCIER P. et al., unpublished results.
48. LUBIN, B., V. PENA, W. C. MENTZER, E. BYMUM, T. B. BRADLEY, and L. PACKER: Dimethyl Adipimidate: A new Antisickling Agent. Proc. Natl. Acad. Sci. USA. 72, 43 (1975).
49. MERSER, C., and P. SINAŸ: Abstr. Symp. Intern. Immunostimulants. Bacteriens. Pasteur Institute, Paris, 1974.
50. MERSER, C., P. SINAŸ, and A. ADAM: Total Synthesis and adjuvant activity of bacterial peptidoglycan derivatives. Biochem. Biophys. Res. Comm. 66, 1316 (1975).
51. MERSER, C.: Synthèse totale du disaccharide de base et de divers fragments du peptidoglycane des parois bactériennes. Activité immunoadjuvante d'un muramyl dipeptide. Thèse de Sciences, Orléans, 1975.
52. MIGLIORE, D., and P. JOLLÈS: A hydrosoluble adjuvant active mycobacterial "poly-saccharide-peptidoglycan". Preparation by a simple extraction technique of the bacterial cells (strain Peurois). FEBS Letters 25, 301 (1972).
52a. NEBELIN, E., and B. C. DAS: Mass spectrometry of Synthetic Glycopeptide analogs of Bacterial Peptidoglycans (MDP and Derivatives) FEBS Letters 107, 254 (1979).
53. OSAWA, T., and R. W. JEANLOZ: An improved, stereoselective synthesis of 2-amino-3-O-(D-1-carboxyethyl)-2-deoxy-D-glucose (Muramic acid). J. Org. Chem. 30, 448 (1965).
54. PARANT, M., C. DAMAIS, F. AUDIBERT, F. PARANT, L. CHEDID, E. SACHE, P. LEFRANCIER, J. CHOAY, and E. LEDERER: In vivo and in vitro stimulation of non specific immunity by the β-D-p-amino-phenyl glycoside of N-acetyl-muramyl-L-alanyl-D-isoglutamine and an oligomer prepared by cross-linking with glutaraldehyde. J. Infec. Dis. 138, 378 (1978).
55. PARANT, M., F. PARANT, L. CHEDID, A. YAPO, J. F. PETIT, and E. LEDERER: Fate of ^{14}C-Labeled muramyl-dipeptide, a synthetic immunoadjuvant, after administration to the mouse. Int. J. Immunopharmacol., 1, 35 (1979).
56. PETIT, J. F.: person. comm.
57. PICHAT, L., J. TOSTAIN, P. LEFRANCIER, P. SINAŸ. and E. LEDERER: Synthèse de la N-acetylmuramyl (oxo-^{14}C-Propyl)-L-alanyl-D-isoglutamine. ["Muramyl (oxo-^{14}C-Propyl) dipeptide": (MDP-^{14}C) J. of labelled compounds and radiopharmaceuticals 17, 153 (1980).

58. POLONSKY, J., E. SOLER, and J. VARENNE: Sur la synthèse du cord-factor et de ses analogues. Carbohyd. Res. **65**, 295 (1978).
59. QURESCHI, N., and K. TAKAYAMA: Characterization of the Purified Components of a new homologous series of α Mycolic Acids from *Mycobacterium tuberculosis*. H 37 Ra. J. Biol. Chem. **253**, 5411 (1978).
60. RAJU, M. S., T. D. J. HALLS, E. WENKERT, M. ZUBER, P. LEFRANCIER, and E. LEDERER: Anomeric configuration of the immunostimulant muramyl dipeptide MDP and some of its derivatives. Carbohyd. Res. **81**, 173 (1980).
61. REICHERT, C. M., C. CARRELI, M. JOLIVET, F. AUDIBERT, P. LEFRANCIER, and L. CHEDID: Synthesis of N-acetyl-muramyl-L-alanyl-D-isoglutamine (MDP) containing conjugates and their use as a hapten-carrier system Mol. Immunol. **17**, 357 (1980).
62. SAITO, H., and I. C. P. SMITH: Carbon-13 Nuclear Magnetic Resonance Studies of poly-amino acids: The Helix-Coil Transition of Poly-Lysine. Arch. Biochem. Biophys. **158**, 154 (1973).
63. SCHLEIFER, K. H., and O. KANDLER: Peptidoglycan types of bacterial Cell Walls and their taxonomic implications. Bacteriol. Rev. **36**, 407 (1972).
64. SELA, M., E. KATCHALSKI, and M. GEHATIA: Multichain Polyaminoacids. J. Amer. Chem. Soc. **78**, 746 (1956).
65. SHARON, N., and S. SEIFTER: A transglycosylation reaction catalyzed by lysozyme. J. Biol. Chem. **239**, P. C. 2398 (1964).
66. SHIBA, T., S. OKADA, S. KUSUMOTO, I. AZUMA, and Y. YAMAMURA: Synthesis of 6-O-mycoloyl-N-acetyl-muramyl-L-alanyl-D-isoglutamine with antitumor activity. Bull. Chem. Soc. Japan **51**, 3307 (1978).
67. SINAŸ, P.: person. comm.
68. VILKAS, E., and E. LEDERER: N-méthylation de peptides par la méthode de HAKOMORI: Structure du mycoside Cb 1. Tetrahedron Lett. 3089 (1968).
69. WALKER, T. E., R. E. LONDON, T. W. WHALEY, R. BARKER, and N. A. MATWIYOFF: Carbon-13 Nuclear Magnetic Resonance Spectroscopy of (1-^{13}C) Enriched monosaccha-rides. Signal Assignments and orientation dependance of geminal and vicinal Carbon-Carbon and Carbon-Hydrogen Spin-Spin coupling constants. J. Amer. Chem. Soc. **98**, 5807 (1976) and references cited therein.
70. WANG, S. S., B. F. GISIN, D. P. WINTER, R. MAKOFSKE, I. D. KULESHA, C. TZOUGRAKI, and J. MEIENHOFER: Facile Synthesis of Amino Acid and Peptide Esters under mild conditions *via* Caesium salts. J. Org. Chem. **42**, 1286 (1977).
71. WHITE, R. G., L. BERNSTOCK, R. G. S. JOHNS, and E. LEDERER: The influence of com-ponents of *Mycobacterium tuberculosis* and other mycobacteria upon antibody pro-duction to ovalbumin. Immunology **1**, 54 (1958).
72. WHITE, R. G.: The adjuvant effect of microbial products on the immune response. Ann. Rev. Microbiol. **30**, 579 (1976).
73. ZAORAL, M., J. JEZEK, R. STRAKA, and K. MASEK: Glycopeptides of Bacterial walls. A novel method of removal of benzyl and benzylidene residues. Coll. Czechoslov. Chem. Commun. **43**, 1797 (1978).
74. ZERVAS, L., and S. KONSTAS: Über Glucosaminide. Chem. Ber. **93**, 435 (1960).
75. ZURABYAN, S. E., T. P. VOLOSYUK, and A. J. KHORLIN: Oxazoline synthesis of 1-2-*trans*-2-acetamido-2-deoxyglycosides. Carbohyd. Res. **9**, 215 (1969).

References of Addendum

76. DUKOR, P., L. TARCSAY, and G. BASCHANG: Annu. Rep. Med. Chem. 14, 146 (1979).
77. LEFRANCIER, P.: First international Conference on Immunopharmacology (1980) Brighton (U. K.).
78. HASEGAWA, A., H. OKUMURA, M. KISO, I. AZUMA, and Y. YAMAMURA: Carbohydr. Res. 79, C. 20 (1980).
79. KISO, M., Y. KANEDA, H. OKUMURA, A. HASEGAWA, I. AZUMA, and Y. YAMAMURA: Carbohydr. Res 79, C. 17 (1980).
80. HASEGAWA, A., H. OKUMURA, M. KISO, I. AZUMA, and Y. YAMAMURA: Agric. Biol. Chem. 44, 1301 (1980).
81. HASEGAWA, A., H. OKUMURA, M. KISO, I. AZUMA, and Y. YAMAMURA: Agric Biol. Chem. 44, 1309 (1980).
82. DURETTE, P. L., E. P. MEITZNER, and T. Y. SHEN: Carbohydr. Res. 77, C. 1 (1979).
83. TSUJIMOTO, M., F. KINOSHITA, T. OKUNAGA, S. KOTANI, S. KUSUMOTO, K. YAMAMOTO, and T. SHIBA: Microbiol. Immunol. 23, 933 (1979).
84. KISO, M., Y. KANEDA, R. SHIMIZU, and A. HASEGAWA: Carbohydr. Res. 86, C. 8 (1980).
85. IVANOV, V. T.: 16th European Peptide Symposium 1980. Helsingør (DK).
86. DURETTE, P. L., E. P. MEITZNER, and T. Y. SHEN: Tetrahedron Lett. 42, 4013 (1979).
87. KOBAYASHI, S., T. FUKUDA, I. IMADA, M. FUJINO, I. AZUMA, and Y. YAMAMURA: Chem. Pharm. Bull. 27, 3193 (1979).
88. AZUMA, I., M. YAMAWAKI, M. UEMIYA, I. SAIKI, Y. TANIO, S. KOBAYADHI, T. FUKUDA, I. IMADA, and Y. YAMAMURA: Gann. 70, 847 (1980).
89. UEMIYA, M., I. SAIKI, T. KUSAMA, I. AZUMA, and Y, YAMAMURA: Microbiol. Immunol. 23, 821 (1979).
90. MASEK, K., M. ZAORAL, J. JEZEK, and V. KRCHNAK: Experientia 35, 1397 (1979).
91. MIGLIORE-SAMOUR, D., J. BOUCHAUDON, F. FLOC'H, A. ZERIAL, L. NINET, G. H. WERNER, and P. JOLLES: C. R. Acad. Sci. Paris 289, 473 (1979).
92. MIGLIORE-SAMOUR, D., J. BOUCHAUDON, F. FLOC'H, A. ZERIAL, L. NINET, G. H. WERNER, and P. JOLLES: Life Sciences 26, 883 (1980).
93. PARANT, M.: Springer Semin. Immunopathol. 2, 101 (1979).
94. LEDERER, E.: J. Med. Chem. 23, 819 (1980).

(Received March 30, 1979)

Appendix

Leading References on Biological Activities of MDP and Derivatives*

For general reviews on the biological properties of MDP and its derivatives see:

JOHNSON, A. G., F. AUDIBERT, and L. CHEDID: Synthetic immunoregulating molecules: a potential bridge between chemotherapy and immunotherapy of cancer. Cancer Immunol. Immunother. 3, 219 (1978).
CHEDID, L., F. AUDIBERT, and A. G. JOHNSON: Biological activities of muramyl dipeptide, a synthetic glycopeptide analogous to bacterial immunoregulating agents. Progress in Allergy 25, 63 (1978).
CHEDID, L., and E. LEDERER: Past, present and future of the synthetic immunoadjuvant MDP and its analogues. Biochemical Pharmacology 27, 2183 (1978).

---

* In chronological order.

PARANT, M.: Biological properties of a new synthetic adjuvant, muramyl dipeptide (MDP). Springer Seminars in Immunopathology **2**, 101 (1979).

CHEDID, L., C. CARELLI, and F. AUDIBERT: Recent developments concerning a synthetic immunoregulating molecule: muramyl dipeptide. J. Reticuloendothelial Soc. **26S**, 631 (1979).

Adjuvant activity:

ELLOUZ, F., A. ADAM, R. CIORBARU, and E. LEDERER: Minimal structural requirements for adjuvant activity of bacterial peptidoglycan derivatives. Biochem. Biophys. Res. Comm. **59**, 1317 (1974).

KOTANI, S., Y. WATANABE, F. KINOSHITA, T. SHIMONO, I. MORISAKI, T. SHIBA, S. KUSUMOTO, Y. TARUMI, and K. IKENAKA: Immunoadjuvant activities of synthetic N-acetyl-muramyl-peptides or -amino acids. Biken J. **18**, 105 (1975).

AZUMA, I., K. SUGIMURA, T. TANIYAMA, M. YAMAWAKI, Y. YAMAMURA, S. KUSUMOTO, S. OKADA, and T. SHIBA: Adjuvant activity of mycobacterial fractions: immunological properties of synthetic N-acetyl-muramyl dipeptide and the related compounds. Infect. Immun. **14**, 18 (1976).

AUDIBERT, F., L. CHEDID, J. LEFRANCIER, and J. CHOAY: Distinctive adjuvanticity of synthetic analogs of mycobacterial water-soluble components. Cell. Immunol. **21**, 243 (1976).

TANAKA, A., R. SAITO, K. SUGIYAMA, I. MORISAKI, S. KOTANI, S. KUSUMOTO, and T. SHIBA: Adjuvant activity of synthetic N-acetyl-muramyl peptides in rats. Infect. Immun. **15**, 332 (1977).

AUDIBERT, F., L. CHEDID, P. LEFRANCIER, J. CHOAY, and E. LEDERER: Relationship between chemical structure and adjuvant activity of some synthetic analogues of N-acetyl-muramyl-L-alanyl-D-isoglutamine (MDP). Ann. Immunol.. (Inst. Pasteur) **128C**, 653 (1977).

KOTANI, S., F. KINOSHITA, I. MORISAKI, T. SHIMONO, T. OKUNAGA, H. TAKADA, M. TSUJI-MOTO, Y. WATANABE, K. KATO, T. SHIBA, S. KUSUMOTO, and S. OKADA: Immunoadjuvant activities of synthetic 6-O-acyl-N-acetyl-muramyl-L-alanyl-D-isoglutamine with special reference to the effect of its administration with liposomes. Biken J. **20**, 95 (1977).

SOUVANNAVONG, V., A. ADAM, and E. LEDERER: Kinetics of the humoral and cellular immune response of guinea pigs after injection of the synthetic adjuvant N-acetyl-muramyl-L-alanyl-D-isoglutamine: comparison with Freund complete adjuvant. Infect. Immun. **19**, 966 (1978).

AZUMA, I., K. SUGIMURA, M. YAMAWAKI, M. UEMIYA, S. KUSUMOTO, S. OKADA, T. SHIBA, and Y. YAMAMURA: Adjuvant activity of synthetic 6-O-"mycoloyl"-N-acetyl-muramyl-L-alanyl-D-isoglutamine and related compounds. Infect. Immun. **20**, 600 (1978).

MASEK, K., M. ZAORAL, J. JEZEK, and R. STRAKA: Immunoadjuvant activity of synthetic N-acetyl-muramyl-dipeptide. Experientia **34**, 1363 (1978).

HEYMER, B., H. FINGER, and C. H. WIRSING: Immunoadjuvant effects of the synthetic muramyl dipeptide (MDP) N-acetylmuramyl-L-alanyl-D-isoglutamine. Zeit.-Immunitätforsch. **155**, 87 (1978).

SOUVANNAVONG, V., and A. ADAM: Opposite effects of the synthetic adjuvant N-acetyl-muramyl-L-alanyl-D-isoglutamine on the immune response in mice depending on experimental conditions. Eur. J. Immunol. **10**, 654 (1980).

AZUMA, I., K. KAMISANGO, I. SAIKI, Y. TANIO, S. KOBAYASHI, and Y. YAMAMURA: Adjuvant activity of N-acetyl-muramyl dipeptides for the induction of delayed-type hypersensitivity to azobenzenearsonate-N-acetyl-L-tyrosine in guinea pigs. Infect. Immun. **29**, 1193 (1980).

Biological activity of diastereoisomers:

ADAM, A., M. DEVYS, V. SOUVANNAVONG, P. LEFRANCIER, J. CHOAY, and E. LEDERER: Correlation of structure and adjuvant activity of N-acetyl-muramyl-L-alanyl-D-iso-glutamine (MDP), its derivatives and analogues. Anti-adjuvant and competition properties of stereoisomers. Biochem. Biophys. Res. Comm. **72,** 339 (1976).

CHEDID, L., F. AUDIBERT, P. LEFRANCIER, J. CHOAY, and E. LEDERER: Modulation of the immune response by a synthetic adjuvant and analogs. Proc. Natl. Acad. Sci. **73,** 2472 (1976).

KOTANI, S., Y. WATANABE, F. KINOSHITA, I. MORISAKI, K. KATO, T. SHIBA, S. KUSUMOTO, Y. TARUMI, and K. IKENAKA: The effect of replacement of L-alanine residue by glycine, L-serine or D-alanine in an N-acetyl muramyl-L- alanyl-D-isoglutamine on immunoadjuvanticities of molecules. Biken J. **20,** 39 (1977).

CHEDID, L., M. PARANT, F. PARANT, F. AUDIBERT, P. LEFRANCIER, J. CHOAY, and M. SELA: Enhancement of certain biological activities of muramyl dipeptide derivatives after conjugation to a multi-poly-(DL-alanyl)-poly-(L-lysine) carrier. Proc. Natl. Acad. Sci. USA **76,** 6557 (1979).

AUDIBERT, F., M. PARANT, C. DAMAIS, P. LEFRANCIER, M. DERRIEN, J. CHOAY, and L. CHEDID: Dissociation of immunostimulant activities of muramyl dipeptide (MDP) by linking amino-acids or peptides to the glutaminyl residue. Biochem. Biophys. Res. Comm. **96,** 915 (1980).

Stimulation of non-specific resistance:

CHEDID, L., M. PARANT, F. PARANT, P. LEFRANCIER, J. CHOAY, and E. LEDERER: Enhancement of non-specific immunity to *Klebsiella pneumoniae* infection by a synthetic immuno-adjuvant (N-acetyl-muramyl-L-alanyl-D-isoglutamine) and several analogs. Proc. Natl. Acad. Sci. **74,** 2089 (1977).

DAMAIS, C., M. PARANT, and L. CHEDID: Non-specific activation of murine spleen cells *in vitro* by a synthetic immunoadjuvant (N-acetyl-muramyl-L-alanyl-D-isoglutamine). Cell Immunol. **34,** 49 (1977).

YAMAMURA, Y., I. AZUMA, K. SUGIMURA, M. YAMAWAKI, M. UEMIYA, S. KUSUMOTO, S. OKADA, and T. SHIBA: Immunological and antitumor activities of synthetic 6-O-mycoloyl-N-acetylmuramyl dipeptides. Proc. Japan Acad. **53,** 63 (1977).

PARANT, M., F. PARANT, and L. CHEDID: Enhancement of the neonate's non-specific immunity to *Klebsiella* infection by muramyl dipeptide, a synthetic immunoadjuvant. Proc. Natl. Acad. Sci. **75,** 3395 (1978).

PARANT, M., C. DAMAIS, F. AUDIBERT, F. PARANT, L. CHEDID, E. SACHE, P. LEFRANCIER, J. CHOAY, and E. LEDERER: *In vivo* and *in vitro* stimulation of non-specific immunity by the β-D-p-aminophenyl glycoside of N-acetyl-muramyl-L-alanyl-D-isoglutamine (MDP) and an oligomer prepared by cross linking with glutaraldehyde. J. Infect. Dis. **138,** 378 (1978).

KIERSZENBAUM, F., and R. W. FERRARESI: Enhancement of host resistance against *Trypanosoma cruzi* infection by the immunoregulatory agent muramyl dipeptide. Infect. Immun. **25,** 273 (1979).

TANIYAMA, T., and H. T. HOLDEN: Direct augmentation of cytolytic activity to tumor-derived macrophages and macrophage cell lines by muramyl dipeptide. Cellular Immunology **48,** 369 (1979).

MCLAUGHLIN, C. A., S. M. SCHWARTZMAN, B. L. HORNER, G. H. JONES, J. G. MOFFATT, J. J. NESTOR JR., and D. TEGG: Regression of tumors in guinea pigs after treatment with synthetic muramyl dipeptides and trehalose dimycolate. Science **208,** 415 (1980).

MATTHEWS, T. R., and E. B. FRASER-SMITH: Protective effect of muramyl dipeptide and analogs against *Pseudomonas aeruginosa* and *Candida albicans* infections of mice. Curr. Chemother. Infect. Diseases, Proc. Int. Cong. Chemother. 11th 1979, **2,** 1734 (1980).

Induction of auto-immune diseases:

Nagai, Y.: Modulation of the immunologic response of myelin basic protein by extrinsic and intrinsic factors. Neurology **26**, 45 (1976).

Toullet, F., F. Audibert, G. A. Voisin, and L. Chedid: Production d'une orchiépididymite aspermatogénétique auto-immune chez le cobaye, à l'aide de différents adjuvants hydrosolubles. Ann. Immunol. (Inst. Pasteur) **128 C**, 267 (1977).

Nagai, Y., K. Akiyama, K. Suzuki, S. Kotani, Y. Watanabe, T. Shimono, T. Shiba, K. Kusumoto, F. Ikuta, and S. Takeda: Minimum structural requirements for encephalitogen and for adjuvant in the induction of experimental allergic encephalomyelitis. Cell. Immunol. **35**, 158 (1978).

Nagai, Y., K. Akiyama, S. Kotani, Y. Watanabe, T. Shimono, T. Shiba, and S. Kusumoto: Structural specificity of synthetic peptide adjuvant for induction of experimental allergic encephalomyelitis. Cell. Immunol. **35**, 168 (1978).

de Kozak, Y., F. Audibert, B. Thillaye, L. Chedid, and J. P. Faure: Effets d'adjuvants hydrosolubles d'origine mycobactérienne sur l'induction et la prévention de l'uvéo-rétinite autoimmune expérimentale chez le cobaye. Ann. Immunol. (Inst. Pasteur) **130 C**, 29 (1979).

Löwy, I., C. Leclerc, and L. Chedid: Induction of antibodies directed against self and altered-self determinants by a synthetic adjuvant, muramyl dipeptide and some of its derivatives. Immunology **39**, 441 (1980).

Nagao, S., and A. Tanaka: Muramyl dipeptide-induced adjuvant arthritis. Infect. Immun. **28**, 624 (1980).

Mechanism of action:

Juy, D., and L. Chedid: Comparison between macrophage activation and enhancement of non-specific resistance to tumors by mycobacterial immunoadjuvants. Proc. Natl. Acad. Sci. **72**, 4105 (1975)

Löwy, I., C. Bona, and L. Chedid: Target cells for the activity of a synthetic adjuvant: muramyl dipeptide. Cell. Immunol. **29**, 195 (1977).

Specter, S., H. Friedman, and L. Chedid: Dissociation between the adjuvant versus mitogenic activity of a synthetic muramyl dipeptide for murine splenocytes. Proc. Soc. Exp. Biol. Med. **155**, 349 (1977).

Takada, H., S. Kotani, S. Kusumoto, Y. Tarumi, K. Ikenaka, and T. Shiba: Mitogenic activity of adjuvant-active N-acetyl-muramyl-L-alanyl-D-isoglutamine and its analogues. Biken J. **20**, 81 (1977).

Tanaka, A., S. Nagao, R. Saito, S. Kotani, S. Kusumoto, and T. Shiba: Correlation of stereochemistry specific structure in muramyl dipeptide between macrophage activation and adjuvant activity. Biochem. Biophys. Res. Comm. **77**, 621 (1977).

Damais, C., M. Parant, L. Chedid, P. Lefrancier, and J. Choay: *In vitro* spleen cell responsiveness to various analogs of MDP(N-acétyl-muramyl-L-alanyl-D-isoglutamine), a synthetic immunoadjuvant, in MDP high-responder mice. Cell. Immunol. **35**, 173 (1978).

Sugimoto, M., R. N. Germain, L. Chedid, and B. Benacerraf: Enhancement of carrier-synthetic helper T-cell function by the synthetic adjuvant, N-acetyl-muramyl-L-alanyl-D-isoglutamine (MDP). J. Immunol. **120**, 980 (1978).

Yamamoto, Y., S. Nagao, A. Tanaka, T. Koga, and K. Onoue: Inhibition of macrophage migration by synthetic muramyl dipeptide. Biochem. Biophys. Res. Comm. **80**, 923 (1978).

Damais, C., M. Parant, and L. Chedid: *In vitro* responsiveness of immunocytes to a synthetic immunoadjuvant, muramyl dipeptide, and different synthetic analogs. Antibiotics Chemother. **24**, 19 (1978).

EMORI, K., and A. TANAKA: Granuloma formation by synthetic bacterial cell wall fragment: muramyl dipeptide. Infect. Immun. **19**, 613 (1978).

LECLERC, C., I. LÖWY, and L. CHEDID: Influence of MDP and of some analogous synthetic glycopeptides on the *in vitro* mouse spleen cell viability and immune response to sheep erythrocytes. Cell. Immunol. **38**, 286 (1978).

FEVRIER, M., J. L. BIRRIEN, C. LECLERC, L. CHEDID, and P. LIACOPOULOS: The macrophage, target cell of the synthetic adjuvant muramyl dipeptide. Eur. J. Immunol. **8**, 558 (1978).

WATSON, J., and C. WHITLOCK: Effect of a synthetic adjuvant on the induction of primary immune responses in T cell-depleted spleen cultures. J. Immunol. **121**, 383 (1978).

LECLERC, C., F. AUDIBERT, and L. CHEDID: Influence of a synthetic adjuvant (MDP) on qualitative and quantitative changes of serum globulins. Immunology **35**, 963 (1978).

ADAM, A., V. SOUVANNAVONG, and E. LEDERER: Non-specific MIF-like activity induced by the synthetic immunoadjuvant: N-acetylmuramyl-L-alanyl-D-isoglutamine (MDP). Biochem. Biophys. Res. Comm. **85**, 684 (1978).

HADDEN, J. W.: Effects of Isoprinosine, Levamisole, Muramyl dipeptide, and SM 1213 on lymphocyte and macrophage function *in vitro*. Cancer Treatment Reports, **62**, 1981 (1978).

PABST, M. J., and R. B. JOHNSTON: Increased production of superoxide anion by macrophages exposed *in vitro* to muramyl dipeptide or lipopolysaccharide. J. Exp. Med. **151**, 101 (1980).

OPPENHEIM, J. J., A. TOGAWA, L. CHEDID, and S. MIZEL: Components of Mycobacteria and muramyl dipeptide with adjuvant activity induce lymphocyte activating factor. Cell. Immunol. **50**, 71 (1980).

TANAKA, A., S. NAGAO, K. IMAI, and R. MORI: Macrophage activation by muramyl dipeptide as measured by macrophage spreading and attachment. Microbiol. Immunol. **24**, 547 (1980).

TENU, J. P., E. LEDERER, and J. F. PETIT: Stimulation of thymocyte mitogenic protein secretion and of cytostatic activity of mouse peritoneal macrophages by trehalose dimycolate and muramyl dipeptide. Eur. J. Immunol. **10**, 647 (1980).

Vaccines:

AUDIBERT, F., L. CHEDID, and C. HANNOUN: Augmentation de la réponse immunitaire au vaccin grippal par un glycopeptide synthétique adjuvant (N-acétyl-muramyl-L)-alanyl-D-isoglutamine). C. R. Acad. Sci. Paris **285**, série D, 467 (1977).

WEBSTER, R. G., W. P. GLEZEN, C. HANNOUN, and W. G. LAVER: Potentiation of the immune response to influenza virus subunit vaccines. J. Immunol. **119**, 2073 (1977).

SIDDIQUI, W. A., D. W. TAYLOR, S. C. KAN, K. KRAMER, S. M. RICHMOND-CRUM, S. KOTANI, T. SHIBA, and S. KUSUMOTO: Vaccination of experimental monkeys against *Plasmodium falciparum:* a possible safe adjuvant. Science **201**, 1237 (1978).

REESE, R. T., W. TRAGER, J. B. JENSEN, D. A. MILLER, and R. TANTRAVAHI: Immunization against malaria with antigen from *Plasmodium falciparum* cultivated *in vitro*. Proc. Natl. Acad. Sci. **75**, 5665 (1978).

MITCHELL, G. H., W. H. G. RICHARDS, A. VOLLER, F. M. DIETRICH, and P. DUKOR: Nor-MDP, saponin, corynebacteria, and pertussis organisms as immunological adjuvants in experimental malaria vaccination of macaques. Bulletin World Health Organization **57**, 189 (1979).

EDELMAN, R.: Vaccine adjuvants. Reviews Infect. Dis. **2**, 370 (1980).

(Received September 30, 1980)

The Chemistry of Longifolene and Its Derivatives

By SUKH DEV, Malti-Chem Research Centre, Nandesari, Vadodara, India

Contents

I. Introduction

With the introduction of modern spectroscopic methods of structure elucidation, the role of chemical reactions in the structure determination of natural products has become minimal. Increasingly, complex organic structures are being elucidated by X-ray crystallography, a technique which essentially by-passes the organic chemist! These advances constitute a water-shed in the development of chemistry of natural products and have enabled the chemists to direct efforts to aspects of natural product chemistry considered unassailable till recently. However, on the other hand, the classical approach involving chemical transformations has, in the past, generated a fund of interesting and unexpected results, often of fundamental importance. This is because many complex natural products have such built-in stereo-electronic features that their chemical transformations, not too infrequently, lead to totally unanticipated results *. It is suggested that chemical transformations of easily available, novel, complex organic molecules deserve to be investigated so that the excitement of the unexpected is not completely lost! The structure of longifolene, a sesquiterpene mono-olefin, was established in 1953, but still, twentyfive years later, continues to attract attention as an unusual substrate for transformations which generate much exciting chemistry. The present article has been written to high-light this aspect of longifolene chemistry, as an example of another facet of natural products chemistry.

The chemistry of longifolene has been reviewed in 1950 (1), 1964 (2) and 1966 (3).

II. Isolation, Occurrence

Longifolene was first isolated in 1920 by SIMONSEN (4) from Indian turpentine oil which is obtained from the oleo-resin of *Pinus roxburghii* Sarg. syn. *P. longifolia* Roxb. It is present in this turpentine to the extent of 5—10% and is the main sesquiterpene constituent co-occurring with minor amounts of other sesquiterpenes: longipinene, longicyclene, caryo-

---

* Viewed with hind-sight, such results offer little difficulty in rationalization!

phyllene, humulene and β-bisabolene (5). Pure longifolene, free from any traces of (−)-caryophyllene (6) has $\{\alpha\}_D$ + 54.06° (CHCl₃).

(+)-Longifolene has since been found to be widely distributed in the family Pinaceae (7). (−)-Longifolene has been isolated from the liverwort *Scapania undulata* (L.) Dum. (8) and from the fungus *Helminthosporium sativum* (142).

(+)-Longifolene is one of the very few sesquiterpenes being produced commercially in hundred ton quantities.

III. Structure

Longifolene was readily recognized as a tricyclic sesquiterpene and its chemistry was extensively investigated (9, 10) by SIMONSEN and his co-workers during the period 1923—1934. As a result of their findings, a summary of which is available (1), these authors proposed a tentative structure (1) which, however, lacked any rigorous proof. For the next, almost two decades, no fresh investigations of this structural problem were reported. However, in 1953, MOFFETT and ROGERS (11) disclosed the complete structure determination of longifolene hydrochloride (2) by X-ray crystallographic analysis. Simultaneously, NAFFA and OURISSON (12) published the results of their detailed chemical investigations which enabled them to extend structure (2) to (3) as the formulation for the parent hydrocarbon, (2) arising from (3) by a Wagner-Meerwein rearrangement during addition of hydrogen chloride. Structure (3) appeared to conflict with some of the previously reported (9, 10) degradations of longifolene, but the major discrepancy was soon resolved when it was found (12) that longifolic and isolongifolic acids *(vide infra)* are C₁₅ acids and not C₁₄ compounds as had been concluded by the earlier investigators.

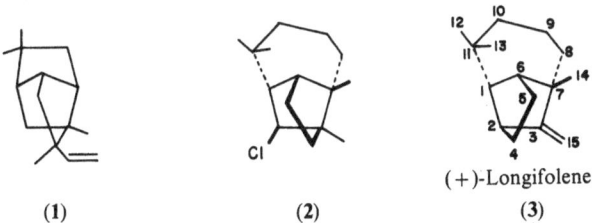

(1)   (2)   (+)-Longifolene
                                              (3)

Longifolene thus proved to be a simple elaboration of camphene (4) and structurally close to β-santalene (5). Molecular rotation considerations (13, 14) further showed that (+)-longifolene is configurationally related

4*

to (+)-camphene (4), and hence (3) represents the absolute stereostructure of (+)-longifolene. This conclusion was further reinforced by optical rotatory dispersion studies carried out subsequently (*15*).

(+)-Camphene   β-Santalene

(4)   (5)

Infrared (*16*), Raman (*16*), n. m. r. (*17*), and mass (*18*) spectra of longifolene have been recorded and discussed. The crystal structure of longifolene hydrochloride (2) has been reinvestigated by 3-dimensional methods and new values for bond lengths and bond angles deduced (*19*).

IV. Synthesis

Though in the context of present-day spectroscopic methods of structure elucidation synthesis can no longer be considered obligatory for the confirmation of a structure, more and more organic chemists have engaged themselves in this field to demonstrate their skill and ingenuity by synthesizing challenging natural product structures. Longifolene (3) is certainly one such molecule and, so far, four syntheses, each ingenious in its own right, have been reported. The first synthesis was achieved in 1961 by COREY and his co-workers (*20, 21*) and the subsequent ones in 1972 (*22*), 1975 (*23*), and 1978 (*24*). The challenge inherent in the longifolene molecule is its intricate tricyclic carbon-framework and each group of workers has tackled this in a unique way.

The first synthesis, that of COREY, OHNO, VATAKENCHERRY, and MITRA (*21*), is schematically shown in Chart 1. The key-feature in the synthesis is the internal Michael addition (7 → 8), which had precedence in the reported (*25*) conversion of santonin (10) into santonic acid (11) under the influence of hot aqueous potassium hydroxide. However, in the present case Michael cyclization proved quite difficult and was finally achieved in a yield of 10—20%, by prolonged exposure of (7) to triethylamine in ethylene glycol at 225°. After methylation of the resulting diketone (8), the less hindered carbonyl function was removed as shown to yield (±)-longicamphenilone (9) which could be readily transformed into (±)-

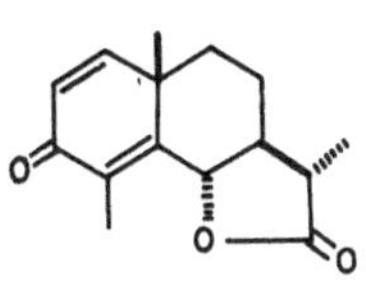

Santonin

(10)

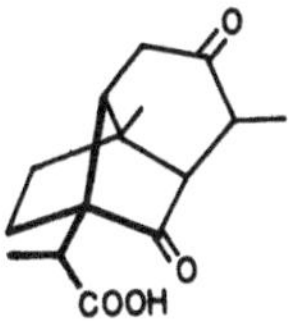

Santonic acid

(11)

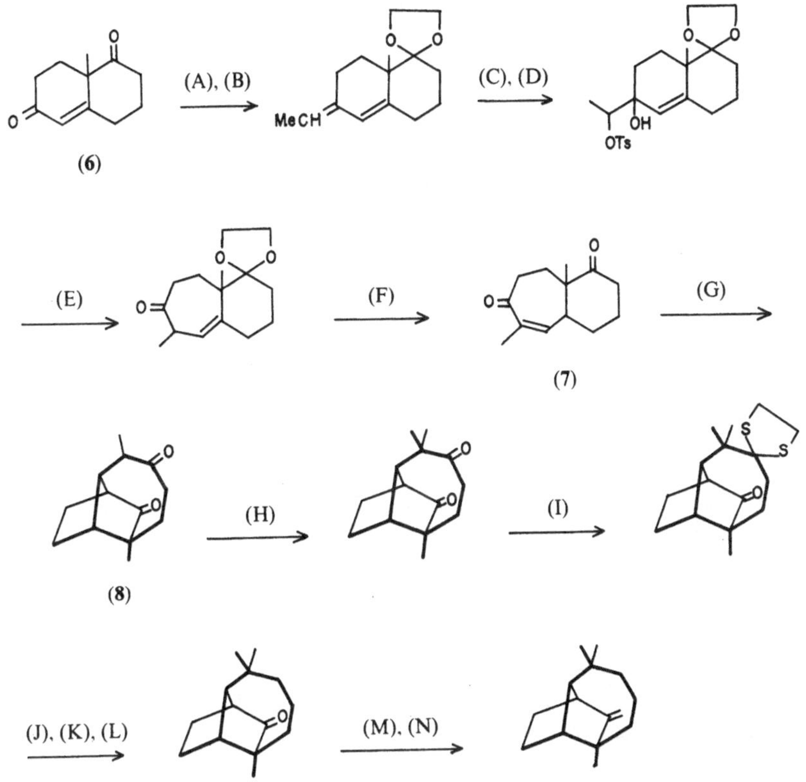

(6) →(A), (B)→ MeCH →(C), (D)→

(E) → →(F)→ (7) →(G)→

(8) →(H)→ →(I)→

(J), (K), (L) → (9) →(M), (N)→ Longifolene (3)

(A) ... (CH₂OH)₂, p–TsOH   (B) :.. Ph₃P=CHMe   (C) ... OsO₄   (D) ... TsCl, py
(E) ... LiClO₄, CaCO₃   (F) ... HCl aq./100°   (G) ... Et₃N, (CH₂OH)₂/225   (H) ...
Ph₃CNa, MeI (I) ... (CH₂SH)₂, BF₃ (J) ... LiAlH₄ (K) ... Wolff-Kishner (L) ... CrO₃
(M) ... MeLi (N) ... SOCl₂, py

Chart 1. Synthesis of longifolene (21)

longifolene (**3**). In an extension of this effort, the racemic diketone (**8**) was resolved using L(+)-2,3-butane-dithiol, which served an additional role in carbonyl protection and reduction, and the product transformed into (+)-longifolene.

The next synthesis, due to MCMURRY and ISSER (*22*), is outlined in Chart 2. The key-intermediate is the keto epoxide (**12**) which was elaborated from the same Wieland-Miescher ketone (**6**) used by COREY, by a fivestep sequence of reactions. The keto epoxide readily underwent intramolecular alkylation to the tricyclic ketol (**13**). Dibromocarbene addition to the olefin derived from (**13**) followed by silver ion assisted solvolytic

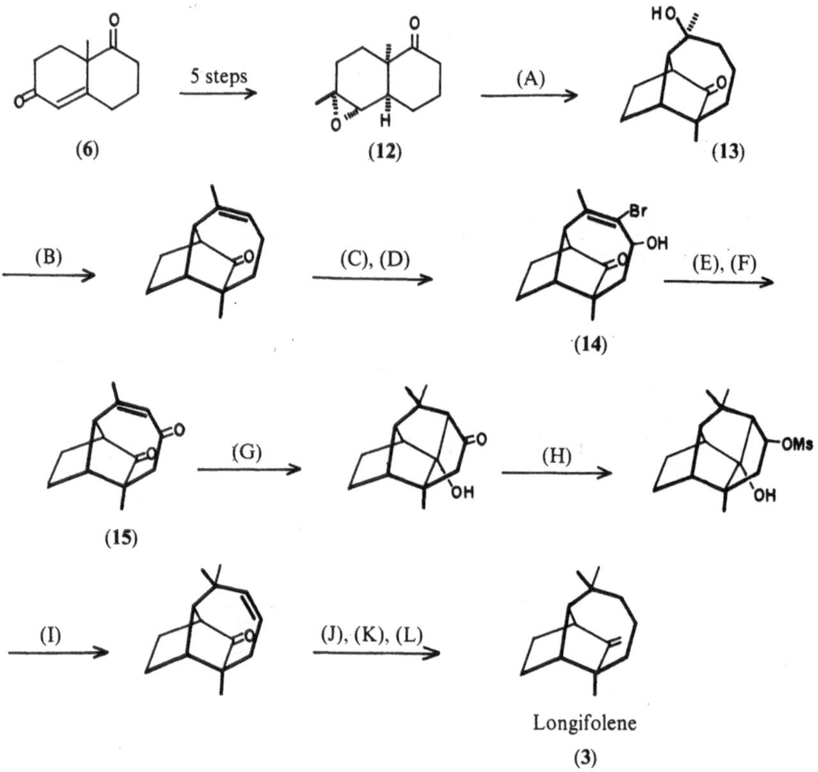

(A) ... MeSOC̄H₂Na⁺, DMSO (B) ... H₂SO₄-aq. (C) ... CHBr₃, KOBuᵗ (D) ... AgClO₄ (E) ... Na, NH₃ (F) ... CrO₃, py. (G) ... Me₂CuLi (H) ... NaBH₄, MeSO₂Cl/NEt₃ (I) ... KOBuᵗ (J) ... (Ph₃P)₃RhCl/H₂ (K) ... MeLi (L) ... SOCl₂, py.

Chart 2. Synthesis of longifolene (*22*)

ring-enlargement, a reaction for which there was ample previous analogy, led to (14). Reductive elimination of bromine followed by oxidation yielded the ene-dione (15), which was transformed into longifolene by the sequence of reactions shown in Chart 2.

The third longifolene synthesis, disclosed by VOLKMANN, ANDREWS, and JOHNSON (23) and depicted in Chart 3, had its genesis in a chance observation. In continuation of their synthetic investigations on steroids

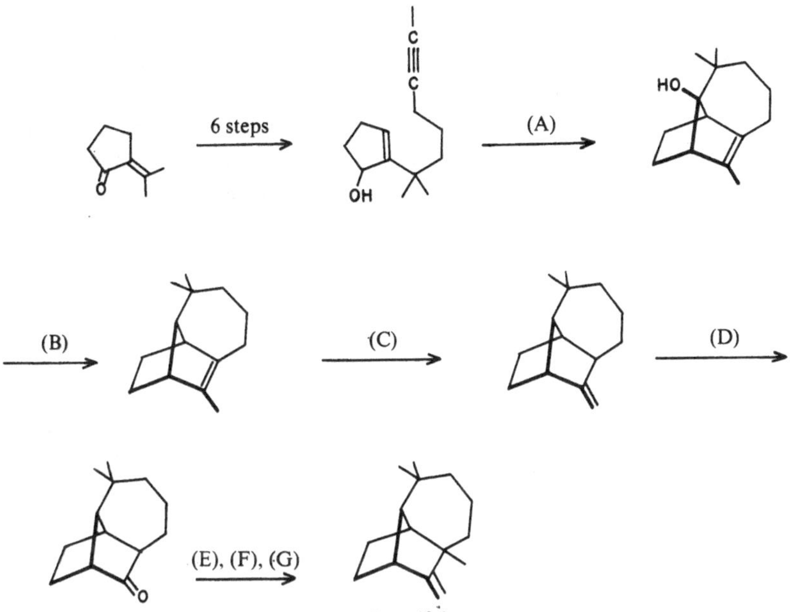

(A) ... CF₃COOH   (B) ... ZnBr₂, NaBH₃CN   (C) ... TsOH   (D) ... RuO₄, HIO₄
(E) ... LiN(iPr)₂, MeI   (F) ... MeLi   (G) ... SOCl₂, py.

Chart 3. Synthesis of longifolene (23)

and triterpenoids, involving participation of triple bonds in directed polyolefin cyclizations, JOHNSON's group examined the stannic chloride-catalyzed cyclization of heptynylmethylcyclopentenol (17), in an effort to construct a hydroazulene system. The anticipated compound (18) was indeed formed, but a side-product recognized as (19) was also isolated. The potential of this unanticipated side reaction for the construction of

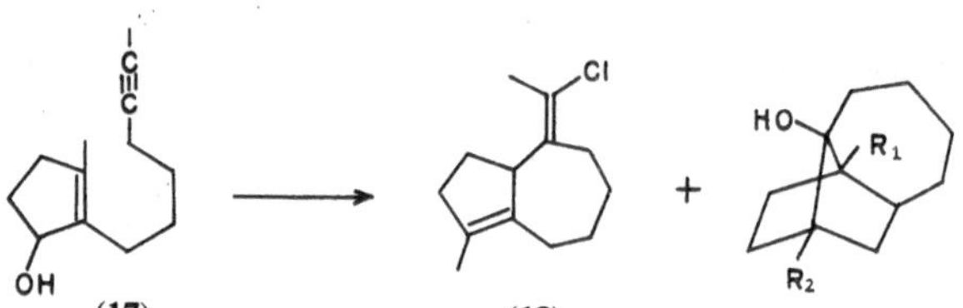

(17)   (18)

R_1 = H or Me
R_2 = Me or H

(19)

the longifolene molecule was at once recognized and the reaction refined and exploited for a novel synthesis of longifolene on the lines summarized in Chart 3.

The latest synthesis (Chart 4) utilizes an intramolecular photoaddition-retroaldol reaction (**20 → 22**) as the key-step in the construction of the tricarbocyclic frame-work. The resulting diketone (**22**) could be easily transformed into the known ketone (**16**; Chart 3) which had been earlier (*23*) converted into longifolene.

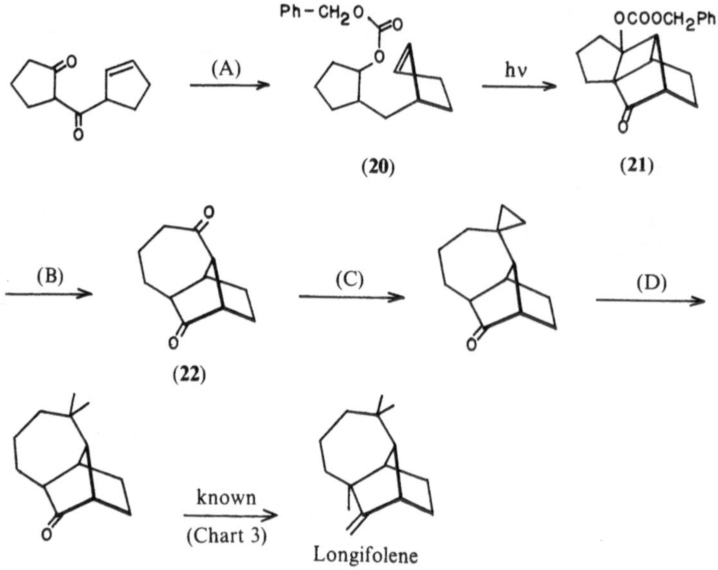

(20)   (21)

(22)

(16)   known (Chart 3)   Longifolene

(3)

(A) ... PhCH₂OCCl, py.
(B) ... Pd – C, H₂

(C) ... CH₂I₂, Zn – Ag
(D) ... PtO₂, H₂

Chart 4. Synthesis of longifolene (*24*)

V. Isolongifolene

Quite at an early stage in the development of chemistry of longifolene its propensity to rearrangement under acid-catalysis to an isomeric tricyclic hydrocarbon, later named isolongifolene, was noted. This rearrangement and the chemistry of isolongifolene itself have attracted considerable attention and hence it appears appropriate to discuss the structure and genesis of isolongifolene at this stage, before proceeding to describe other reactions of longifolene and, where appropriate, those of isolongifolene.

The formation of an isomeric hydrocarbon of varying optical rotation (α_D $-24°$ to $-80°$) was observed during the reaction of longifolene with hydrogen chloride (26) and, during its "hydration" with Bertram-Walbaum reagent (26, 27) or on exposure to sulphuric acid-acetic acid (28). Subsequently, it was shown (29) that all these preparations are nothing but isolongifolene racemized to varying degrees. The α_D of optically pure isolongifolene has been computed to be $\sim -125°$ (30). Isolongifolene can now be prepared conveniently by the action of $BF_3 \cdot Et_2O$ (31) or Amberlyst-15 (32) or acid-treated silica gel (32) on longifolene.

1. Structure

The structure (23) of isolongifolene has been elucidated (29, 30) by a combination of spectroscopic methods and incisive chemical degradations. The key-compounds involved in the structure determination were (24), (25), (26) and (27). The i.r. carbonyl absorption of (24) (two epimers:

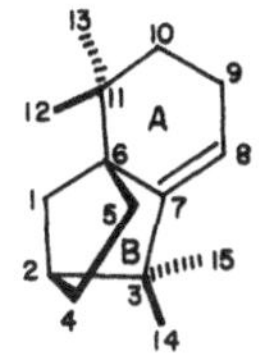

Isolongifolene

(23)   (24) 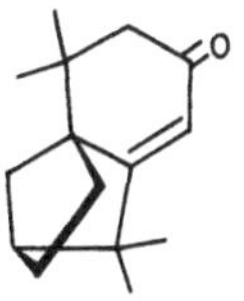   (25)

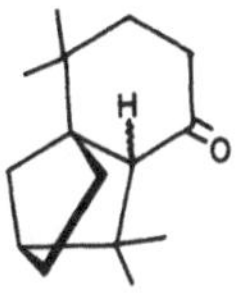

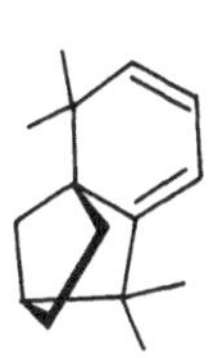

(26) 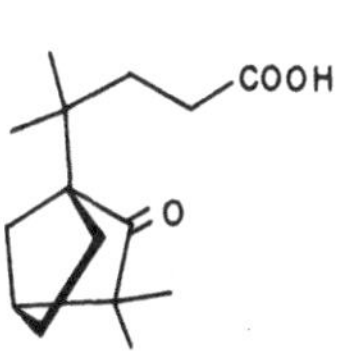   (27) 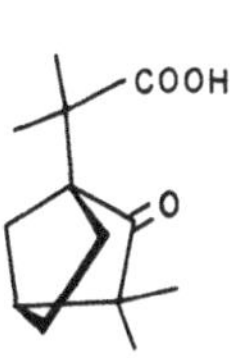   (28)

$v_{C=O}$ 1715, 1698 cm^{-1}), as well as the u. v. absorption of the diene (26) (λ_{max}^{EtOH} 266.5 nm, = 7470) helped establish the size of ring A. From the C=O stretching frequency of the keto acid (27) ($v_{C=O}$ 1735, 1710 cm^{-1}) and its methyl ester ($v_{C=O}$ 1732 cm^{-1}), the ring size of cycle B could be established as 5-membered. By systematic degradation the keto acid (27) was converted into the bis-nor keto acid (28) (33) which was later synthesized (34). All these transformations coupled with the n. m. r., i. r. and u. v. characteristics of isolongifolene as well as those of derived compounds helped define unambiguously the structure of isolongifolene as (23).

2. Synthesis

Starting with camphene-1-carboxylic acid (29), isolongifolene has been synthesized (34) along the lines depicted in Chart 5. The crucial step

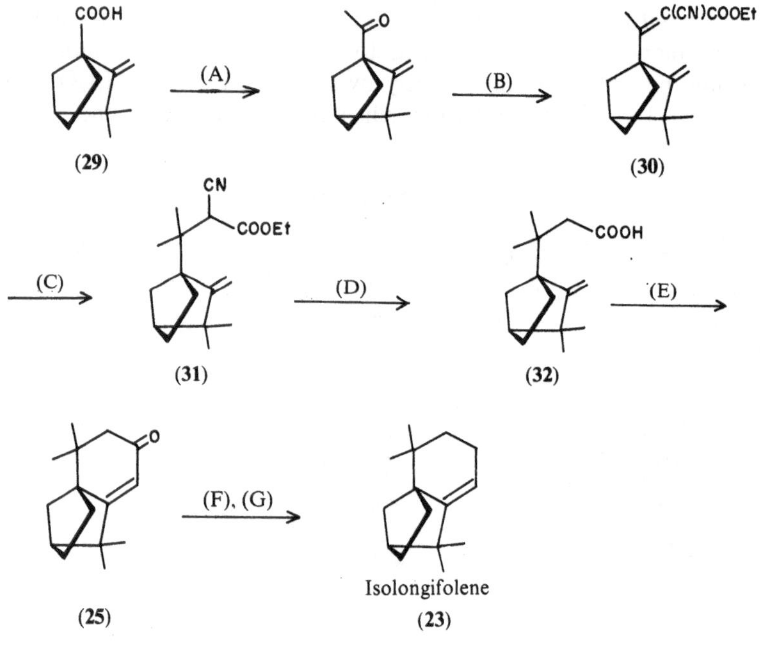

(A) ... MeLi   (B) ... CNCH$_2$COOEt, NH$_4$OAc   (C) ... LiCuMe$_2$ (D) ... KOH, (CH$_2$OH)$_2$ (E) ... (COCl)$_2$, SnCl$_4$ (F) ... (CH$_2$SH)$_2$, BF$_3$ · Et$_2$O   (G) ... Raney Ni

Chart 5. Synthesis of isolongifolene (34)

in the envisaged route was the addition of a methyl group to (30) and this was achieved by utilizing the then recently discovered reagent, lithium dimethyl copper (35). Conjugate addition of the reagent to (30) proceeded in 80% yield to furnish the required (**31**; diastereoisomeric mixture). The intramolecular Friedel-Crafts acylation of (**32**) proceeded smoothly in over 85% yield to furnish the known unsaturated ketone (**25**) which could be readily transformed into isolongifolene.

3. Mechanism of Rearrangement

The deep-seated rearrangement of longifolene to isolongifolene with concomitant partial racemization of the product, was rationalized (*2, 30*) in terms of a series of 1,2-shifts, as outlined in Chart 6. Racemization of

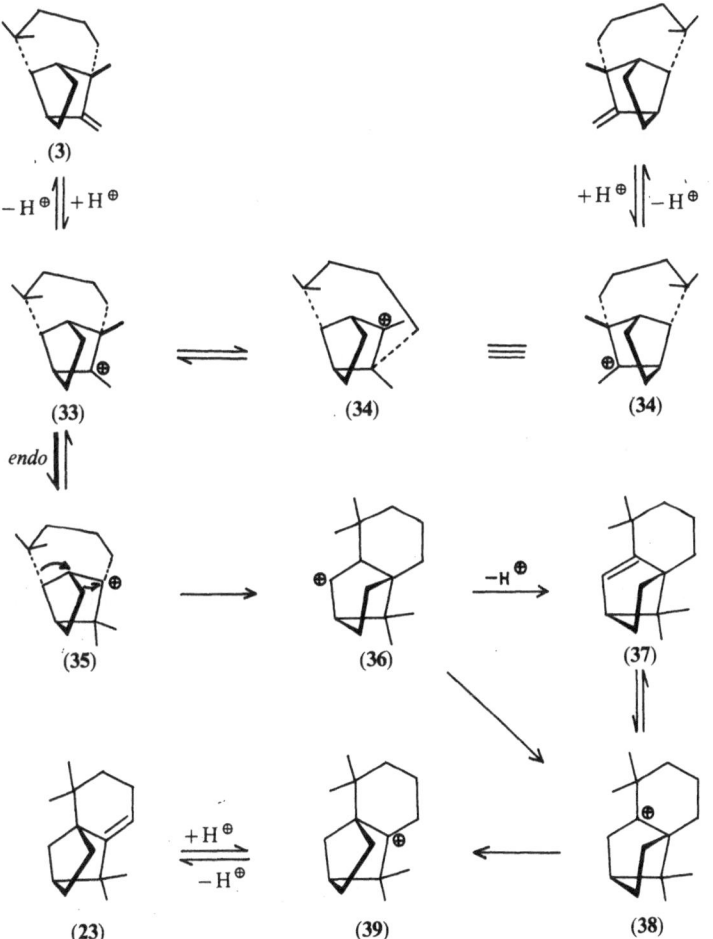

Chart 6. Rearrangement of longifolene to isolongifolene (*2, 30*)

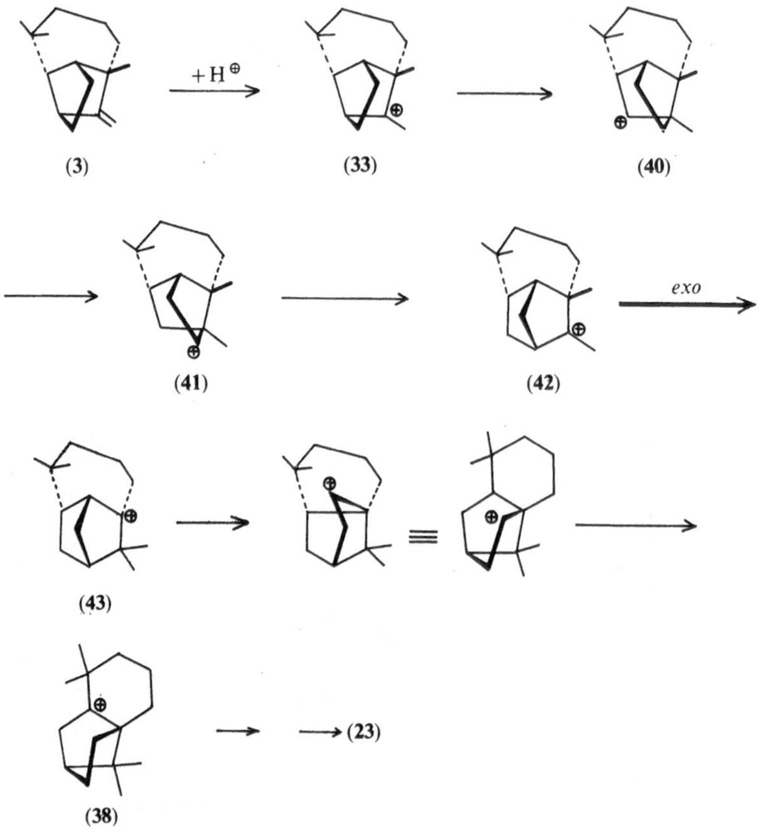

Chart 7. Rearrangement of longifolene to isolongifolene (37)

(−)-isolongifolene is explained by initial racemization of longifolene (*via* **33 ⇌ 34**), which must intercept the isomerization, as it is difficult to visualize a suitable mode for the racemization of isolongifolene, once it has been formed. The remaining steps were considered unexceptional; the formation of the bridgehead carbocation (**35**) has earlier analogy (*36*). However, subsequently, BERSON *et al.* (*37*), based on their extensive studies on methylnorbornyl cations, pointed out that the proposed *endo, endo* methyl shift (**33 → 35**; Chart 6) should be energetically unfavourable and proposed a modified pathway (Chart 7), wherein the much more precedented *exo, exo* (**42 → 43**) shift occurs. Still later, MCMURRY (*38*) suggested the intermediacy of longicyclene (**44**) (*39*), in an effort to simplify (**33 → 42**; Chart 8) the pathway, earlier suggested by BERSON and co-workers.

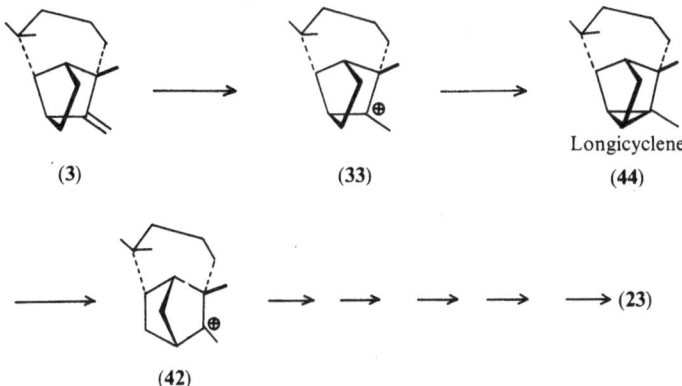

Chart 8. Rearrangement of longifolene to isolongifolene (*38*)

In a recent investigation (*40, 41*), SUKH DEV and co-workers have clarified the mechanism of rearrangement. Isomerization of longifolene-4,4,5,5-d_4 (**45**) would, according to the preferred pathway (Chart 6/Chart 7), lead to either (**46**) or (**47**) (Chart 9). In actual practice, $BF_3 \cdot Et_2O$-catalyzed rearrangement of (**45**) gave isolongifolene-d_4 in which deuterium label was shown to be distributed as in (**47**). These results fully support the *exo, exo* migration pathway (Chart 7). The problem of the intermediacy of longicyclene (Chart 8) was sorted out in favour of the non-involvement of longicyclene by investigating the isomerization of longifolene with $BF_3 \cdot Et_2O$-AcOD. Had longicyclene (**44**) been an obligatory intermediate, one would have expected incorporation of deuterium at C-1/C-2

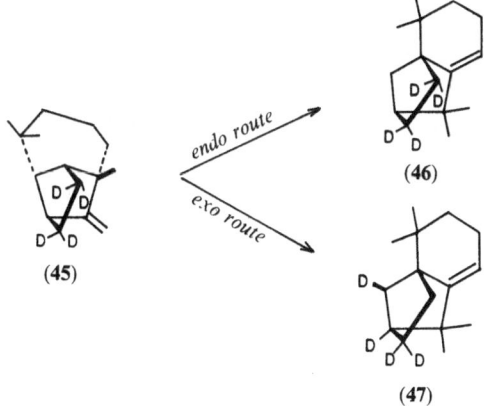

Chart 9. Possible products from rearrangement of longifolene-4,4,5,5-d_4

(Chart 10). Experimentally, it was found that such incorporation of label does not occur. Furthermore, it was shown that cleavage of the cyclopropane ring in longicyclene (**44**) to give (**42**), as envisaged by McMurry (Chart 8), is unimportant in the acid-catalyzed rearrangement of longicyclene to isolongifolene.

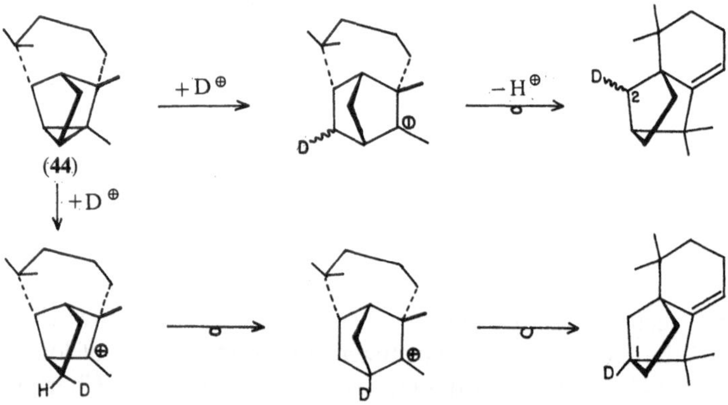

Chart 10. Possible cleavage and rearrangement pathways of longicyclene

VI. Reactions of Longifolene

Many reactions of longifolene have been investigated since elucidation of its structure, while others which were previously studied could be clarified only after this event. Much interesting chemistry has emerged from these studies by virtue of certain special features of the molecule. With the kind of information now available, it is possible to classify the reactions of longifolene essentially into four categories:

1. "Normal" addition, substitution reactions.
2. Reactions involving skeletal rearrangements.
3. Reactions leading to products from steric diversion.
4. Transannular reactions.

Which of these pathways predominates is governed by the nature of the reagent and the reaction conditions. In general, in the case of electrophilic additions, it is conceded (*e.g. 42*) that many reactions proceed through a cyclic activated complex, for example (**48**), which may be strongly bridged (**49**), weakly bridged (**50**) or may lead to a fully developed carbonium ion at the more substituted carbon atom (**51**). The importance of one or other of these ions may be expected to determine, in the present context, which of the four pathways delineated above dominate product development. That in many reactions of longifolene products arising from more than one of these pathways have been encountered can be readily understood from the above considerations.

$$-\overset{|}{\underset{\underset{X}{\overset{\delta+}{\cdots}}}{C}}\overset{\delta+}{\cdots}\overset{|}{\underset{}{C}}-$$

(48)

$$-\overset{|}{\underset{\underset{X}{\overset{+}{\diagdown}}}{C}}\overset{|}{\underset{}{C}}-$$

(49)

$$-\overset{|}{\underset{\underset{X}{\overset{\delta+}{\diagdown}}}{C}}\overset{|}{\underset{}{C}}-$$

(50)

$$-\overset{|}{\underset{+}{C}}\overset{|}{\underset{X}{\diagup}}\overset{|}{\underset{}{C}}-$$

(51)

It appears worthwhile to discuss the reactions of longifolene in terms of the above four categories, even though some reactions of longifolene are known to lead to products of more than one type. In such cases, the reaction has been discussed under only one head or both the heads, depending on whether one pathway is dominant or not. Besides these four categories of reactions, some other pertinent transformations of longifolene and its derivatives are discussed under two additional heads: (i) Conversions into other sesquiterpene skeletons, and (ii) Miscellaneous transformations.

1. "Normal" Reactions

There are some reactions of longifolene which proceed along pathways generally established for olefins. A comparison of some of these reactions with those of comphene (4) serves to emphasize the fact that the olefinic linkage in longifolene is much less accessible than that in camphene. In camphene (52) the side most accessible to a reagent is the *exo*-face. However, in longifolene (53), this very face is essentially fully screened by the bridge forming the third ring and *endo*-side (in terms of the camphene part-structure of longifolene) is now *relatively* less hindered. Thus, in comparison to camphene, the same reactions often proceed sluggishly

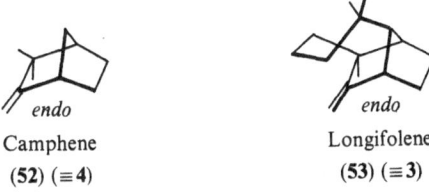

Camphene
(52) (≡4)

Longifolene
(53) (≡3)

or in poor yield with longifolene and attack from the *endo*-face is reflected in the stereochemistry at the new chiral centre. There are other consequences of the "larger bridge" (3) in the longifolene framework, which will be discussed in subsequent subsections.

a) Addition Reactions

Longifolene could not be hydrogenated over a Ni or Pd catalyst in neutral medium (*43*). However, it can be reduced with Pd catalyst in acetic acid (*43*) or with Adam's PtO_2 catalyst in either ethyl acetate or acetic acid (*44*). The product of hydrogenation consists of an approximately 1 : 1 mixture of two longifolanes (**54**) and (**55**) (*44*). In contrast, camphene is readily hydrogenated to isocamphane (**56**); the other epimer arising from *endo* addition is formed in trace quantities only (*45*).

		Isocamphane
(54)	**(55)**	**(56)**

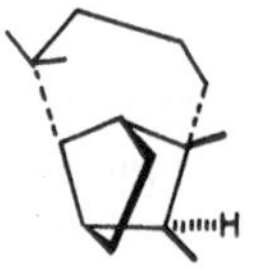

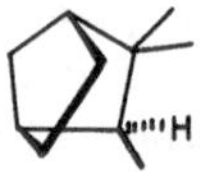

Longifolene on hydroboration followed by oxidation with alkaline hydrogen peroxide furnishes, in 74% yield, essentially pure longifolol (**57**) as a result of *endo* attack (*46*). On the other hand, camphene, under the same treatment, is known (*47*) to give almost exclusively *endo*-isocamphanol (**58**) by way of *exo* approach of the reagent. Oxidation of longifolyl-borane with air or silver oxide results in transannular products which are discussed in a later section.

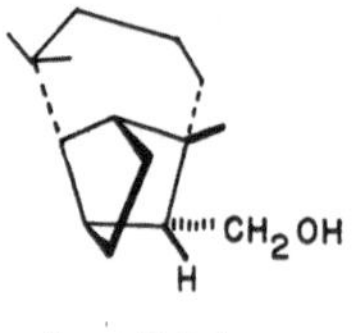

Longifolol	*endo*-Isocamphanol
(57)	**(58)**

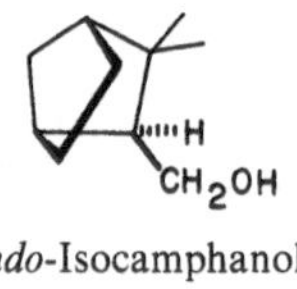

b) Substitution Reactions

Acetoxymethylation (Prins reaction) of longifolene proceeds smoothly to yield ω-acetoxymethyl longifolene which was shown to possess the *E*-configuration (**59**) (*48*). The product was converted into a number of related compounds (**60—64**).

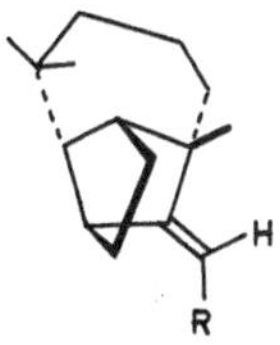

(59): R = CH₂OAc
(60): R = CH₂OH
(61): R = CH₂OMe
(62): R = CHO
(63): R = COOH
(64): R = COOMe
(65): R = COMe

Acylation of longifolene with $BF_3 \cdot Et_2O$ and acetic anhydride gives 30—40% of ω-acetyllongifolene (65), the stereochemistry of which rests on its hypochlorite oxidation to the known (63) (31). Isolongifolene (23) is the major by-product of the reaction.

Exposure of longifolene to manganic acetate in refluxing acetic acid-acetic anhydride furnished in poor yield (9%) the product (66), the stereochemistry of which remains unresolved. Under the same conditions, camphene gave in 30% yield γ-lactone mixture (67), in which one isomer predominated (49). A carboxymethyl radical arising from the thermolysis of Mn (III) acetate is considered to be the reactive species in such reactions (50).

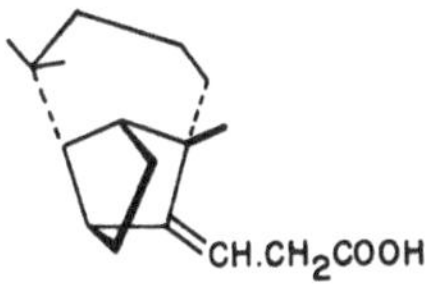

CH.CH₂COOH

(66)

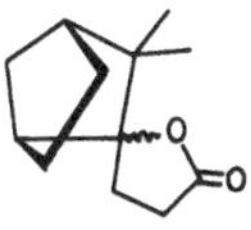

(67)

2. Skeletal Rearrangements

a) Simple Wagner-Meerwein Rearrangements

As a first approximation, the camphene part-structure in longifolene may be expected to mimic rearrangements typical of camphene.

Camphene (4) undergoes two types of 1,2-shifts. On exposure to hydrogen chloride camphene (4) gives an unstable tertiary chloride (68) which smoothly rearranges to isobornyl chloride (71). This rearrangement, which was investigated in detail by MEERWEIN and VAN EMSTER (51), was interpreted in terms of ions (69) and (70) (Chart 11); it may be worthwhile to recall that it was this reaction which led to the first suggestion that carbocations were intermediates in such molecular rearrangements. The reaction proceeds with a high degree of stereoselectivity, though isobornyl chloride can slowly epimerize to bornyl chloride (72) under certain conditions. This high stereoselectivity may be explained in terms of a σ-bridged ion (73) or carbonium ion-chloride ion pairs (69, 70). At one time, the so-called non-classical ion (73) was considered to account uniquely for the stereochemical outcome of the reaction. However, during the past fifteen years, the concept of σ-bridged ions itself has come under considerable scrutiny and debate (52, 53), and as yet no clear picture has emerged regarding secondary norbornyl derivatives, which are relevant to the present discussion (73 vs 74). The second type of 1,2-alkyl shift characteristic of camphene is the degenerate rearrangement (75) to (76) (the so-called Nametkin change), demonstrated (54) to be in part responsible for the racemization of optically active camphene, under acid-catalysis.

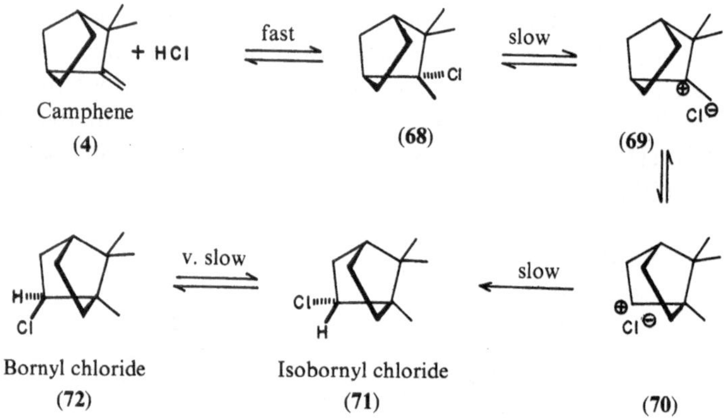

Chart 11. Transformation of camphene to isobornyl chloride

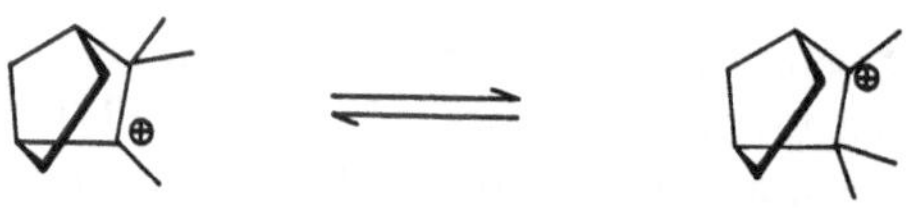

Longifolene on treatment with hydrogen halides undergoes the expected 1,2-shift of the bridge to furnish longibornyl halide (77), the structure of which (X = Cl), as mentioned earlier, was established by X-ray crystallographic analysis (11). Viewed in the present context, the stereo-chemistry of the new C-X bond, though in apparent contrast to the behaviour of camphene, is in accord with approach of the halide from the less hindered *endo*-face. At one time, however, this transformation generated interest as an example of a rearrangement where a non-classical carbonium ion was considered unimportant (55). Another contrasting feature is the failure to demonstrate, so far, the formation of the un-rearranged tertiary halide of longifolene (**78**) corresponding to (**68**) in the camphene series. Undoubtedly, this is due to steric compression at C-3 (*vide* discussion under steric diversion). In the section on isolongi-folene, it was demonstrated that racemization of longifolene (Chart 6) proceeding by way of (**33** ⇌ **34**) must intercept the isomerization process; this racemization pathway is the longifolene counterpart of camphene racemization process (**75** ⇌ **76**).

(77)   (78)

Solvolysis of longibornyl halides regenerates longifolene under suit-able conditions. Thus, exposure of longibornyl chloride (**77**, X = Cl) to 10% sodium hydroxide in refluxing ethylene glycol (**26**) or silver acetate in acetic acid (**26**) or 80% ethanol buffered with calcium carbonate (**55**) furnishes longifolene. The rate of solvolysis of the chloride in 80% ethanol has been studied and found to be 100 times faster than that of bornyl chloride, though still a thousand times smaller than for isobornyl chloride (55). The acceleration has been explained (55) by the relief of steric strain in going from (**79**) to (**80**), a factor absent in the case of bornyl chloride. Though the larger bridge in longifolene can possibly have several low energy conformations, detailed analysis (55) has shown that (**81**) clearly has the least non-bonded interactions, a conclusion borne out by the X-ray analysis of longibornyl chloride (11, 19).

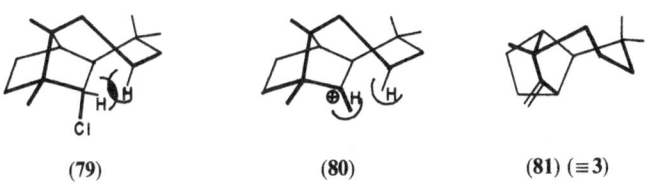

(79)   (80)   (81) (≡ 3)

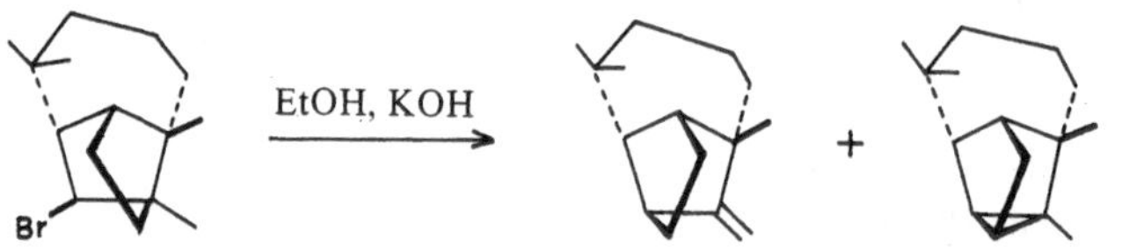

Longibornyl bromide (82) on treatment with ethanolic potassium hydroxide at reflux gives a 1 : 3 mixture of longicyclene (44) and longifolene (39). The formation of longicyclene reflects an alternative pathway for stabilization open to the ion (80) (and 33) and may be compared with the formation of tricyclene (84) during solvolysis of certain isobornyl esters (83) (56).

| Longibornyl bromide | Longifolene | Longicyclene |
| (82) | (3) | (44) |

| | | Tricyclene |
| (83) | | (84) |

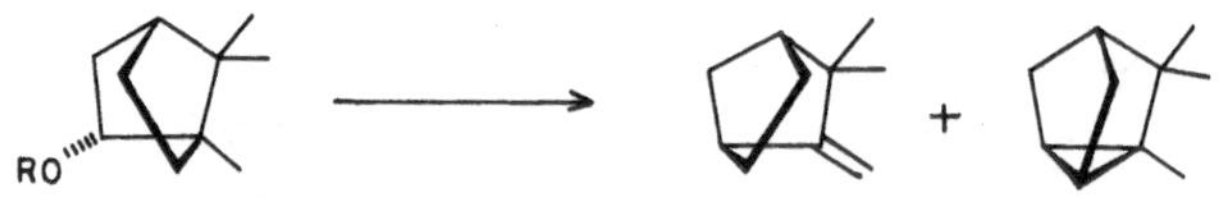

Photo-irradiation of longibornyl iodide (85) in heptane containing some triethylamine yields longifolene (40%), longicyclene (44%) and a transannular product, longiborn-8-ene (86; 14%). It has been suggested (57) that the reaction proceeds by way of homolytic photodissociation of the iodide, rapidly followed by a fast electron-transfer process in the solvent cage thus leading to a "hot" carbocation, which generates the observed products.

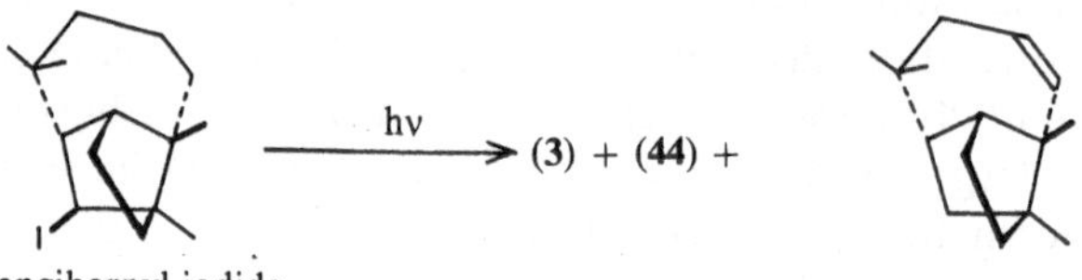

| Longibornyl iodide | |
| (85) | (86) |

Under the influence of cupric acetate in refluxing acetic acid, longifolene and longicyclene equilibrate to a 2 : 1 mixture (3 : 44), along with formation of some isolongifolene (23) (39).

On exposure to hot acetic acid containing sulphuric acid, camphene gives an excellent conversion to isobornyl acetate (83, R = Ac). Under similar conditions, longifolene yields isolongifolene (23) as the major (~ 60%) product and an approximately 1 : 3 mixture of acetates of longiborneol (**87**; R=H) and the transannular alcohol (**88**; R=H) discussed in some detail in Section VI 4 (*6, 58, 59*). The formation of longiborneol, rather than the epimer longiisoborneol, is in accord with the behaviour of longifolene towards halogen acids discussed earlier.

Longiborneol had been earlier obtained from longibornyl bromide *via* its Grignard derivative (*26*) and was also shown to be identical with the natural products juniperol (and macrocarpol, and kuromatsuol) (*60*). Oxidation of longiborneol gives a ketone (longi-camphor) in which the carbonyl function is highly unreactive as might have been anticipated from its sterically shielded location.

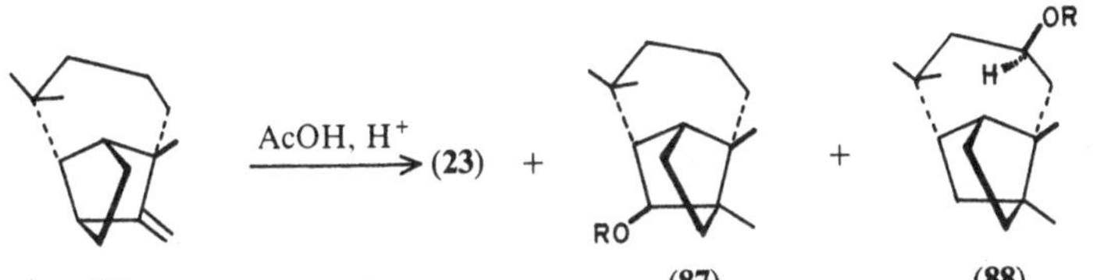

$$ (3) \xrightarrow{\text{AcOH, H}^+} (23) \ + \ (87) \ + \ (88) $$

b) Deep-Seated Rearrangements

The rearrangement of longifolene (**3**) to isolongifolene (**23**) through a series of alkyl/hydride shifts has been described earlier in some detail. Isolongifolene is still a high-energy molecule due to the presence of the strained bicyclo{2,2,1}heptane system and hence it is not surprising that the molecule reorganizes further under appropriate conditions. The same

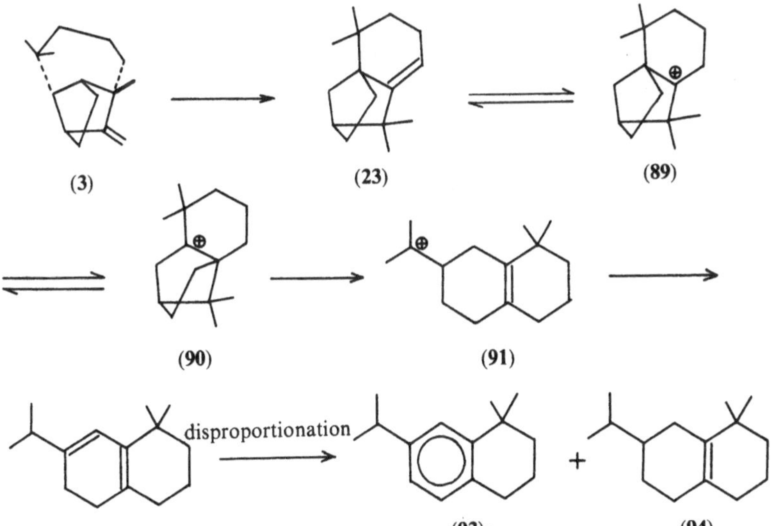

Chart 12. Further rearrangements of isolongifolene (*32, 61*)

product composition is reached from either longifolene or isolongifolene; however, the reaction with longifolene can be stopped at the isolongifolene stage by conducting the reaction under milder conditions.

Longifolene/isolongifolene on being refluxed with Amberlyst-15 or phosphoric acid-silica gel gives, besides a polymer (25—30%), the tetralin **(93)** and the octalin **(94)**. This transformation can be effected more conveniently with BF$_3$ · Et$_2$O: addition of this reagent (3%) to longifolene at room temp. (27°) results in an exothermic reaction leading to the above products. The possible pathway for this rearrangement is depicted in Chart 12 (*32, 61*). Tetralin **(93)** has also been prepared from isolongifolene by the action of zinc chloride (*62*) or trifluoroacetic acid (*63*).

As a corollary to the mechanism shown in Chart 12, one would expect facilitation of the rearrangement in structures in which the change **(90)** to **(89)** is inhibited. This was fully borne out by the rearrangement of 8-

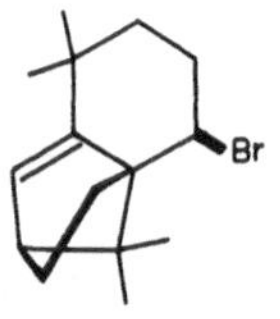

(95) 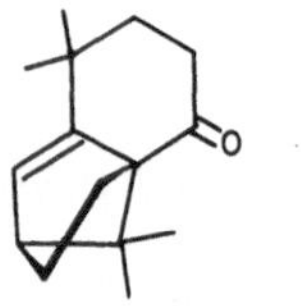   **(96)** 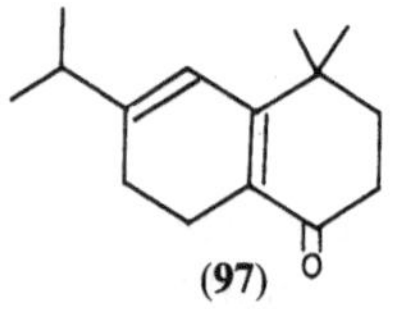   **(97)**

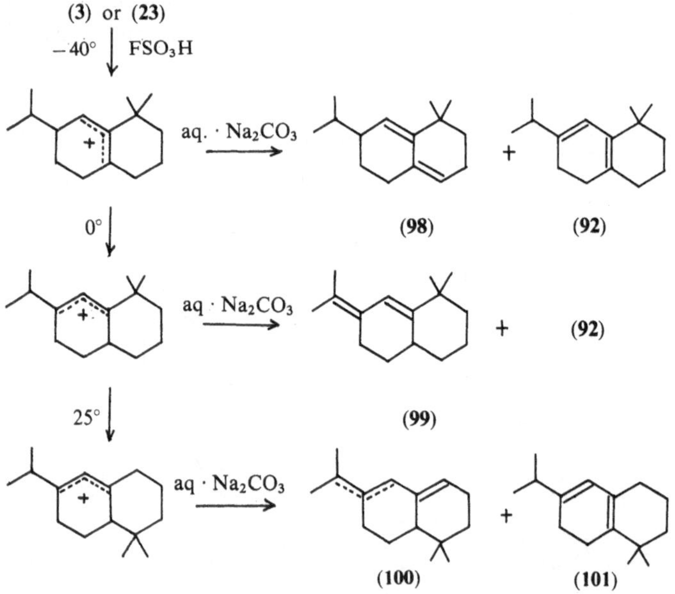

Chart 13. Carbocations and quenching products derived from longifolene in fluorosulfonic acid (*64*)

bromo-neoisolongifolene (**95**) and the ketone (**96**). Solution of 8-bromo-neoisolongifolene in 90% H_2SO_4 at 0° results in rapid generation of hydrogen bromide and eventually furnish tetralin (**93**) in good yield. Similarly, ketone (**96**) on exposure to 90% H_2SO_4 at $\sim 0°$ readily rearranged to the anticipated dienone (**97**) (*61*).

The generation of stable carbonium ions from longifolene in fluorosulfonic acid has been investigated (*64*). The structure of the carbonium ion and consequently that of the end product was shown to be temperature dependent (Chart 13). The activation energies for the rearrangement of cations involved have been estimated to be in the range 15—20 Kcal/mol.

In a recent study, SURYAWANSHI and NAYAK (*65*) investigated the rearrangement of methyl- (**102**) and dimethyl-longifolene (**103**) with $BF_3 \cdot Et_2O$ to find the anticipated transformations quite facile.

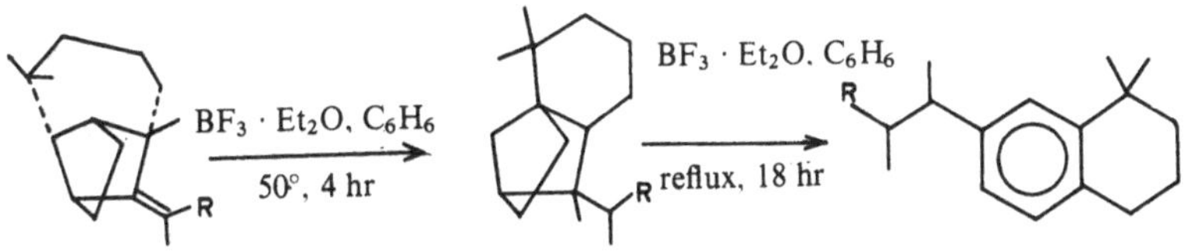

(**102**: R = H)
(**103**: R = Me)

In a less complex rearrangement, COATES and CHEN (*66*) found that acetolysis of longicamphenilyl tosylate (**104**) affords, in addition to norlongicyclene (**105**) and acetate products, the novel tricyclic olefins (**106**) and (**107**). The rearrangement pathway from (**104**) to (**106**) is pictured as a C-3 to C-4 hydride shift, followed by methylene migration and then proton elimination. In formic acid, (**106**) goes over to (**107**).

Another interesting rearrangement of longifolene skeleton is the BF_3-catalyzed transformation of (**108**) to (**109**) which is considered to proceed as shown in Chart 14 (*67*).

(**104**)   Norlongicyclene (**105**)   (**106**)   (**107**)

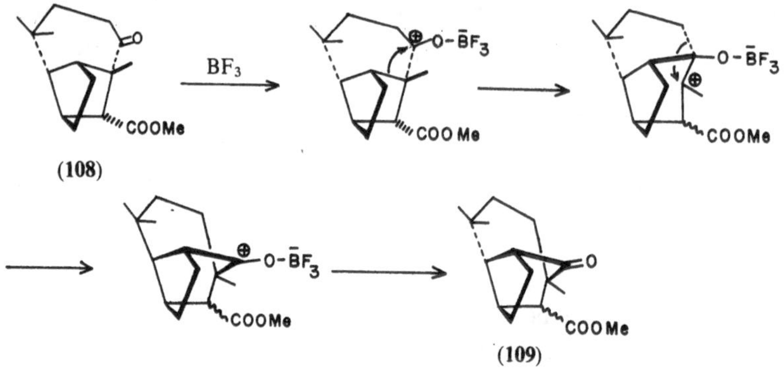

Chart 14. (67)

3. Steric Diversion

Conceivably, in an olefin addition reaction, if the more substituted end of the ethylenic linkage is sterically shielded such that the approach of the nucleophile or the radical is essentially blocked, the product cannot be expected to be the result of a simple addition reaction, but would always be complicated by the intervention of other pathways, such as elimination/rearrangement. To illustrate this point, the reaction of chlorine-free hypochlorous acid with *unsym*-dineopentylethylene (110) and 2,4,4-trimethyl-1-pentene (112) may be cited: (110) gives a complex reaction product in which (111) predominates (48%) and no oxygen-containing material was detected in the total product, while (112) which is less hindered furnishes 34% of the "normal" product (113) and 46% of elimination product (114) (68).

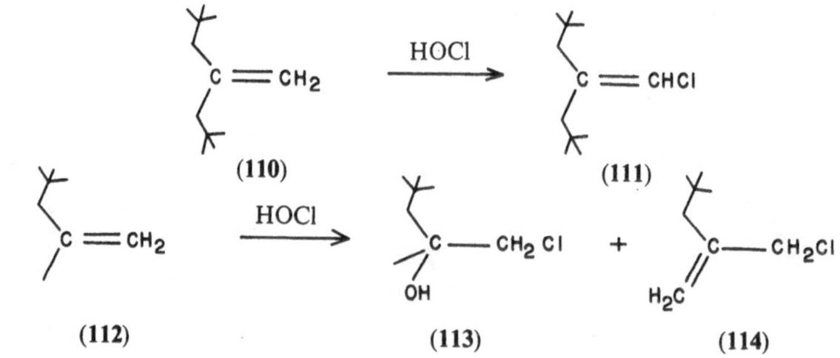

Another consequence of such steric crowding would be that, even if ordinary addition does occur under favorable circumstances or is made to occur under special reaction conditions, the product, due to steric compression, would be especially prone to undergo reactions in which, in the transition state, there is a change from sp^3 to sp^2 hybridization at the new fully substituted carbon (69, 70).

Thus, sterically crowded situations can divert the "normal" reaction pathway and the term *steric diversion* has been proposed (71) to describe this switch-over from the "normal" route. In longifolene (115), the ethylenic linkage is well-shielded and crowding at the more substituted end of the olefinic bond is, at least, as severe as in *unsym*-dineopentylethylene (110) (72, 73). Hence it is not surprising that the chemistry of longifolene is replete with examples of "abnormal products", more appropriately called sterically diverted products.

Longifolene
(115) (≡3)

a) Electrophilic Additions

Though the reaction of longifolene (3) with bromine was investigated quite early (16), a reinvestigation has been carried out recently (74). As might have been anticipated, no normal dibromide could be obtained, the products being the elimination products (116, 15%) and (117, 18%), the Wagner-Meerwein product (118, 15%), and the transannular product (119, 10%), besides some 22% longibornyl bromide (82).

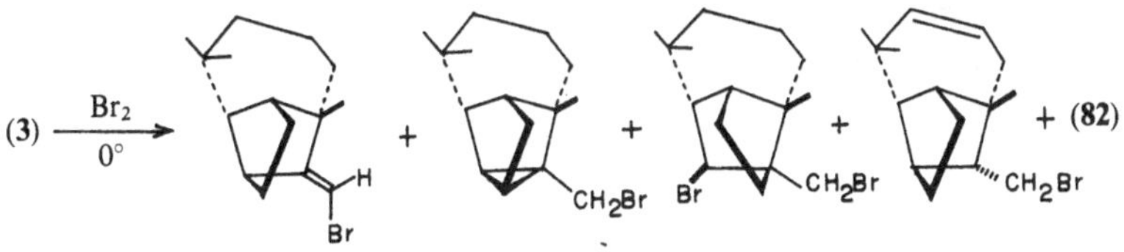

(116)   (117)   (118)   (119)

Reaction with iodine monochloride or phenylsulphenyl chloride also results in ω-substituted products (120, 121) (74). Longifolene reacts with oxides of nitrogen to furnish, like camphene, the ω-nitrolongifolene

(122) (*16*). Chlorosulphonyl isocyanate, which normally reacts with olefins in a cryptoionic reaction to furnish as the *major* product a β-lactam-N-sulphonyl chloride (*75, 75 a*), gives with longifolene *only* the unsaturated N-chlorosulphonylamide **(123)** (*76*). Reaction of longifolene with mercuric acetate-acetic acid followed by treatment with sodium chloride gives the vinyl derivative **(124)**, besides other products (*77*).

(**120**): X = I
(**121**): X = S – C₆H₅
(**122**): X = NO₂
(**123**): X = CONHSO₂Cl
(**124**): X = HgCl

ω-Bromolongifolene **(116)**, which is best obtained by the action of N-bromosuccinimide on longifolene (*78*), on fusion with KOH furnished, in low yield, longihomocamphenilone **(125a)** and longi-isohomocamphenilone **(125b)** besides a dimeric dilongifolenyl ether (*79*). The mechanism of this reaction, along the lines of a similar investigation on ω-bromocamphene (*80*), has been studied (*78*).

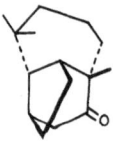

(125a)   (125b)

b) Oxidations

Much interesting and complex chemistry has resulted from oxidation of longifolene with various oxidizing agents. Invariably, the products have arisen from such "abnormal" reactions as:
(a) terminal oxidation to acids (CrO₃, peracids, O₃),
(b) cyclopropane ring formation (CrO₃),
(c) over-oxidation and bond cleavage by peracids,
(d) epoxidation and formation of lactones by ozone,
(e) ring expansion (CrO₃, peracids, O₃, lead tetraacetate, RuO₄).

Chromic Acid Oxidation

Oxidation of longifolene with chromic acid was first investigated by SIMONSEN and his co-workers (*9, 10*), who reported isolation of "longifolic" and isolongifolic acids. A third acid, "α-longifolic acid" was obtained by the ozonation of longifolene (10). SIMONSEN regarded these acids as having the molecular formula, $C_{14}H_{22}O_2$. It was left for NAFFA and OURISSON (81) to establish that these acids are not C_{14} acids, but are in fact $C_{15}H_{24}O_2$ acids arising from terminal oxidation of the olefinic linkage. Isolongifolic and "longifolic" acids were considered by these authors to be epimeric, while "α-longifolic acid" was regarded as a molecular compound of these acids. Support for the epimeric character of these two acids was forthcoming from two other independent investigations (28, 82).

However, in a later study, NAYAK and SUKH DEV (83) demonstrated that there are in fact three C_{15} monocarboxylic acids derived from longifolene and not two. The so-called "longifolic acid" of the previous authors was shown (84) to be tetracyclic, having a cyclopropane ring, and was termed ψ-longifolic acid. The "α-longifolic acid" was indeed found to be an epimeric mixture, from which the thermodynamically less stable epimer, for which the name longifolic acid was retained, was obtained pure for the first time. These three acids thus have structures (126), (127) and (128). These authors further showed that all three acids are formed during the chromic acid oxidation of longifolene, the yields being in the ratio (126 : 127 : 128) :: (2 : 15 : 3). Besides these acids there are at least two neutral products of oxidation, longicamphenilone (9) (81) and longidione (129) (9, 10, 81), obtainable in yields of 12% and 8% respectively (81). Structures of these two products were established by suitable chemical transformations (3).

Though the terminal oxidation of a vinylidene group in an hindered olefin to a carboxyl function by chromic acid had prior analogy (73, 81), the formation of ψ-longifolic acid (128) and the ring-expanded dione

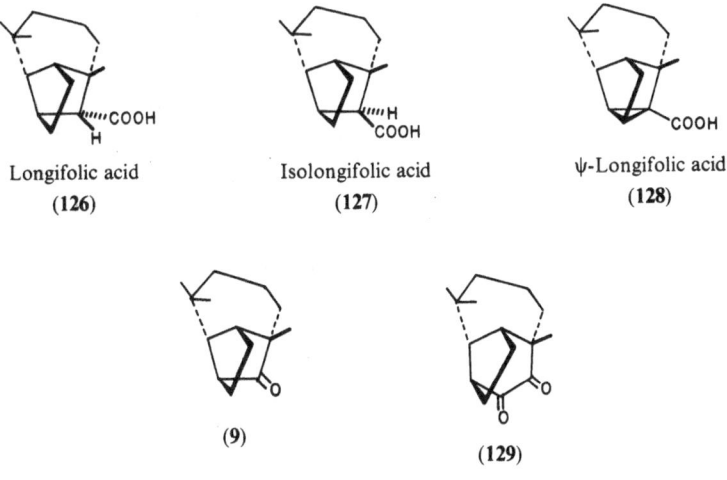

Longifolic acid   Isolongifolic acid   ψ-Longifolic acid
(126)   (127)   (128)

(9)

(129)

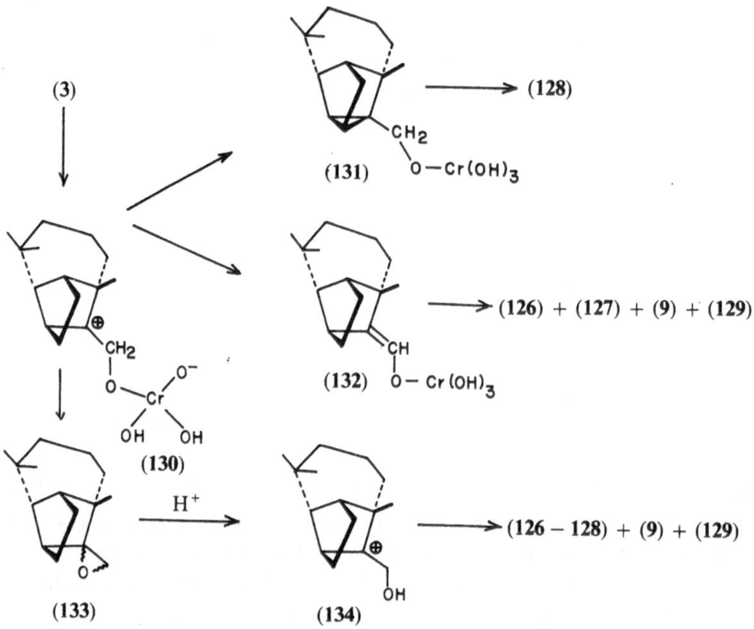

Chart 15. Oxidation of longifolene with chromic acid

(129) under these conditions is quite unusual. It is conceivable that species (130), arising from the elctrophilic attack of Cr^{VI} on the longifolene double bond (e. g. see 85) collapses by 1,3-proton elimination to (131) or by 1,2-elimination to the longifolaldehyde enol derivative (132), and that these intermediates next generate the products observed, including (129). Alternately, longifolene epoxide (133) which can arise from (130) can be implicated as the intermediate leading to the final observed product mixture; it should be noted that due to steric compression (86), the epoxide would be highly susceptible to ring opening to (134).

By way of comparison it may be pointed out that camphene (4) with chromic acid in aqueous acid gives, as the chief product, camphor (135) and not camphenilanic acid (136), apparently due to fast prior hydration of camphene to isoborneol (83, R = H) (85). This reaction has not been observed with longifolene. Low conversions of camphene to the acid (136) can be, however, achieved by carrying out this oxidation in acetic anhydride-carbon tetrachloride (87), and under these conditions camphene epoxide has been demonstrated (88) as the primary oxidation product.

Camphor
(135)

Camphenilanic acid
(136)

Peracid Oxidation

The effect of steric compression at C-3 in leading to sterically diverted products has been found to be more profound in the oxidation of longifolene by peracids.

NAFFA and OURISSON (81) found that longifolene, which has only one olefinic linkage, consumes almost two mole equivalents of perbenzoic acid (in chloroform, at ~0°) and generates completely unexpected products: the norketone longicamphenilone (9, 68%) and the ring-expanded dione (129) but none of the anticipated epoxide.

In a later investigation, NAYAK and SUKH DEV (86) demonstrated that longifolene epoxide is indeed the primary product of oxidation and can be isolated in an excellent yield by carrying out the oxidation with stoichiometric amounts of perbenzoic acid in *benzene* solution. Since approach of the peracid would be preferred from the *endo*-face, the epoxide was assigned configuration (137). It was further shown that the epoxide slowly consumed one mole equivalent of perbenzoic acid (in chloroform, at 0°) to furnish a mixture of compounds, qualitatively similar to the products of uncontrolled oxidation of longifolene by per-benzoic acid, which in a reinvestigation were shown to consist of, besides longicamphenilone (9) and traces of longidione (129), an α-ketol (giving longidione on oxidation with cupric acetate), C_{14}-alcohols (corresponding to 9), longifolic (126) and isolongifolic acids (127). The reaction of longifolaldehydes *(vide infra)* with perbenzoic acid was also investigated.

The above information was exploited to rationalize the peracid oxidation of longifolene. The key suggestion (86) made was that the susceptibility of longifolene epoxide to further peracid attack has its origin in steric compression at C-3 in the epoxide which is relieved by its isomerization to the enol (138). Thus, isomerization of the epoxide to an aldehyde would be exceptionally facile (under acid or base catalysis) due to a decreased energy barrier between the epoxide and the enol. As a matter of fact, epoxide (137) is readily isomerized to an epimeric mixture of aldehydes (139, 140) by mere rapid filtration of its hexane solution through a bed of silica gel (83). Protonation of the enol should preferentially occur from the *endo*-face to give the thermodynamically less stable longifolaldehyde (139) which evidently rapidly epimerizes to isolongifol-aldehyde (140). This reaction proved valuable in establishing the stereo-

Longifolene epoxide (137)

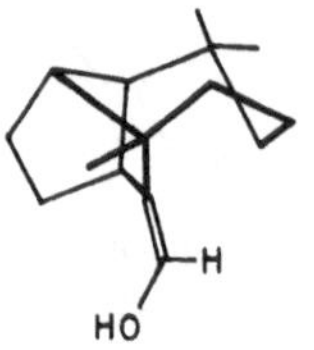

(138)

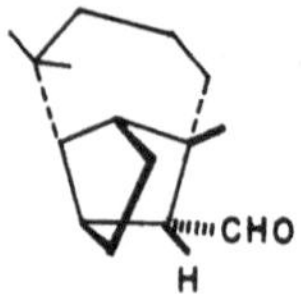

Longifolaldehyde (139)

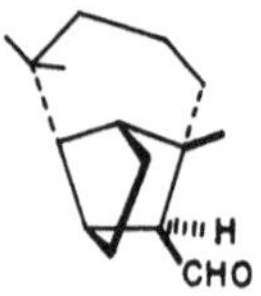

Isolongifolaldehyde (140)

chemistry of longifolic and isolongifolic acids already described earlier. It was further suggested (86) that the products of uncontrolled peracid oxidation of longifolene are essentially summation of the products

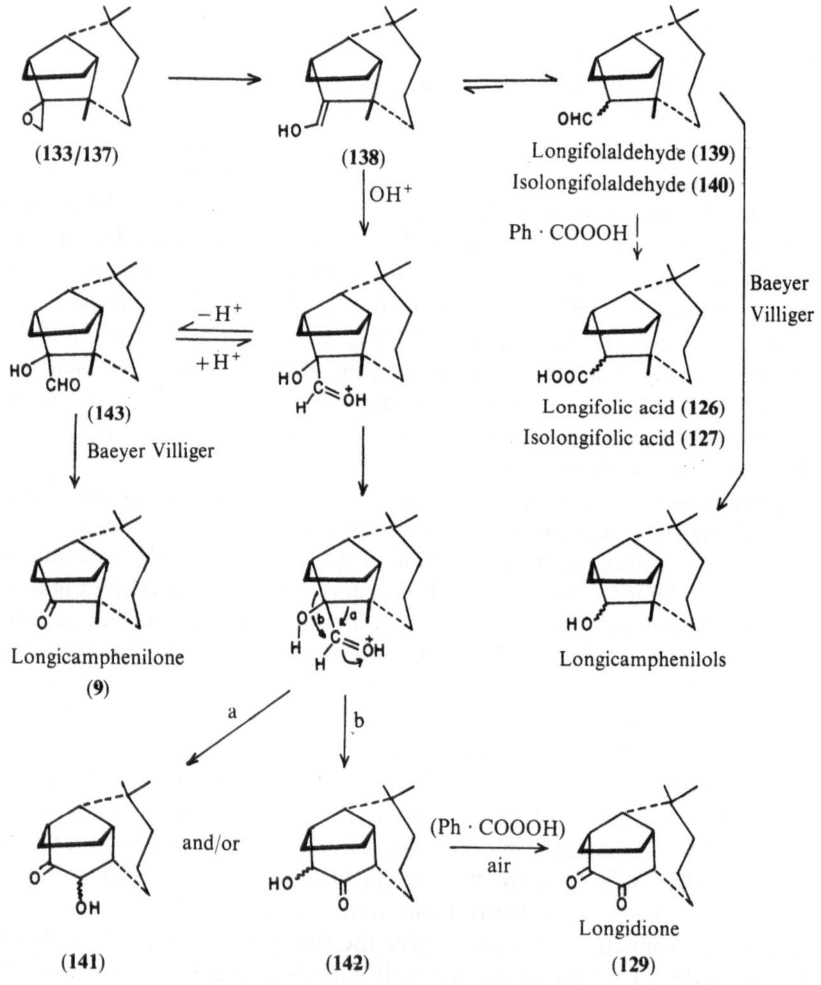

(133/137) → (138) ⇌ Longifolaldehyde (139) Isolongifolaldehyde (140)

OH⁺

Ph · COOOH

Baeyer Villiger

(143) ⇌ −H⁺ / +H⁺

Longifolic acid (126) Isolongifolic acid (127)

Baeyer Villiger

Longicamphenilone (9)

Longicamphenilols

a

b

(141)   and/or   (142)   (Ph · COOOH) / air   Longidione (129)

Chart 16. Peracid oxidation of longifolene

resulting from attack of the peracid on the enol and the aldehyde, the latter arising from the epoxide under acid-catalysis (benzoic acid in $CHCl_3$?) (Chart 16). Though the presence of α-ketols (141/142) as required by the mechanism outlined (Chart 16), was only inferred from the results of the further oxidation of the total peracid oxidation product with cupric acetate in acetic acid (86), a more direct proof was forthcoming, when LHOMME and OURISSON (89) succeeded in isolating (142) as the acetate from the oxidation of longifolene with monoperphthalic acid.

From an examination of the mechanistic interpretation outlined in Chart 16 it becomes clear that the 3-hydroxylongifolaldehyde (143) occupies a key position. α-Hydroxy aldehyde (145) is the main product of peracid oxidation of the olefin (144) (90). However, till recently (91), all attempts to isolate (143) or even obtain evidence for its presence were unsuccessful. It has now been demonstrated that by using a mixture of carefully purified chloroform and ethanol, oxidation of longifolene with two mole equivalents of perbenzoic acid gives, in good yield, the elusive 3-hydroxylongifolaldehyde, which has been shown to possess the stereochemistry (146). As expected, (146) readily furnished the C_{14}-ketone (9) on further peracid oxidation and smoothly rearranged to ketol (141) (rather than the isomeric 142!) on short exposure to p-toluenesulfonic acid in chloroform (at 25°). All these results fully support the gross mechanistic features of this "abnormal" peracid oxidation.

(144)   (145)

(146)   (147)

It may also be noted that LAH reduction of (146), besides giving the expected diol, furnished in 40% yield the ring-expanded α-ketol (142). Conceivably, this rearrangement can occur through a complex such as (147) in which the aldehydic carbon is well-shielded and the driving force for the ring-enlargement would be steric compression at C-3; Lewis acid character of LAH has been recognized (92).

Reaction of longifolene with peracetic and performic acids has also been investigated (81): lesser yields of longicamphenilone (9), higher amounts of acids (126, 127) and longidione (129), and some amounts of α-longiforic acid (148) result.

COOH
H
H
COOH

α-Longiforic acid
(148)

=O
O

(149)

COOH
H
H
HOOC
H

β-Longiforic acid
(150)

α-Longiforic acid was first described by SIMONSEN (9, 10), as a cleavage product of longidione (129) with a mixture of nitric and sulphuric acids. The relationship of this acid with the dione became clear, when NAFFA and OURISSON (81) found that the dione on oxidation with alkaline hydrogen peroxide gives the anhydride (149) of the acid. In a further investigation, these authors, demonstrated that β-longiforic acid, obtained by SIMONSEN by heating the α-acid with hydrogen bromide in a sealed tube, is the thermodynamically more stable epimer (150). This epimerization can be conveniently effected by heating the α-acid with a trace of p-toluenesulfonic acid at 240—250° (93). In connection with structural investigations on longifolene, the dehydrogenation of longiforic acids has been studied (94, 95).

Though the overconsumption of a peracid by certain mono-olefins is known (96), and the example of olefin (144) has already been cited, the complex of products generated by longifolene is truly remarkable. In contrast, the main product from the reaction of camphene with per-benzoic acid is the expected epoxide (97).

Oxidation with Ozone, Other Oxidants

Longifolene under the usual ozonolysis conditions funishes, besides a poor yield of the expected cleavage product, longicamphenilone (9), longifolic and isolongifolic acids (126, 127) (10, 28, 81); formation of small amounts of two δ-lactones (151) has also been reported (98). When ozonation was carried out in presence of tetracyanoethylene in an effort to improve the yield of (9), though a 30% yield of (9) could be obtained,

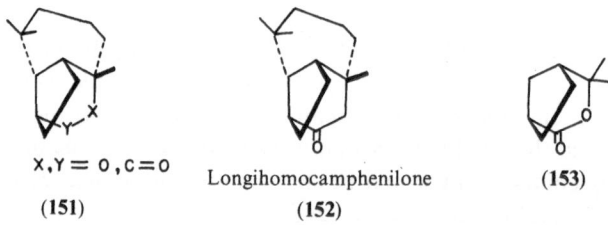

X,Y = O, C=O

(151)

Longihomocamphenilone

(152)

(153)

an unexpected reaction leading to the ring-expanded longidione (129) also occurred (99). Ozonation in acetic acid gives besides (9) and (129) small amounts of longihomocamphenilone (152) (99) whose origin is discussed later in this section.

Terminal oxidation and ring-enlargement observed during the above reactions, can be readily rationalized by mechanistic reasoning similar to that developed for the peracid oxidation, and by implicating longifolene epoxide during ozonation. It is known (100) that with hindered olefins epoxide formation effectively competes with ozonolysis. As a matter of fact, by carrying out the ozonation of longifolene in the presence of pyridine longifolene epoxide (133) has been isolated in 70% yield (99).

Camphene, under usual ozonolysis conditions gives largely the lactone (153) and products derived from this (101). As already noted, this is a minor pathway for longifolene.

As a matter of convenience, the results of oxidation of longifolene with lead tetraacetate, and ruthenium tetraoxide will also be summarized here; steric diversion may have little role in forming the products of these reactions. One of the important reactions of lead tetraacetate with olefins is allylic substitution/rearrangement (102). Since, this pathway is blocked for longifolene, the major product of this reaction is the ring-expanded enol acetate (154) (Chart 17), exactly parallel to what happens with camphene (81, 99).

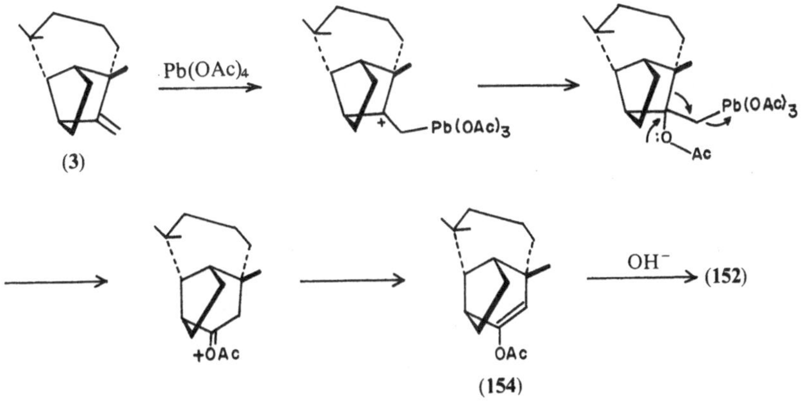

Chart 17. Oxidation of longifolene with lead tetraacetate

Saponification of (154) gives longihomocamphenilone (152) in which the position of the carbonyl function was established from a study of the spectral characteristics of the derived monobromoketone. Oxidation of (152) with selenium dioxide furnishes an excellent yield of longidione (129). This sequence of reactions was utilized by NAFFA and OURISSON (81) to establish that the yellow diketone (longidione) is not derived from any impurity present in longifolene, as was contemplated by SIMONSEN.

Longifolene on oxidation with RuO$_4$ gives a 30% yield of longi-camphenilone and a small yield of longidione (*99*).

4. Transannular Reactions

An important characteristic of medium-sized alicyclic rings is the ease with which transannular reactions arising from proximity of atoms occur (*e. g.* see *103*). In the preferred conformation of the four-carbon bridge in the longifolane (**155**) and longibornane (**156**) systems, the proximity of the C-3, C-10 and C-2, C-9 positions, respectively has been responsible for interesting transannular reactions, both of homolytic and heterolytic types.

(155) 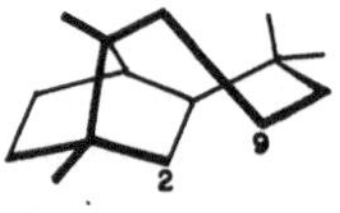   (156)

a) Radical Reactions

Reaction of camphene with polyhalomethanes, for example bromotrichloromethane, in the presence of peroxides furnishes the adduct (**157**) arising from addition of CCl$_3$ and Br radicals, in the expected manner (*e.g.* see *104*). Longifolene, under the same reaction conditions, also reacts

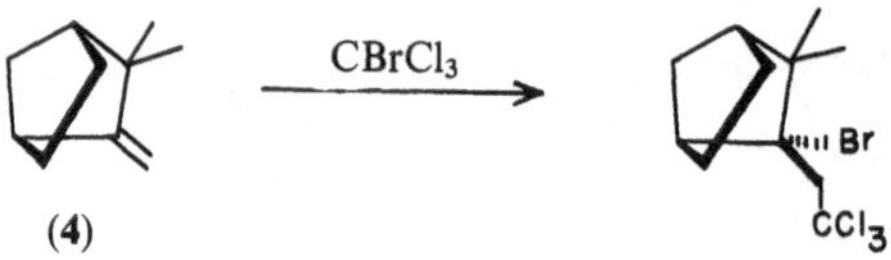

(4)   (157)

to give an adduct which, however, was shown to possess the unexpected structure (**158**), a consequence of a transannular 1,5-hydrogen atom shift in the cyclooctane ring of the longifolene system (Chart 18); it may be noted that in this step the migrating hydrogen atom barely changes its position and all that happens is a different pairing of one bonding electron (*105*).

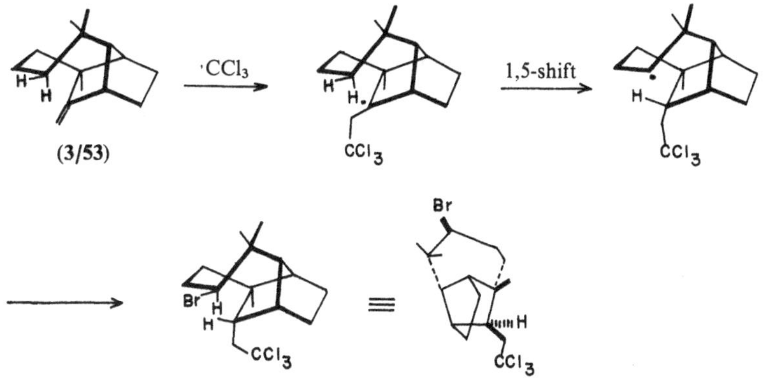

(158)

Chart 18. Addition of bromotrichloromethane to longifolene

The photosensitized addition of formamide to olefins is essentially an addition of H˙ and ˙CONH₂ to the double bond (see *e.g. 106*); trans-annular participation of a second suitably located olefinic linkage has also been observed (*106*). Longifolene, however, gives only the unsaturated amides (**159, 160**), with no evidence of a 1,5-H shift product possibly because of fast abstraction by triplet acetone of a hydrogen α to the radical centre (*107*).

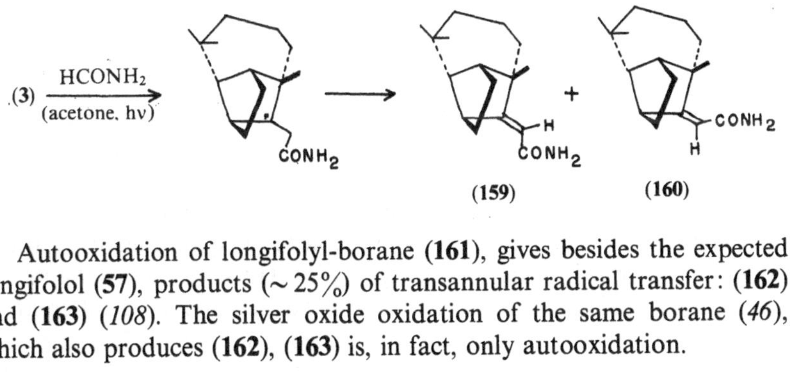

(159)   (160)

Autooxidation of longifolyl-borane (**161**), gives besides the expected longifolol (**57**), products (~ 25%) of transannular radical transfer: (**162**) and (**163**) (*108*). The silver oxide oxidation of the same borane (*46*), which also produces (**162**), (**163**) is, in fact, only autooxidation.

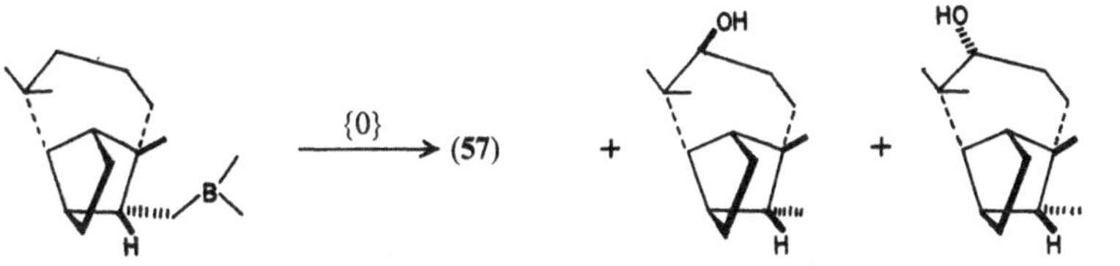

(161)   (162)   (163)

6*

b) Ionic Reactions

The first example of a transannular substitution reaction in the longibornane series was reported by PRAHLAD, NAYAK and SUKH DEV (59, 109) while investigating the products of "hydration" of longifolene, a reaction which has already been discussed earlier (Section: VI 2a). The ion (164) resulting from protonation followed by Wagner-Meerwein rearrangement is all set for a 1,5-hydride shift (Chart 19: path b), to furnish finally the transannular product (88: R = Ac). A similar 1,5-hydride shift was also observed during the photolysis of longibornyl iodide (85) which led to the formation of longiborn-8-ene (86) as already discussed.

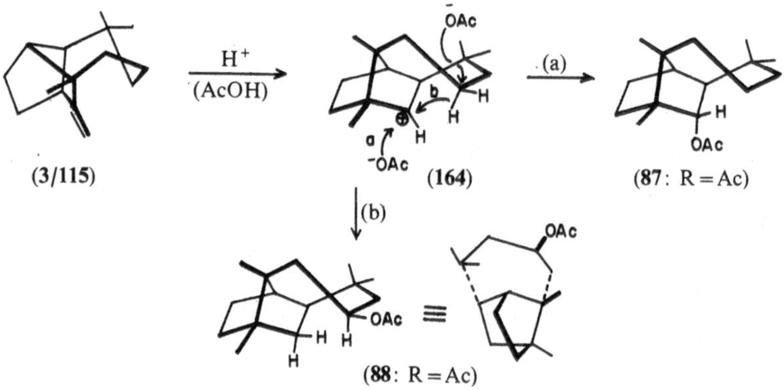

Chart 19. "Hydration" of longifolene (59)

Products available from the transannular reactions described so far have been exploited as substrates for further transformations, some involving 1,5-hydride shifts in "reverse".

The bromide (165: R = Me), obtainable from the adduct (158) by a degradative sequence, regenerates longifolene (3) quantitatively on mild solvolysis or on exposure to silica gel (105). However, when the methyl ester (166), again obtainable from (158), is solvolyzed, a different rearrangement leading to the ring-contracted products (167, and related compounds) occurs because the methoxycarbonyl function at C-3 would now inhibit generation of positive charge at this carbon, thus blocking the "reverse" hydride shift. Solvolysis of tosylates (168, 169) derived from (162), (163) have also been investigated (110): only ring-contracted products (170) were obtained, but no longifolene. The mechanism of solvolysis has been discussed and, in the case of (169), ring-inversion has been invoked to rationalize the product development. From a detailed kinetic and product analysis study of solvolysis of bromides of type (165) having substituents with differing electronic demands (R: CN, COOMe, $C \equiv C - Cl$, CH_2Br, CH_2I, CH_2OMe, CH_2OH, CH_3, CH_2CH_3). OURISSON et al. (111) concluded that whenever 1,5-hydride migration occurs, there is rate acceleration.

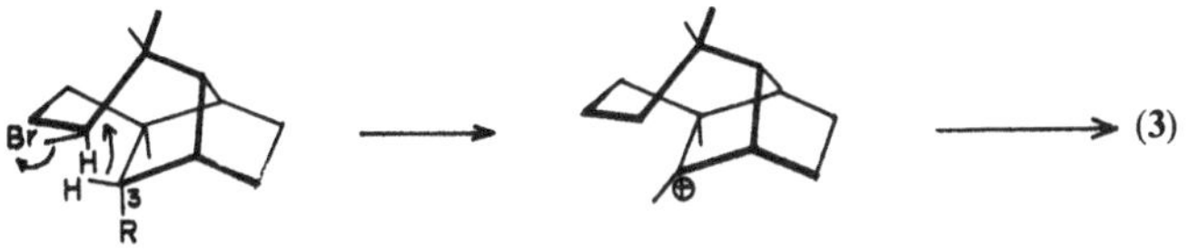

(165)

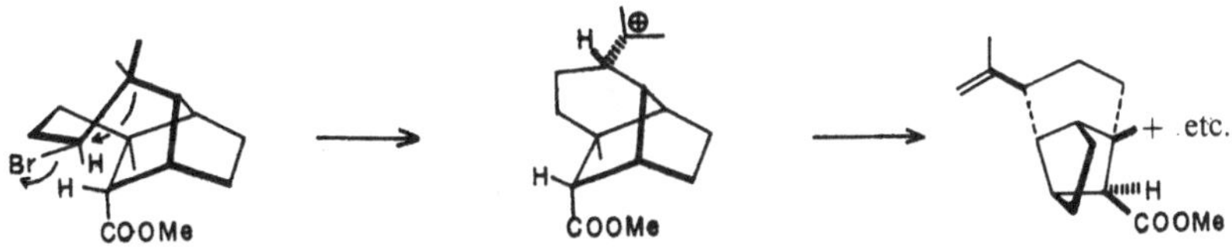

(166)   (167)

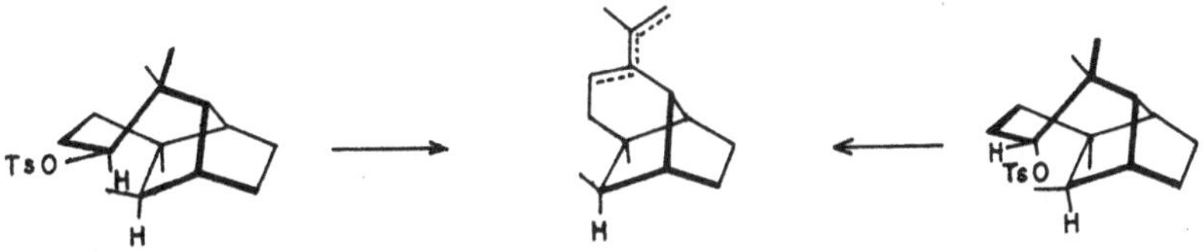

(168)   (170)   (169)

An interesting transannular cyclization was uncovered by HELMLIN-GER and OURISSON (112) who found that mere chromatography of bromo olefin (171) over silica gel (but not alumina) gives the cyclized product (172). Acetolysis of (171) gives both (172) and (173). Structure (172) is fully secured by an X-ray crystallographic analysis study (113).

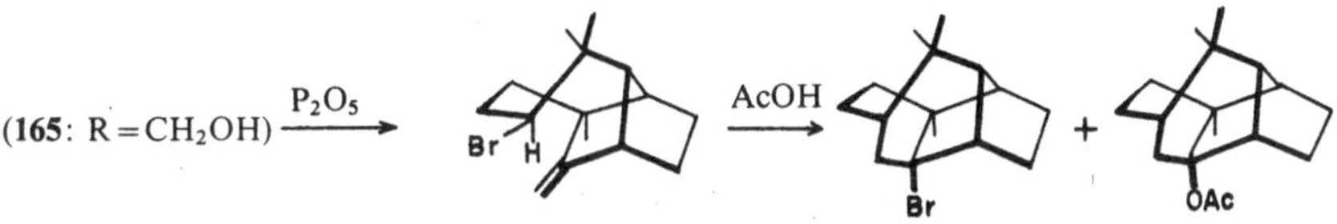

(165: R = CH₂OH) $\xrightarrow{P_2O_5}$ (171) $\xrightarrow{AcOH}$ (172)   +   (173)

In the longibornane series exposure of longiborn-9-yl mesylate (88: R = MeSO₂) to alumina or nitrous acid deamination of 9-amino-longibornane (174) generates hydrocarbons consisting of, besides longiborn-8-ene (86) and longiborn-9-ene (175), longifolene (3), longicyclene (44) and isolongifolene (23). In the case of the deamination reaction longifolene is formed to the extent of 30% (114).

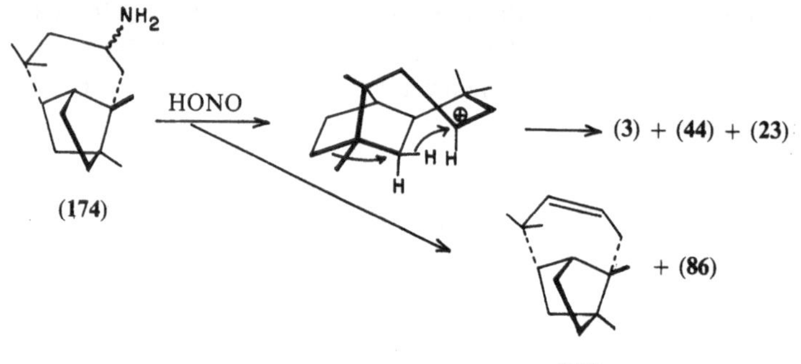

(174)   HONO →   ⟶ (3) + (44) + (23)

           + (86)

(175)

c) Lead Tetraacetate Oxidation of Longifolols

An important reaction of suitable alcohols with lead tetraacetate in apolar solvents is intramolecular cyclization leading to cyclic ethers (115). This reaction has been investigated with a view to effecting functionalization of appropriate carbon atoms in the essentially rigid framework of longifolane by employing longifolol (57) and isolongifolol (181) as substrates.

Longifolol (57), when exposed to the reagent in refluxing benzene, gave the corresponding acetate (20%) and the anticipated oxide (20—30%) shown to be (176); besides these, norlongicyclene (105, 5—6%) and ψ-longifolaldehyde (178, 8—10%) were also formed (109). In another study employing hypoiodite {Pb(OAc)$_4$ − I$_2$}, Jadhav and Nayak (117) found that transannular ether formation occurs in excellent yield to furnish, besides (176), a second oxide (177) which was anticipated from a con-

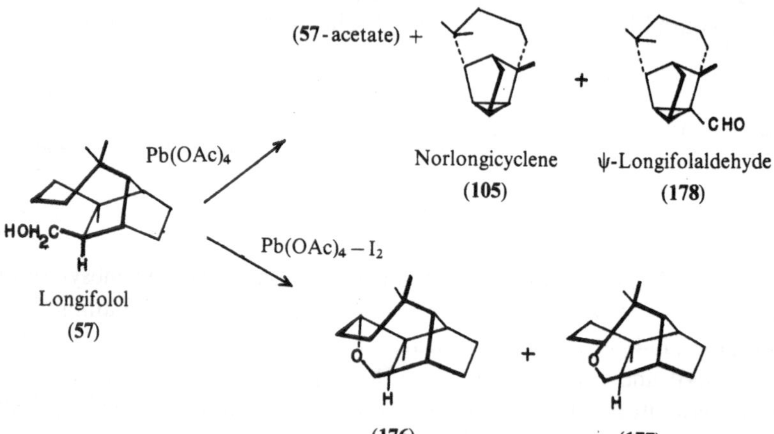

(57-acetate) +

Norlongicyclene   ψ-Longifolaldehyde
(105)   (178)

Pb(OAc)$_4$

HOH$_2$C

Longifolol
(57)

Pb(OAc)$_4$ − I$_2$

(176)   (177)

sideration of proximity effects in the medium-ring cycle of longifolane. These two ethers have been exploited by these authors for the synthesis of two naturally occurring longifolene derivatives (179) and (180), isolated by a Japanese group from the heartwood of *Juniperus conferta* Parl. (*117*).

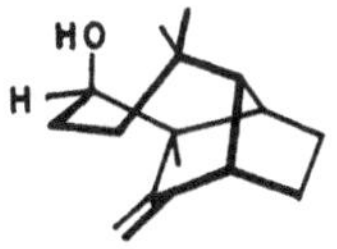

(179)    (180)

Isolongifolol (181) under similar conditions (lead tetraacetate) gives a 20—30% yield of the oxide (182) (*109, 118*) besides much acetate (181-acetate; 60%) and some norlongicyclene (105, 6%), but no ψ-longifol-aldehyde (*109*). The combination Pb(OAc)$_4$—I$_2$, on the other hand, furnishes (182) in almost quantitative yield (*40*). This ether has been exploited (*40*) for a synthesis of longifolene-4,4,5,5-d$_4$ (45), required in connection with studies on the mechanism of isomerization of longifolene to isolongifolene.

Isolongifolol
(181)   (182)

The formation of cyclopropane compounds in the lead tetraacetate oxidations is rather unexpected. LHOMME and OURISSON (*109;119*) explain the formation of norlongicyclene (105) by extending the mechanism usually considered for the oxidation of primary alcohols: C—C bond cleavage with loss of formaldehyde and formation of radical (183); the radical on further oxidation with Ph(OAc)$_4$ gives the carbocation (184), which by 1,3-proton elimination furnishes (105). These conclusions find support from experiments specially designed to generate (183) or (184). The formation of ψ-longifolaldehyde (178) from longifolol, but not from isolongifolol, has been rationalized in terms of initial oxidation of the alcohol to the aldehyde, rapid enolization of longifolaldehyde (139) *alone* (steric compression) to the enol (138), which is then oxidized to the hybrid cation (185) *via* the radical and final cyclization to (178) by β-proton elimination.

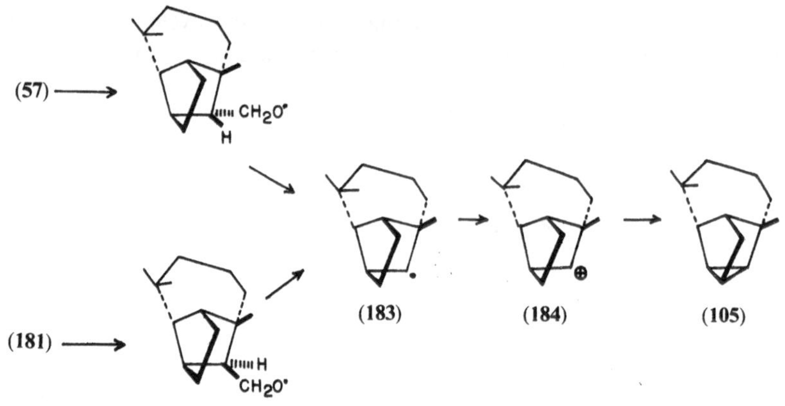

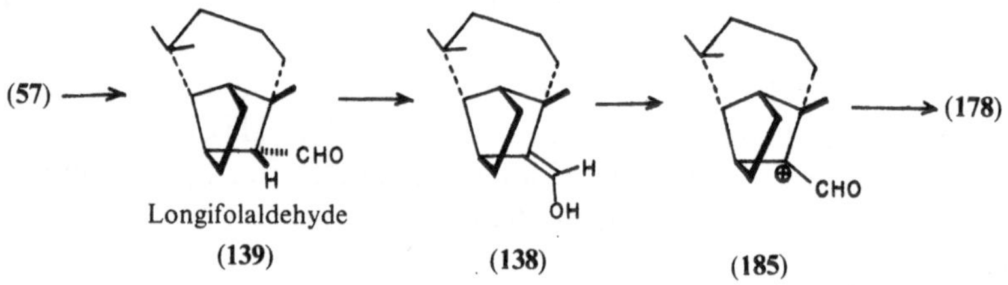

(57) ⟶

(181) ⟶

(183) → (184) → (105)

(57) ⟶

Longifolaldehyde
(139)   (138)   (185)   ⟶ (178)

5. Conversions into Other Sesquiterpene Skeletons

Skeletal transformations within a group of natural products either along biogenetic concepts or by devising exotic reaction sequences have attracted considerable attention (e. g. see: *120, 121; 122*, note 4). The key reactions usually employed are: cyclization, cationic rearrangements and fragmentation. It is not surprising that in this exercise longifolene has not failed to get attention.

Conversion of longifolene into the tetracyclic longicyclene (**44**) has already been discussed earlier. Transformation of longifolene into the naturally occurring longiborneol (**87**: R = H), a transformation which was incidental to the investigations on the chemistry of longifolene, may also be noted. DE MAYO and WILLIAMS (*123*) have utilized OURISSON's bromo ester (**166**) and have exploited its propensity to ring-contraction, discussed in the preceding section, for its transformation into (+)-sativene (**187**), a metabolite of *Helminthosporium sativum*. Distillation of (**166**) with iron powder furnished the ring-contracted product (**186**) which could be converted into (**177**) along conventional lines.

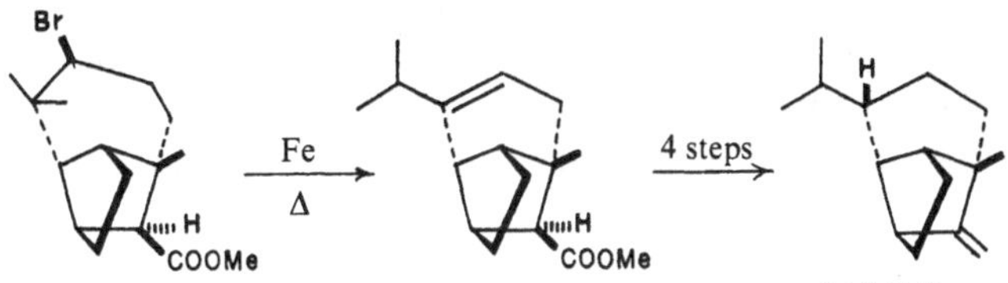

(166)   (186)   (+)-Sativene
                              (187)

Longifolene has also been transformed in a seven-step sequence into (+)-himachalene hydrochloride (190) (122), the parent hydrocarbon of which has biogenetic kinship with longifolene (124). The key-step in the process is the base-catalyzed fragmentation of the bromoketone (188) whereby the longibornane skeleton of (188) is converted into the bicyclic himachalane frame-work (189). The required bromoketone was prepared from ω-bromolongifolene (116) by an acid-catalyzed "hydration" involving an intramolecular 1,5-hydride shift, along the known conversion of longifolene into the transannular alcohol (88: R = OH). Transformation of (189) into (190) was straightforward.

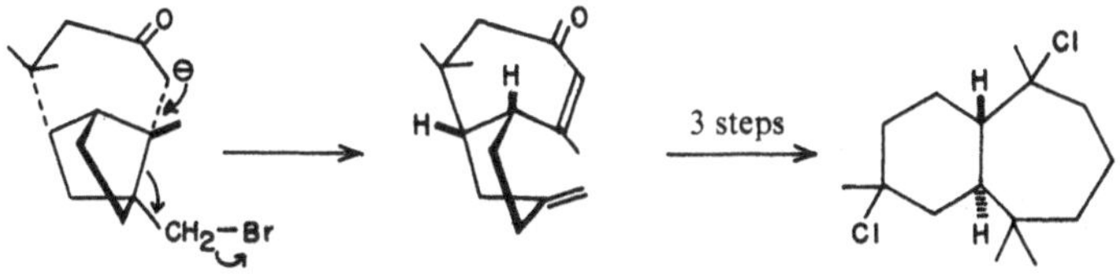

Himachalene hydrochloride

(188)   (189)   (190)

(+)-Secolongifolene diol is one of the metabolites of the fungus *Helminthosporium sativum* (125). Recently, a simple synthesis of its optical antipode (193) has been achieved by exploiting the electrophilic addition-fragmentation reaction (126) of homoallylic alcohols. The homoallylic alcohol (191), which can be obtained from longicyclene (44) by cyclo-

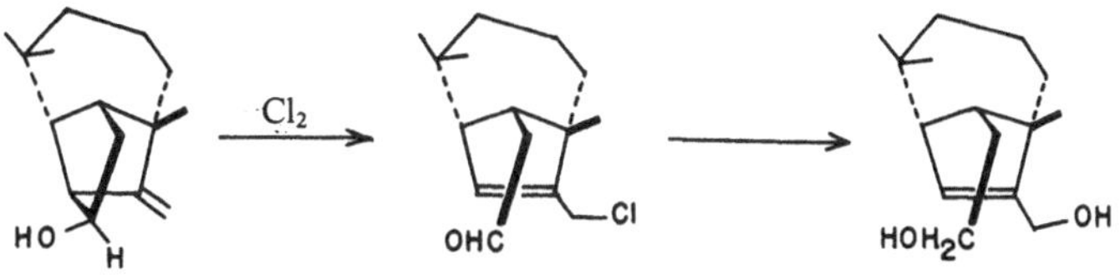

(−)-Secolongifolene diol

(191)   (192)   (193)

propane ring cleavage by NBS (*127*) followed by hydrolysis, or from ψ-longifolic acid (**128**) *via* acid-catalyzed equilibration of the derived alcohol, is the key-intermediate. Exposure of (**191**) to chlorine furnished (**192**) in an excellent yield, which was readily transformed into the targetted (**193**) by unexceptional steps (*128*).

6. Miscellaneous Transformations

Longifolene and selenium dioxide in refluxing acetic anhydride give in some 55% yield a crystalline compound formulated as the divinyl selenide (**194**)* (*129*).

Irradiation of a solution of longicamphenilone (**9**) in cyclohexane furnishes a complex mixture of products, in which a crystalline oxetane (**196**) predominates to the extent of 30%. The reaction proceeds by way of an initial Norrish type-I cleavage to (**195**) which is followed by a secondary Paterno-Büchi photocycloaddition (*130*) to give the tetracyclic oxetane (*131*).

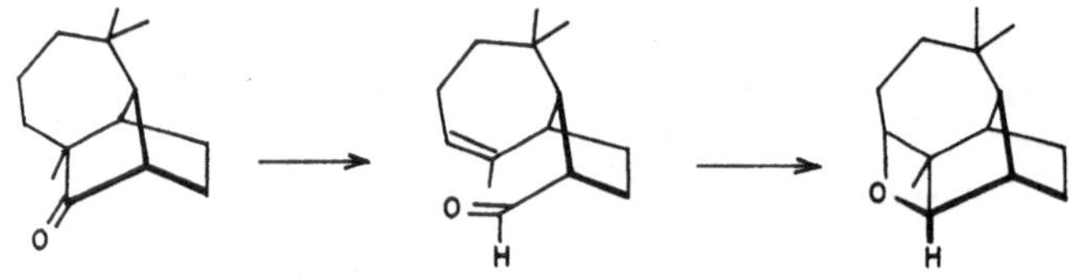

(**9**)   (**195**)   (**196**)

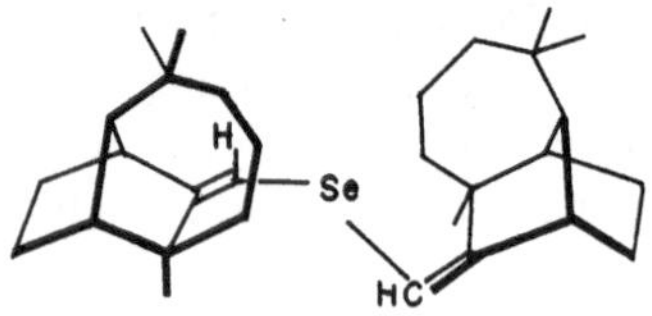

(**194**)

---

* In the original formulation of the authors, the two longifolene moieties have an object-mirror image relationship, which, obviously, is an error.

VII. Reactions of Isolongifolene

While describing the chemistry of longifolene it is only appropriate to draw attention to some reactions of isolongifolene, which have some novel features. Till the realization that isolongifolene, as obtained from acid-catalyzed isomerization of longifolene is varyingly racemized, the experimental work involving preparation of pure products proved often misleading because of side by side formation of optically active and racemic compounds which had different solubility characteristics, and melting points.

The steric crowding at the more substituted end of the ethylenic bond, in isolongifolene (23) also is at least as severe as in *unsym*-dineopentyl-ethylene (110), hence it is not surprising to find that many of its reactions involving additions to the olefinic linkage are dominated by steric diversion.

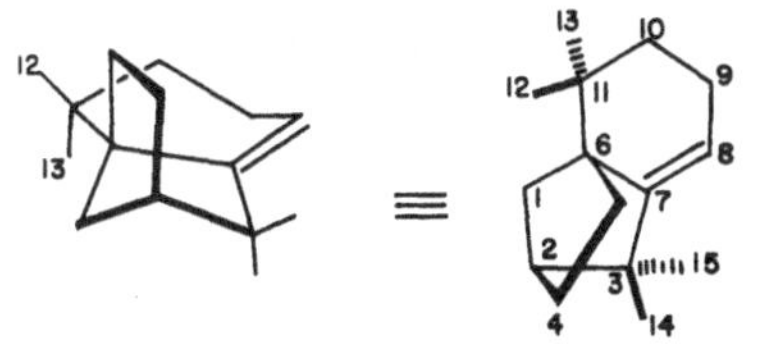

(23)

1. Epoxidation and Reactions of Epoxide

If oxidation of isolongifolene with perbenzoic acid is carried out in chloroform solution, almost two mole equivalents of the reagent are consumed, an experience reminiscent of the "uncontrolled" peracid oxidation of longifolene. The main product of the reaction was shown to be the lactone (197) while interruption of the reaction at one mole equivalent consumption led, as expected, to the isolation of the ketone (198) (33). Extrapolating the longifolene-peracid chemistry to the present case, SUKH DEV et al. (30) prepared the primary product of oxidation, the epoxide (199), by carrying out the reaction in benzene solution. In this case also, it is the facility with which isomerization of the oxide to the carbonyl derivative occurs as the result of steric compression, which causes further uptake of peracid.

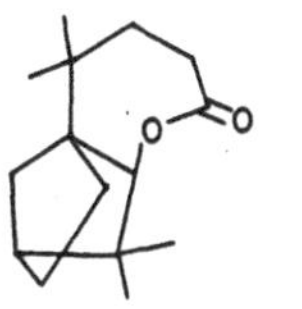

(197) 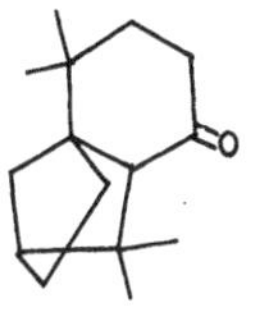   (198) 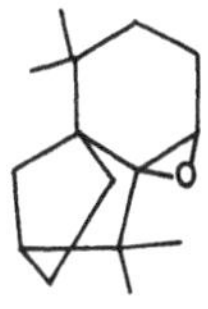   (199)

The stereochemistry of the oxide (199) has been the subject of some controversy, but was finally settled in favour of the β-structure (200) by X-ray crystallographic analysis (132). Thus, as in longifolene, the olefinic linkage in isolongifolene is preferentially attacked from the *endo*-face (of the camphene part-structure of the molecule). By a study of model compounds, Greengrass and Ramage (133) were able to conclude that it is the axial C-13 methyl which directs the approach of the reagent from the *endo*-face of the bicycloheptyl moiety.

Some reactions of isolongifolene oxide (200) deserve special mention. The oxide on exposure to chloroform containing 1% hydrogen chloride gives, besides the ketone (201) which can be more efficiently obtained by the action of $BF_3 \cdot Et_2O$ (30), some 20% yield of the homoallylic alcohol (202) (134). It was this transformation (200 → 202), thought to be proceeding in a concerted fashion, that led to the initial wrong assignment of stereochemistry to the oxide (30, 134). It is now clear that the rearrangement proceeds in a *syn* fashion and this points to the importance of ion (51) in the transition state. Several examples of *syn*-migrations in epoxide opening are on record (e. g. see 135). The transformation (200 → 202) can be carried out more effectively by just heating the oxide in dimethylsulphoxide (136). A rather novel rearrangement (134) of the oxide (200) in an alumina matrix generates a tetracyclic hydrocarbon (203, 48%) and a tetracyclic alcohol (204, 42%), besides some (201). Structures (203, 204) rest on sound chemical and spectroscopic evidence. The closure of a cyclopropane ring under such conditions is rather unprecedented. Silica gel, on the other hand, furnishes the same alcohol (50%) but no hydrocarbon and besides gives large amounts (42%) of ketone (201). This isomerization to the tetracyclic alcohol can also be carried out by the action of aluminium isopropoxide (137).

(200)   1% HCl in CHCl₃   (201)   (202)

Al₂O₃

(201) +   (203)   +   (204)

2. Addition of Halogens and Pseudo-Halogens

Addition of halogens (Cl_2, Br_2) and pseudo-halogens (ICl, halogen azides, NOCl) to isolongifolene does not yield any "normal" addition products due to severe steric hindrance to approach of counter ion at C-7. The initially formed halogenonium ion undergoes elimination/rearrangement to furnish one or more of the products (205, 206, 207) (71). Thus, exposure of isolongifolene to one mole equivalent of chlorine, gives (205, 55%, and 207, 45%; X = Cl); bromine, on the other hand, leads mostly to (205, 95%; X = Br), while ICl gives over 90% of the elimination compound (206; X = I). These results have been rationalized in terms of the bridging capacity (42) of halogens and, consequently, in terms of the nature of the ions (49/50/51) involved.

(23) 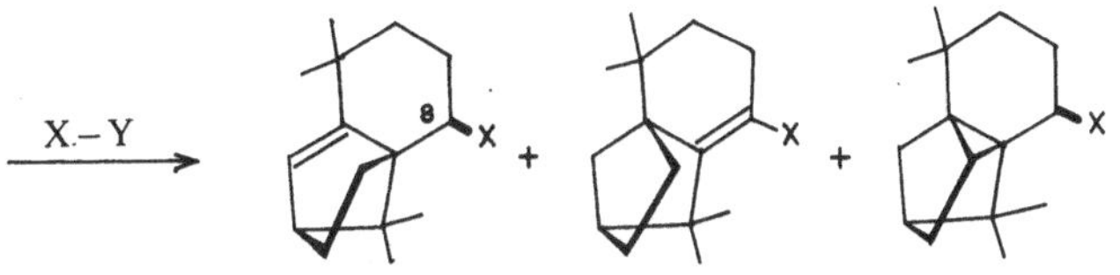 $\xrightarrow{\text{X–Y}}$

8-Haloneoisolongifolene

(205)   (206)   (207)

With two mole equivalents of bromine, isolongifolene gives chiefly (208), whereas two mole equivalents of chlorine leads to the products (209) and (210) in a ratio of 1 : 12 (71).

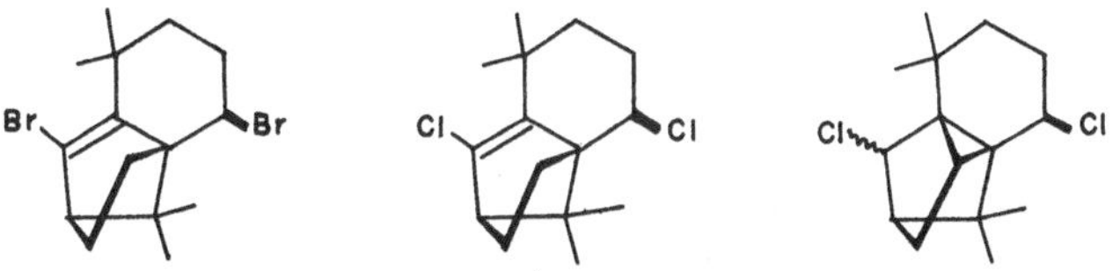

(208)   (209)   (210)

Hydrolysis (lime-water) of the chloride (207; X = Cl) proceeds with retention of configuration to give alcohol (204); this would require that σ-bridged ions (138) such as bicyclobutonium cation, are unimportant in this reaction (71).

Acetolysis of 8-bromo- or 8-chloroneoisolongifolene (205; X = Br/Cl) generates, besides two normal products (of elimination and S_N^2 displacement) which are minor, one rearranged elimination product (212) and the tertiary acetate (213). The diene (212) results from acetic acid-catalyzed rearrangement of the normal elimination product (211) while the tertiary acetate (213) is a Wagner substitution product resulting from a *syn*-migration (*139*). In the rearrangement (205 → 213), the strained bicyclo-{2,2,1}heptene part-structure expands, thus overcoming strain due to bond angle deformation. The migration of the bond actually observed,

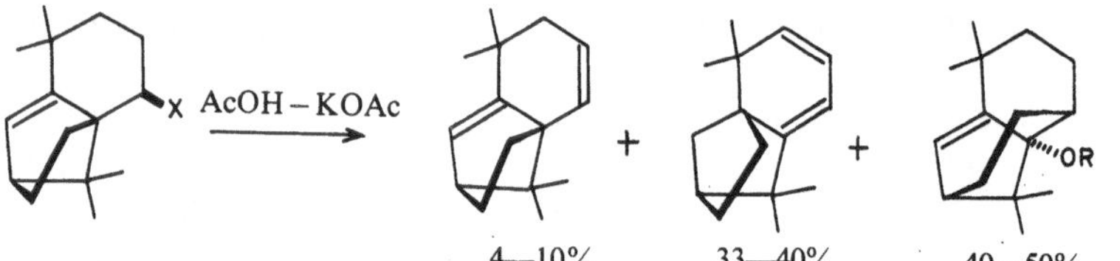

	4—10%	33—40%	40—50%
(205)	**(211)**	**(212)**	**(213)**; R = Ac
			(214); R = H

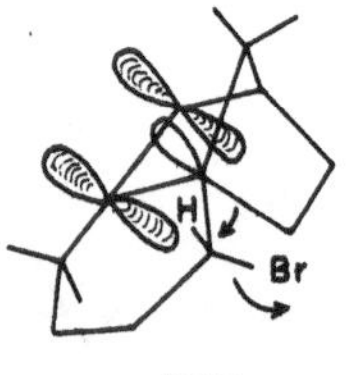

(215)

besides being electronically favoured, possibly derives further impetus for *syn*-migration from the fact that the developing vacant orbital has the correct geometry for lateral overlap with the π-orbitals (215), thus deriving additional stabilization. The alcohol (214) on exposure to aqueous sulphuric acid-dioxane, readily rearranges to the ketone (216). Similarly, the epoxide (217) derived from (214) on treatment with

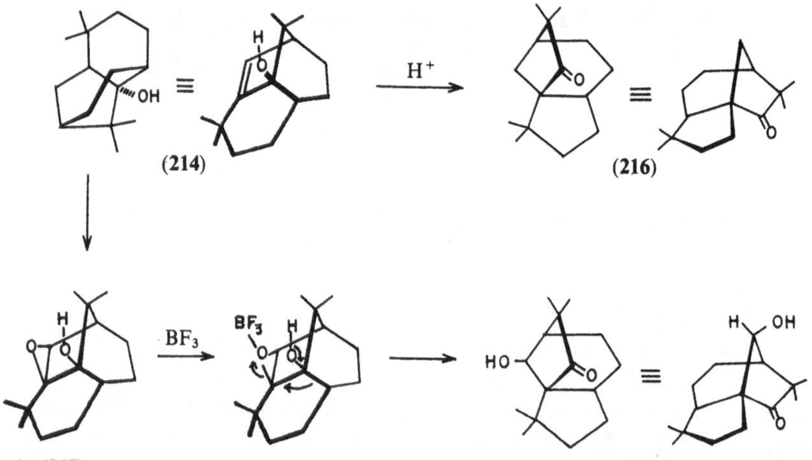

(214)   (216)

(217)   (218)

$BF_3 \cdot Et_2O$, smoothly passes into (218); it may be noted that isomerization of the oxirane ring to the carbonyl function under the influence of $BF_3 \cdot Et_2O$, the usual reaction, is blocked in the present case as the migration of the *endo*-hydrogen from C-1 to C-6 required by such an isomerization will result in inversion at C-6, which in the present case is sterically prohibitive (*139*).

VIII. Ultraviolet Absorption of Some Longifolene Derivatives

A comparison of the λ_{max} values for ω-formyl-longifolene (220; X = CHO) with that of ω-formyl-camphene (219; X = CHO) revealed a bathochromic shift for the longifolene derivative. To see if this trend was general, NAYAK, SANTHANAKRISHNAN, and SUKH DEV (*48*) examined several other derivatives and found a consistent positive shift in going from camphene to the longifolene series (Table 1). It has been suggested that this red-shift may have its origin in a raised ground state for the longifolene system which arises from a slight twisting of the C=C bond in longifolene.

Table 1. *Ultraviolet Light Absorption of Some Longifolene and Camphene Derivatives*

Substituent (X)	Camphene Derivative  (219)		Longifolene Derivative 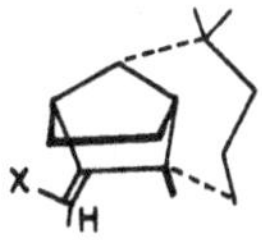 (220)		$\Delta\lambda_{max}$ (nm)
	λ_{max}^{EtOH} (nm)	10^{-3}	λ_{max}^{EtOH} (nm)	10^{-3}	
CHO	244	17.43	253	15.63	9
COOH	219	14.45	230	11.88	11
COOMe	226	14.36	235	12.43	9
	210	215	220	210	215	220
H	1255	410	251	4455	1932	429
CH$_2$OH	4849	1881	896	8660	5712	2580
CH$_2$OAc	7207	3913	2781	9040	6328	3390

IX. Biosynthesis

OURISSON (*14*) suggested that biogenesis of longifolene from the usual sesquiterpene precursor farnesol is likely to proceed *via* the longibornyl cation (**225**). Later, HENDRICKSON (*140*), while examining the stereo-chemical implications in sesquiterpene biogenesis, suggested that the *cis*-cation (**222**) arising from 1,11-cyclization of *cis, trans*-farnesol (**221**) would, by a 1,3-hydride transfer, generate the more stable cation (**223**) which is singularly well-set for leading to longifolene *via* the longibornyl cation (**225**) (Chart 20). Gross features of this scheme received support from the finding (*141*) that in longicamphenilone (**9**) from degradation of labelled longifolene, obtained after feeding {1-^{14}C} acetate to *Pinus longifolia* Roxb., none of the original activity had been lost.

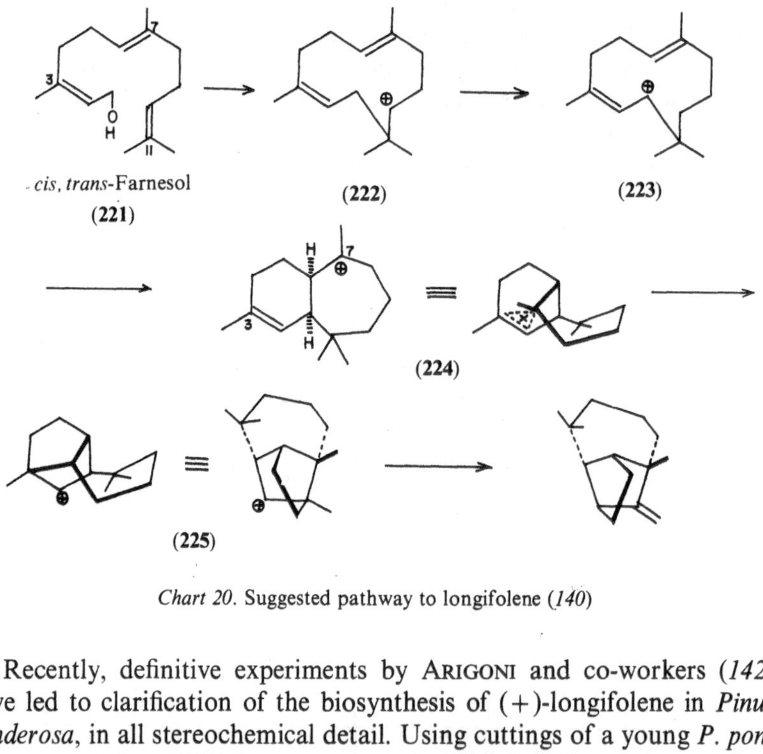

- *cis, trans*-Farnesol   (**222**)   (**223**)
(**221**)

(**224**)

(**225**)

Chart 20. Suggested pathway to longifolene (*140*)

Recently, definitive experiments by ARIGONI and co-workers (*142*) have led to clarification of the biosynthesis of (+)-longifolene in *Pinus ponderosa*, in all stereochemical detail. Using cuttings of a young *P. ponderosa* tree 0.1 to 0.2% incorporation of radiolabelled mevalonates into (+)-longifolene could be achieved. By employing appropriately labelled mevalonates and by carrying out suitable degradations to locate the label these authors were able to establish that: (i) a 1,3-hydride shift occurs, as had been suggested by Hendrickson; the migrating hydrogen is derived from the pro-5R position in mevalonate and is located in an *exo*-position at C-10 (in **225**), (ii) these results uniquely define the *si*-direction of attack on the more substituted end of the isopropylidene double bond from an anti-conformation of *cis,trans*-farnesol-PP (**226**) and this enforces the subsequent migration of H_R.

'(**226**)

Fusarium culmorum elaborates culmorin (**227**) (*143*), a sesquiterpenoid closely related to (−)-longiborneol (**228**). Biosynthetic work by HANSON and NYFELER (*144, 145*) using doubly-labelled mevalonates has led to elucidation of the stereochemistry of culmorin biosynthesis and the demonstration that migration of pro-5S H (in mevalonate) occurs after initial cyclization to the eleven-membered ring.

.Culmorin   (−)-Longiborneol
(**227**)   (**228**)

X. Longifolene in Industry

No account of the chemistry of longifolene would be complete without a brief reference to its use in industry. Several essential oils rich in oxygenated sesquiterpenoids find use in perfumery. Therefore, with longifolene available in commercial quantities, it is not surprising that efforts have been directed to evaluate its derivatives in perfumery compositions. A number of compounds have been found to possess desirable characteristics and their preparation as well as their use in perfumery have been covered by patents (see *e.g. 146, 147*). It should suffice to mention here that acetyl-longifolene (**65**), ω-hydroxymethyl-longifolene (**60**), and the isolongifolene ketones (**198**, both epimers) have already reached wide acceptance and are being manufactured by different companies.

98 SUKH DEV:

Acknowledgement. The author wishes to thank Dr. S. C. BISARYA for updating the literature references on the subject.

References

1. SIMONSEN, J., and D. H. R. BARTON: The Terpenes, Vol. III, p. 92—98. Cambridge: University Press. 1951.

2. OURISSON, G.: Molecular Rearrangements of Terpenes. Proc. Chem. Soc. (London) **1964,** 274.

3. LHOMME, J., and G. OURISSON: Longifolene. Recherches **15,** 15 (1966).

4. SIMONSEN, J. L.: The Constituents of Indian Turpentine from *Pinus longifolia* Roxb. J. Chem. Soc. (London) **117,** 570 (1920).

5. BISARYA, S. C.: Terpenoids. Ph. D. Thesis, Agra Univ. 1965.

6. NAYAK, U. R., and SUKH DEV: Erratum. Tetrahedron **28,** 4466 (1972).

7. OURISSON, G., S. MUNAVALLI, and C. EHRET: Data relative to Sesquiterpenoids, p. 36. Oxford: Pergamon Press. 1966.

8. HUNECK, S., and E. KLEIN: Inhaltsstoffe der Moose. III. Über die Vergleichende Gas- und Dünnschicht-Chromatographische Untersuchung der Ätherischen Öle einiger Lebermoose und die Isolierung von (−)-Longifolen und (−)-Longiborneol aus *Scapania undulata* (L.) Dum. Phytochem. **6,** 383 (1967).

9. SIMONSEN, J. L.: The Constituents of Indian Turpentine from *Pinus longifolia* Roxb. Part III. J. Chem. Soc. (London) **123,** 2642T (1923).

10. BRADFIELD, A. E., E. M. FRANCIS, and J. L. SIMONSEN: Constituents of Indian Turpentine from *Pinus longifolia* Roxb. III (continued). J. Chem. Soc. (London) **1934,** 188.

11. MOFFETT, R. H., and D. ROGERS: The Molecular Configuration of Longifolene Hydrochloride. Chem. and Ind. **1953,** 916.

12. NAFFA, P., and G. OURISSON: Chemical Approach to the Structure of Longifolene. Chem. and Ind. **1953,** 917.

13. OURISSON, G.: Molecular Rotations in the Series of Longifolene and β-Santalene. Chem. and Ind. **1953,** 918.

14. — Le Longifolène. (V). Structure absolue du longifolène, stéréochimie de ses dérivés. Stéréochimie des santalènes. Bull. soc. chim. France **1955,** 895.

15. JACOB, G., G. OURISSON, and A. RASSAT: Le longifolène. (VI). Dispersion rotatoire et configuration absolue du longifolène. Dispersion rotatoire de cétones bicycliques pontées. Bull. soc. chim. France **1959,** 1374.

16. DUPONT, G., R. DULOU, P. NAFFA, and G. OURISSON: Le longifolène. Introduction générale. Isolement et caractères physiques. Étude Chimique Préliminaire. Bull. soc. chim France **1954,** 1075.

17. SUKH DEV: Studies in Sesquiterpenes. XVIII. The Proton Magnetic Resonance Spectra of Some Sesquiterpenes and the Structure of Humulene. Tetrahedron **9,** 1 (1960).

18. HILL, H. C., R. I. REED, and M. T. ROBERT-LOPES: Mass Spectra and Molecular Structure. Part I. Correlation Studies and Metastable Transitions. J. Chem. Soc. (London) (C) **1968,** 93.

19. CESUR, A. F., and D. F. GRANT: The Crystal Structure of longifolene hydrochloride. Acta Crystallogr. **18,** 55 (1965).

20. COREY, E. J., M. OHNO, P. A. VATAKENCHERRY, and R. B. MITRA: Total Synthesis of d,1-Longifolene. J. Amer. Chem. Soc. **83,** 1251 (1961).

21. COREY, E. J., M. OHNO, R. B. MITRA, and P. A. VATAKENCHERRY: Total Synthesis of Longifolene. J. Amer. Chem. Soc. **86,** 478 (1964).

22. McMurry, J. E., and S. J. Isser: Total Synthesis of Longifolene. J. Amer. Chem. Soc. **94**, 7132 (1972).

23. Volkmann, R. A., G. C. Andrews, and W. S. Johnson: A Novel Synthesis of Longifolene. J. Amer. Chem. Soc. **97**, 4777 (1975).

24. Oppolzer, W., and T. Godel: A New and Efficient Total Synthesis of ($\pm$)-Longifolene. J. Amer. Chem. Soc. **100**, 2583 (1978).

25. Woodward, R. B., E. J. Brutschy, and H. Baer: The Structure of Santonic Acid. J. Amer. Chem. Soc. **70**, 4216 (1948).

26. Naffa, P., and G. Ourisson: Le longifolène. (III). Addition des hydracides halogenes sur le longifolène. Les halogenures de longibornyle. Produits d'isomerisation du longifolène. Bull. soc. chim. France **1954**, 1410.

27. Nayak, U. R., and Sukh Dev: Structure of Longifolene. Chem. and Ind. **1954**, 989.

28. Zeiss, H. H., and M. Arakawa: The Structure of Longifolene. The Longifolic Acids. J. Amer. Chem. Soc. **76**, 1653 (1954).

29. Prahlad, J. R., R. Ranganathan, U. R. Nayak, T. S. Santhanakrishnan, and Sukh Dev: On the Structure of Isolongifolene. Tetrahedron Letters **1964**, 417.

30. Ranganathan, R., U. R. Nayak, T. S. Santhanakrishnan, and Sukh Dev: Studies in Sesquiterpenes. XL. Isolongifolene (Part 1): Structure. Tetrahedron **26**, 621 (1970).

31. Beyler, R. E., and G. Ourisson: ω-Acetyllongifolene. J. Org. Chem. (USA) **30**, 2838 (1965).

32. Bisarya, S. C., U. R. Nayak, and Sukh Dev: Further Rearrangement of Isolongifolene. Tetrahedron Letters **1969**, 2323.

33. Santhanakrishnan, T. S., U. R. Nayak, and Sukh Dev: Studies in Sesquiterpenes. XLII. Isolongifolene (Part 3): Systematic Degradation. Tetrahedron **26**, 641 (1970).

34. Sobti, R. R., and Sukh Dev: Studies in Sesquiterpenes. XLIII. Isolongifolene (Part 4): Synthesis. Tetrahedron **26**, 649 (1970).

35. House, H. O., W. L. Respess, and G. M. Whitesides: The Chemistry of Carbanions. XII. The Role of Copper in the Conjugate Addition of Organometallic Reagents. J. Org. Chem. (USA) **31**, 3128 (1966).

36. Clunie, J. S., and J. M. Robertson: The Structure of Isoclovene. Proc. Chem. Soc. (London) **1960**, 82.

37. Berson, J. A., J. H. Hammons, A. W. McRowe, R. G. Bergman, A. Remanick, and D. Houston: The Chemistry of Methylnorbornyl Cations. VI. The Stereochemistry of Vicinal Hydride Shift. Evidence for the Nonclassical Structure of 3-Methyl-2-norbornyl Cations. J. Amer. Chem. Soc. **89**, 2590 (1967).

38. McMurry, J. E.: The Total Synthesis of Copacamphene and its Acid-catalyzed Interconversion with Sativene. J. Org. Chem. (USA) **36**, 2826 (1971).

39. Nayak, U. R., and Sukh Dev: Studies in Sesquiterpenes. XXXV. Longicyclene, the First Tetracyclic Sesquiterpene. Tetrahedron **24**, 4099 (1968).

40. Yadav, J. S., U. R. Nayak, and Sukh Dev: Studies in Sesquiterpenes. LV. Isolongifolene (Part 6): Mechanism of Rearrangement of Longifolene to Isolongifolene — I. Tetrahedron, in press.

41. Yadav, J. S., R. Soman, R. R. Sobti, U. R. Nayak, and Sukh Dev: Studies in Sesquiterpenes. LVI. Isolongifolene (Part 7): Mechanism of Rearrangement of Longifolene to Isolongifolene — II. Tetrahedron, in press.

42. Freeman, F.: Possible Criteria for Distinguishing between Cyclic and Acyclic Activated Complexes and among Cyclic Activated Complexes in Addition Reactions. Chem. Rev. **75**, 439 (1975).

43. Ogura, I., N. Taniguchi, C. Izulani, and M. Tangita: Isoprenoids. I. Longifolane. J. Pharm. Soc. Japan **76**, 1085 (1956).

44. Carnduff, J., and G. Ourisson: Le longifolène (XIII) (1). Hydrogenation du longifolène. Bull. soc. chim. France **1965**, 3297.

45. ALDER, K., G. STEIN, S. SCHNEIDER, M. LIEBMANN, E. ROLLAND, and G. SCHULZE: The Steric Course of Addition and Substitution Reactions. VI. The Exo-addition of Catalytically added Hydrogen to the Bicyclo{1,2,2}heptene Linkage. Liebigs Ann. Chem. **525**, 183, 203 (1936).

46. LHOMME, J., and G. OURISSON: Le longifolène. XIII. Hydroboration du longifolène. Tetrahedron **24**, 3167 (1968).

47. DULOU, R., and Y. CHRETIEN-BESSIERE: Reactions avec le diborane. Bull. soc. chim. France **1959**, 1362.

48. NAYAK, U. R., T. S. SANTHANAKRISHNAN, and SUKH DEV: Studies in Sesquiterpenes. XX. Acetoxymethylation of Longifolene. Tetrahedron **19**, 2281 (1963).

49. NAMBUDIRY, M. E. N., and G. S. KRISHNA RAO: Studies in Terpenoids: Part XXXVI. Manganic Acetate Oxidation of Camphene, Longifolene and β-Pinene. Indian J. Chem. **13**, 633 (1975).

50. HEIBA, E. I., R. M. DESSAU, and W. J. KOEHL, Jr.: Oxidation by Metal Salts. IV. A New Method for the Preparation of γ-lactones by the reaction of Manganic Acetate with Olefins. J. Amer. Chem. Soc. **90**, 5905 (1968).

51. MEERWEIN, H., and K. VAN EMSTER: Über die Gleichgewichts-Isomerie zwischen Bornyl chlorid, Isobornyl chlorid und Camphen-Chlorhydrat. Ber. dtsch. chem. Ges. **55**, 2500 (1922).

52. OLAH, G. A.: The σ-Bridged 2-Norbornyl Cation and its Significance to Chemistry. Accounts Chem. Res. **9**, 41 (1976).

53. BROWN, H. C.: The Nonclassical Ion Problem. New York: Plenum Press. 1977.

54. ROBERTt, J. D., and J. A. YANCEY: Mechanisms of Racemization of Camphene-8-C[14]. J. Amer. Chem. Soc. **75**, 3165 (1953).

55. OURISSON, P., and G. OURISSON: Le longifolène (IV). Mecanisme de l'addition des hydracides halogenes sur le longifolène. Mecanisme de la solvolyse des halogenures de longibornyle. Bull. soc. chim. France **1954**, 1415.

56. GREAM, G. E., and D. WEGE: Conversion of (+)-Camphor to the Enantiomeric Hydrocamphenyl-Isobornyl Cations by the σ- and π-routes of Solvolysis. Tetrahedron Letters **1964**, 535.

57. GOKHALE, P. D., A. P. JOSHI, R. SAHNI, V. G. NAIK, N. P. DAMODARAN, U. R. NAYAK, and SUKH DEV: Photochemical Transformations. I. Reactions of some Terpene Iodides. Tetrahedron **32**, 1391 (1976).

58. NAYAK, U. R., and SUKH DEV: Studies in Sesquiterpenes. XVII. Hydration of Longifolene. Tetrahedron **8**, 42 (1960).

59. PRAHLAD, J. R., U. R. NAYAK, and SUKH DEV: Studies in Sesquiterpenes. XLV. Structure of an alcohol from Hydration of Longifolene. Tetrahedron **26**, 663 (1970).

60. AKIYOSHI, S., H. ERDTMAN, and T. KUBOTA: Chemistry of the Natural Order Cupressales. XXVI. The Identity of Junipene, Kuromatsuene and Longifolene and of Juniperol, Kuromatsuol, Macrocarpol and Longiborneol. Tetrahedron **9**, 237 (1960).

61. BISARYA, S. C., U. R. NAYAK, SUKH DEV, B. S. PANDE, J. S. YADAV, and H. P. S. CHAWLA: Studies in Sesquiterpenes. LIX. Isolongifolene (Part 9): Further Rearrangement. J. Indian Chem. Soc. **55**, 1138 (1978).

62. KETTENS, D. K., J. B. H. VAN LIEROP, B. VAN DER WAL, and G. SIMPA: Musk Odorants. VI. Formation of 1,1-Dimethyl-7-isopropyltetralin *via* a Novel Rearrangement-Dehydrogenation of Longifolene. Rec. trav. chim. Pays-Bas **88**, 313 (1969).

63. MEHTA, G.: An Efficacious Aromatisation of Isolongifolene in Trifluoroacetic Acid. Chem. and Ind. **1972**, 766.

64. FARNUM, D. G., R. A. MADER, and G. MEHTA: Rearrangements of Longifolene through Stable Carbonium Ions. Preparation of Novel C_{15} Hexahydronaphthalenes. J. Amer. Chem. Soc. **95**, 8692 (1973).

65. SURYAWANSHI, S. N., and U. R. NAYAK: Private Communication.

66. COATES, R. M., and J. P. CHEN: Tricyclic Olefins from Solvolysis of Longicamphenilyl Tosylate. Chem. Commun. **1970**, 1481.e

67. LHOMME, J., and G. OURISSON: Le longifolène. XVI. Rearrangements dans la serie du longifolène. Obtention de squelettes tricycliques nouveaux. Bull. soc. chim. France **1970**, 3935.

68. MARMOR, S., and J. G. MAROSKI: Reactions of Hypochlorous Acid with Hindered Olefins. J. Org. Chem. (USA) **31**, 4278 (1966).

69. BROWN, H. C.: Chemical Effects of Steric Strains. J. Chem. Soc. (London) **1956**, 1248.

70. TIDWELL, T. T.: Sterically Crowded Organic Molecules: Synthesis, Structure and Properties. Tetrahedron **34**, 1855 (1978).

71. YADAV, J. S., H. P. S. CHAWLA, SUKH DEV, A. S. C. PRAKASA RAO, and U. R. NAYAK: Steric Diversion. I. Addition of Halogens and Pseudo-halogens to Isolongifolene. Tetrahedron **33**, 2441 (1977).

72. BARTLETT, P. D., G. L. FRASER, and R. B. WOODWARD: The Isolation and Properties of 1,1-Dineopentylethylene, a Component of Triisobutylene. J. Amer. Chem. Soc. **63**, 495 (1941).

73. WHITMORE, F. C., and J. D. SURMATIS: Polymerisation of Olefins. VII. The Isolation and Oxidation of 1,1-Dineopentylethylene. J. Amer. Chem. Soc. **63**, 2200 (1941).

74. MEHTA, G., S. K. KAPOOR, and B. G. B. GUPTA: Electrophilic Additions to Longifolene: Addition of Bromine, Iodine Monochloride and Phenylsulphenyl Chloride. Indian J. Chem. **14 B**, 364 (1976).

75. GRAF, R.: New Methods of Preparative Organic Chemistry. VI. Reactions with N-Carbonylsulfamoyl Chloride. Angew. Chem. internat. Edit. **7**, 172 (1968).

75 a. SASAKI, T., S. EGUCHI, and H. YAMADA: Studies on Reactions of Isoprenoids. XVIII. Reactions of Chlorosulfonyl Isocyanate with Bicyclic Monoterpene Olefins. J. Org. Chem. (USA) **38**, 679 (1973).

76. MEHTA, G., D. N. DHAR, and S. C. SURI: Addition of Chlorosulphonyl Isocyanate to Some Polycyclic Sesquiterpenes, Longifolene, Isolongifolene, α-Cedrene, Caryophyllene and Thujopsene. Indian. J. Chem. **16 B**, 87 (1978).

77. SURYAWANSHI, S. N., and U. R. NAYAK: Unique Mercuration-Halogenation of Isoprenoid Ectocyclic Olefins: Novel ωω-Dihalides from Longifolene and Camphene *via* ωω-Dimercurichlorides. Tetrahedron Letters **1978**, 4425.

78. MEHTA, G.: Base-Catalyzed Rearrangement of ώ-Bromolongifolene. J. Org. Chem. (USA) **36**, 3455 (1971).

79. MEHRA, M. M., B. B. GHATGE, and S. C. BHATTACHARYYA: Terpenoids. LXVI. Ring Enlargement Produced by the Alkaline Fusion of ω-Bromolongifolene. Tetrahedron **21**, 637 (1965).

80. ERICKSON, K. E., and J. WOLINSKY: Rearrangement of Bromomethylenecycloalkanes with Potassium t-Butoxide. J. Amer. Chem. Soc. **87**, 1142 (1965).

81. NAFFA, P., and G. OURISSON: Le longifolène (II). Produits d'oxydation du longifolène. Bull. soc. chim. France. **1954**, 1115.

82. OGURA, I.: The Relationship between Longifolic and Isolongifolic Acids. Bull. Chem. Soc. Japan **29**, 363 (1956).

83. NAYAK, U. R., and SUKH DEV: Studies in Sesquiterpenes. XXI. Longifolic Acids. Tetrahedron **19**, 2293 (1963).

84. MEHTA, G., U. R. NAYAK, and SUKH DEV: Studies in Sesquiterpenes. XXXVI. Structure of ψ-Longifolic Acid. Tetrahedron **24**, 4105 (1968).

85. WIBERG, K. B.: Oxidation by Chromic Acid and Chromyl Compounds. In: Oxidation in Organic Chemistry, Part A (WIBERG, K. B., ed.), pp. 125—131. New York: Academic Press. 1965.

86. NAYAK, U. R., and SUKH DEV: Studies in Sesquiterpenes. XIX. Action of Perbenzoic Acid on Longifolene. Tetrahedron **19**, 2269 (1963).

 87. TREIBS, W., and H. SCHMIDT: Oxidation of Reactive Methylene Groups. Ber. dtsch.
     chem. Ges. **61,** 459 (1928).
 88. HICKINBOTTOM, W. J., and D. G. M. WOOD: Reactions of Unsaturated Compounds.
     Part X. The Oxidation of Camphene by Chromic Acid. J. Chem. Soc. (London) **1953,** 1906.
 89. LHOMME, J., and G. OURISSON: Le longifolène. (VIII). Mechanisme de l'oxydation du
     longifolène par les peracids. Bull. soc. chim. France **1964,** 1888.
 90. WEISENBORN, F. L., and D. TAUB: The Reaction of Perbenzoic Acid with certain
     Olefins. J. Amer. Chem. Soc. **74,** 1329 (1952).
 91. JOSHI, A. P., U. R. NAYAK, and SUKH DEV: Studies in Sesquiterpenes. L. 3-Hydroxy-
     longifolaldehyde, the Elusive Intermediate in the Abnormal Perbenzoic Acid Oxidation
     of Longifolene. Tetrahedron **32,** 1423 (1976).
 92. CHEN, SHI-CHOW: Molecular Rearrangements in Lithium Aluminum Hydride Reduc-
     tion. Synthesis **1974,** 691.
 93. NAYAK, U. R., and SUKH DEV: Epimerisation of α-Longiforic Acid. Chem. and Ind.
     1959, 1157.
 94. KUBOTA, T., and J. OGURA: 4:7-Dimethylazulene from Longifolene. Chem. and Ind.
     1958, 951.
 95. MUNAVALLI, S., and G. OURISSON: Le longifolène. IX. Deshydrogenation de derives
     bicycliques du longifolène. Bull. soc. chim. France **1964,** 2822.
 96. WRAGG, W. R.: Dosage des doubles liaisons par les peracids. Bull. soc. chim. France
     1952, 911.
 97. SIMONSEN, J. L., and L. N. OWEN: The Terpenes, Vol. II, p. 280. Cambridge: Uni-
     versity Press. 1949.
 98. GHATGEY, B. B., and S. C. BHATTACHARYYA: Longifolene Fraction of Indian Turpentine
     Oil. I. New Derivatives of Longifolene and the Isolation of a New Hydrocarbon β-
     Longifolene. Perf. Essent. Oil Rec. **47,** 122 (1956).
 99. MUNAVALLI, S., and G. OURISSON: Le longifolène. VII. Structure de la longihomo-
     camphenylone. Produits d'oxydation du longifolene. Bull. soc. chim. France **1964,** 729.
100. BAILEY, P. S.: Ozonation in Organic Chemistry, Vol. I, p. 197. New York: Academic
     Press. 1978.
101. — Ozonation in Organic Chemistry, Vol. I, p. 170. New York: Academic Press. 1978.
102. MORIARTY, R. M.: The Lead Tetraacetate Oxidation of Olefins. In: Selective Organic
     Transformations, Vol. 2 (THYAGARAJAN, B. S., ed.), p. 183. New York: Wiley-Inter-
     science. 1972.
103. BAIRD, M. S.: Aspects of the Chemistry of Seven- to Eleven-membered Rings. In:
     MTP International Review of Science, Organic Chemistry, Series One, Vol. 5 (PARKER,
     W., ed.), p. 222. London: Butterworths. 1973.
104. WALLING, C., and E. S. HUYSER: Free Radical Additions to Olefins to Form Carbon-
     Carbon Bonds. In: Organic Reactions, **13,** 91 (1963).
105. HELMLINGER, D., and G. OURISSON: Über Longifolen, XI. Homolytische und Hetero-
     lytische Transannulare Wasserstoffverschiebungen in der Longifolenreihe. Liebigs
     Ann. Chem. **686,** 19 (1965).
106. DOV ELAD: Photochemical Additions to Multiple Bonds. In: Organic Photochemistry,
     Vol. 2 (CHAPMAN. O. L., ed.). p. 168. New York: Marcel Dekker. 1969.
107. FISCH, M., and G. OURISSON: Le longifolène XIV. Reaction photochimique du form-
     amide avec le longifolène. Bull. soc. chim. France **1966,** 1325.
108. TANAHASHI, Y., J. LHOMME, and G. OURISSON: Le longifolène. XVII. Autoxydation de
     longifolyl-borane. Tetrahedron **28,** 2655 (1972).
109. LHOMME, J., and G. OURISSON: Le longifolène. XIV. Oxydation des longifolols par le
     tetracetate de plomb. Tetrahedron **24,** 3177 (1968).
110. TANAHASHI, Y., J. LHOMME, and G. OURISSON: Le longifolène. XVIII. Solvolyse des
     tosylates de (7αH)-longifolanyle-3α et 3β. Tetrahedron **28,** 2663 (1972).

111. STEHELIN, L., J. Lhomme, and G. OURISSON: Transannular Hydride Shifts. A Mechanistic Study in the Longifolene Series. J. Amer. Chem. Soc. **93**, 1650 (1971).
112. HELMLINGER, D., and G. OURISSON: Le longifolène. XV. Cyclisation transannulaire du bromo-3α longifolene. Tetrahedron **25**, 4895 (1969).
113. THIERRY, J.-C., and R. WEISS: Études cristallographiques en serie Sesquiterpenique. I. Structures cristallines et moleculaires du bromo-3α longifolene, du bromo-3α (7βH) longifolane et du bromo-7 cyclo (3:15) longifolane. Tetrahedron Letters **1969**, 2663.
114. GUPTA, A. S., and SUKH DEV: Chemistry of Longibornan-9 Cation. Eighth I.U.P.A.C. Symposium on the Chemistry of Natural Products, Abstracts Book, p. 202. New Delhi. 1972.
115. MIHAILOVIC, M. L., and R. E. PARTCH: Alcohol Oxidation by Lead Tetraacetate. In: Selective Organic Transformations, Vol. 2 (THYAGARAJAN, B. S., ed.), p. 97. New York: Wiley-Interscience. 1972.
116. JADHAV, P. K., and U. R. NAYAK: Novel Hypoiodite Functionalization of Longifolol: Synthesis of Longifol-7 (15)-en-5β-ol and Longifolan-3α,7α-oxide Sesquiterpenes from *Juniperus conferta*. Tetrahedron Letters **1976**, 4855.
117. DOI, K., T. SHIBUYA, T. MATSUO, and S. MIKI: Longifol-7(15)-en-5β-ol and Longifolan-3α,7α-oxide: New Sesquiterpenes from *Juniperus conferta* Parl. Tetrahedron Letters **1971**, 4003.
118. PATNEKAR, S. G., and S. C. BHATTACHARYYA: Terpenoids. XCIV. Synthesis of Novel Longifolane Derivatives *via* Oxidation with Lead Tetraacetate. Tetrahedron **23**, 919 (1967).
119. LHOMME, J., and G. OURISSON: Lead Tetra-acetate Oxidation of the Longifolols. Formation of Cyclopropane Derivatives. Chem. Commun. **1967**, 436.
120. SCOTT, A. I.: Biogenetic Type Syntheses. In: Techniques of Chemistry, Vol. X, Part II (WEISSBERGER, A., ed.), p. 555. New York: Wiley-Interscience. 1976.
121. COATES, R. M.: Biogenetic-Type Rearrangements of Terpene. In: Progress in the Chemistry of Organic Natural Products, Vol. 33 (HERZ, W., H. GRISEBACH, and G. W. KIRBY, eds.), p. 73. Wien-New York: Springer. 1976.
122. MEHTA, G., and S. K. KAPOOR: Terpenes and Related Systems. IX. A Synthesis of (+)-Himachalene Dihydrochloride and (+)-*ar*-Himachalene. J. Org. Chem. (USA) **39**, 2618 (1974).
123. DE MAYO, P., and R. E. WILLIAMS: Sativene, Parent of the Toxin from *Helminthosporium sativum*. J. Amer. Chem. Soc. **87**, 3275 (1965).
124. JOSEPH, T. C., and SUKH DEV: Studies in Sesquiterpenes. XXIX. Structure of Himachalenes. Tetrahedron **24**, 3809 (1968).
125. DORN, F., and D. ARIGONI: Ein bicyclischer Abkömmling von (−)-Longifolen aus *Helminthosporium sativum* und *H. victoriae*. Experientia **30**, 851 (1974).
126. YADAV, J. S., H. P. S. CHAWLA, and SUKH DEV: Cleavage of Homoallylic Alcohols. A Novel Fragmentation Reaction. Tetrahedron Letters **1977**, 201.
127. GAITONDE, M., P. A. VATAKENCHERRY, and SUKH DEV: Cyclopropane Ring Cleavage with N-Bromosuccinimide. Tetrahedron Letters **1964**, 2007.
128. YADAV, J. S., H. P. S. CHAWLA, and SUKH DEV: Synthesis of (−)-Secolongifolene Diol. Tetrahedron Letters **1977**, 1749.
129. MEHTA, G., and U. R. NAYAK: Selenium Dioxide Oxidation of Ectocyclic Olefins: Formation and Structures of Selenides from Longifolene and Camphene. Indian J. Chem. **15B**, 419 (1977).
130. COWAN, D. O., and R. L. DRISKO: Elements of Organic Photochemistry, p. 181. New York: Plenum Press. 1976.
131. JOSHI, A. P.: Ph. D. Thesis. Poona: University. 1972.
132. McMILLAN, J. A., I. C. PAUL, U. R. NAYAK, and SUKH DEV: Molecular Structure of Isolongifolene Epoxide. Tetrahedron Letters **1974**, 419.

133. Greengrass, C. W., and R. Ramage: Stereochemical Studies of Tricyclo{6.2.1.0.^{1,6}} undecanes. II. Stereochemistry of Isolongifolene epoxide. Tetrahedron 31, 689 (1975).
134. Santhanakrishnan, T. S., R. R. Sobti, U. R. Nayak, and Sukh Dev: Studies in Sesquiterpenes. XLIV. Isolongifolene (Part 5): Rearrangement of Isolongifolene Epoxide. Tetrahedron 26, 657 (1970).
135. Coates, R. M.: Biogenetic-Type Rearrangements of Terpenes. In: Progress in the Chemistry of Organic Natural Products, Vol. 33 (Herz, W., H. Grisebach, and G. W. Kirby, eds.), p. 82. Wien-New York: Springer. 1976.
136. Bisarya, S. C., and Sukh Dev: unpublished results.
137. Eschinasi, E. H., G. W. Shaffer, and A. P. Bartels. The Aluminium Alkoxide Rearrangement of Epoxides. Part III. Rearrangement of Isolongifolene Epoxide. Tetrahedron Letters 1970, 3523.
138. Wiberg, K. B., B. A. Hess, Jr., and A. J. Ashe: Cyclopropyl-carbinyl and Cyclobutyl Cations. In: Carbonium Ions, Vol. III (Olah, G. A., and P. von R. Schleyer, eds.), p. 1295. New York: Wiley-Interscience. 1972.
139. Yadav, J. S., H. P. S. Chawla, and Sukh Dev: Studies in Sesquiterpenes. LVII. Isolongifolene (Part 8): Solvolysis of 8-Bromoneoisolongifolene. Tetrahedron 34, 475 (1978).
140. Hendrickson, J. B.: Stereochemical Implications in Sesquiterpene Biogenesis. Tetrahedron 7, 82 (1959).
141. Sandermann, W., and K. Bruns: Über die Biogenese von Longifolen in *Pinus longifolia* Roxb. Tetrahedron Letters 1962, 261.
142. Arigoni, D.: Stereochemical Aspects of Sesquiterpene Biosynthesis. Pure Appl. Chem. 41, 219 (1974).
143. Barton, D. H. R., and N. H. Werstiuk: The Constitution and Stereochemistry of Culmorin. Chem. Commun. 1967, 30.
144. Hanson, J. R., and R. Nyfeler: The Mevalonoid Origin of the Hydrogen Atoms of Culmorin. Chem. Commun. 1975, 171.
145. — — Stereochemistry of Culmorin Biosynthesis. Chem. Commun. 1975, 824.
146. Ansari, H. R., and A. J. Curtis: Sesquiterpenes in the Perfumery Industry. J. Soc. Cosmet. Chem. 25, 203 (1974).
147. James, R. W.: Fragrance Technology, Synthetic and Natural Perfumes, p. 27. New Jersey: Noyes Data Corporation. 1975.

(Received June 11, 1979)

Homoisoflavanones and Biogenetically Related Compounds

By W. HELLER and CH. TAMM, Institut für Organische Chemie, Universität Basel, Switzerland

With 3 Figures

Contents

1. Introduction

The homoisoflavanones belong to a small family of natural products whose first member was isolated by BOEHLER and TAMM in 1967 from bulbs of *Eucomis bicolor* Bak. (*9*). Their discovery resulted from a systematic chemical analysis of *Liliaceae* for cardiac glycosides. However, unlike the botanically closely related *Urginea maritima* (L.) Bak. (Squill) and some species of *Scilla, Ornithogalum* and *Dipcadi* (*33*), *Eucomis* plants did not contain even traces of these compounds.

The genus *Eucomis* is indigenous to southern and eastern Africa. The widespread *E. autumnalis* (Mill.) Chitt. (Syn. *E. undulata* Ait.) has been used by native medical doctors against rheumatism, veneric, mental and many other diseases. A cardiotonic action has not been observed (*73*). A botanical survey revising this genus had been published by REYNEKE in 1972 (*61*).

Eucomin and eucomol are the main metabolites in *E. bicolor* (*9*). Later 3,9-dihydroeucomin, the three 7-O-methyl as well as the 4'-demethyl derivatives and 4'-demethyl-5-O-methyl-3,9-dihydroeucomin were isolated from this plant as minor constituents (*38, 64*). These compounds which show a 4',5,7-oxygenation pattern, are called "eucomins". "Eucomnalins" and "punctatins", bearing an additional methoxy group at C-6 and C-8 respectively, have been isolated from *E. comosa* (Houtt.) Wehrh. (Syn. *E. punctata* L'Hérit.) (*27*) and *E. autumnalis* (*65*) together with some metabolites which belong to the eucomin series (*26, 37*) (Table 1)[1].

The three benzylidene-type derivatives were usually characterized as their (*E*)-isomers. FINCKH and TAMM (*27*) observed the simultaneous occurrence of (*E*)- and (*Z*)-4'-O-methylpunctatin. As was demonstrated later by chromatographic methods this compound as well as eucomin occur exclusively as (*Z*)-isomers in the living plant (*38*).

The isolation of the homoisoflavanones was simplified to a great extent by the observation that these phenolic compounds are accumulated specifically in a waxy layer between the storage leaves of the bulb.

---

[1] Eucomnalin is identical with the former autumnalin (*65*), a name which had already been conferred on an alkaloid from *Colchicum autumnale* L.

Other metabolites like 5,7-dihydroxy-8-methoxychroman-4-one (*37*) and possibly a coumarin derivative (*35*) have also been found in the bulb wax of *E. comosa*. From the bulb tissue of different *Eucomis* spp. a series of sterol derivatives, dibenzo-α-pyrones and a dicarboxylic acid have been isolated (*69, 82*).

Another source of homoisoflavanones is *Scilla scilloides* Druce. KOUNO *et al.* (*44*) have described the isolation and identification of 3,9-dihydro-eucomnalin along with the structurally related compounds scillascillin and 2-hydroxy-7-O-methylscillascillin. The new structural feature of these substances is a 3-spirocyclobutene ring system. The name "scil-lascillins" was adopted for this group of compounds.

DEWICK elucidated the biosynthesis of eucomin (*18, 19*). He could demonstrate a relationship to the flavanoids by administering various biogenetic precursors including suitable chalcone intermediates to the plants. The additional carbon atom which is present in the skeleton of the homoisoflavanones was shown to be C-2. It originates from the methyl group of L-methionine. Their possible role as precursors for the scillascillins as well as brazilin and hematoxylin was also discussed (*19*).

Brazilin and hematoxylin were first isolated a long time ago from the heart wood of *Caesalpinia* spp. and *Haematoxylon campechianum* L. *(Leguminosae)* respectively. Aqueous plant extracts have been widely used as mordant dyes especially on silk and furs (*63*). The extract of the Mexican species *H. braziletto* was found to exhibit antibacterial activity (*59*). It was assumed that these compounds belong to the class of the neoflavonoids (*20, 31*).

The chemical and physical properties of the homoisoflavanones and of further metabolites isolated from *Eucomis* spp. have been summarized earlier by TAMM (*69*) and ZIEGLER and TAMM (*82*).

2. Isolation and Identification

2.1. Isolation

Early in the investigations the bulbs were stored in ethanol and were macerated and extracted with mixtures of ethanol containing increasing amounts of water. After filtration over hyflo super cel or celite and concentration *in vacuo* the different metabolites were extracted subsequently with petroleum ether (b. p. 50—70°), diethyl ether, chloro-form, and chloroform-methanol (2 : 1). Emulsions, frequently occurring with petroleum ether, are prevented by careful stirring or addition of a

W. Heller and Ch. Tamm:

Table 1. *Natural Homoisoflavanones*

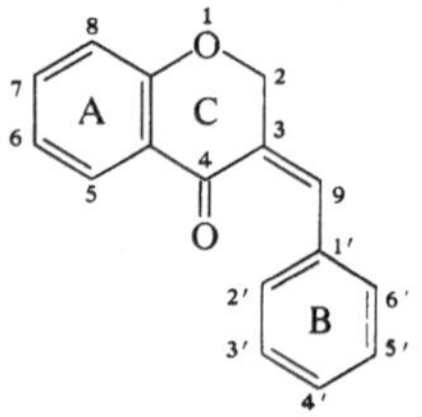

I (Z)-, *cis-*

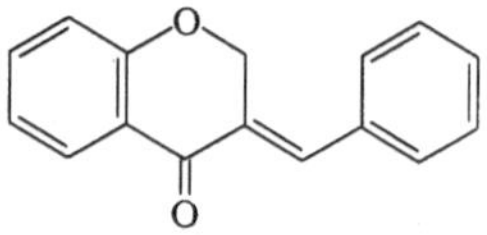

II (E)-, *trans-*

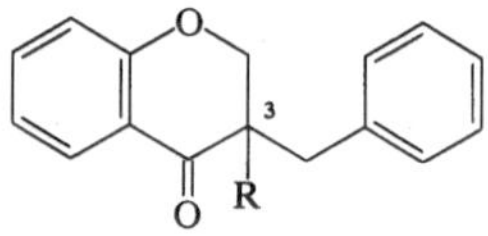

III R = H
IV R = OH

Com-pound	Name (Series)	3	5	6	7	8	4'	Structural Type
(1)		—[2]	OH [3]		OH		OH	II
(2), (3)	Eucomin	—	OH		OH		OCH$_3$	I, II
(4)		—	OH		OCH$_3$		OCH$_3$	II
(5)	3,9-Dihydroeucomin		OH		OH		OH	III
(6)			OH		OH		OCH$_3$	
(7)			OH		OCH$_3$		OCH$_3$	
(8)			OCH$_3$		OH		OH	
(9)	Eucomol	OH	OH		OH		OH	IV
(10)		OH	OH		OH		OCH$_3$	'
(11)		OH	OH		OCH$_3$		OCH$_3$	
(12)	Eucomnalin	—	OH	OCH$_3$	OH		OH	II
(13)	3,9-Dihydro-eucomnalin		OH	OCH$_3$	OH		OH	III
(14)	Punctatin	—	OH		OH	OCH$_3$	OH	II
(15), (16)		—	OH		OH	OCH$_3$	OCH$_3$	I, II
(17)	3,9-Dihydro-punctatin		OH		OH	OCH$_3$	OH	III
(18)			OH		OH	OCH$_3$	OCH$_3$	

References, pp. 149—152

Melting Point (C)	$[\alpha]_D$	(solvent)	Source[1]	References
209—213°	0°		(b)	(27)
143—145°/199—201°	0°		(a) (b) (c)	(9, 26, 38, 65)
147—148°	0°		(b)	(38)
			(b)	(64)
161—163°	+38°	(chloroform)	(b) (c)	(37, 38)
	+23°	(chloroform)[4]	(b)	(38)
196—197°	−38°	(dioxane)	(b) (c)	(27, 64)
105—107°[5]	−90°	(acetone)[5]	(b)	(64)
133—134°	−32°/−26°	(chloroform)	(b)	(9, 38)
83— 84°	−27°	(chloroform)	(b)	(38)
244—247°	0°		(a)	(65)
207—209°	−10°	(dioxane)	(a) (d)	(44, 65)
189—190°	0°		(c)	(27)
213—214°[6]	0°		(a) (c)	(26, 65)
204—206°	−37°	(dioxane)	(c)	(27)
			(c)	(27)

Table 1 (continued)

Com-pound	Name (Series)		Structural Type
(19)	Scillascillin	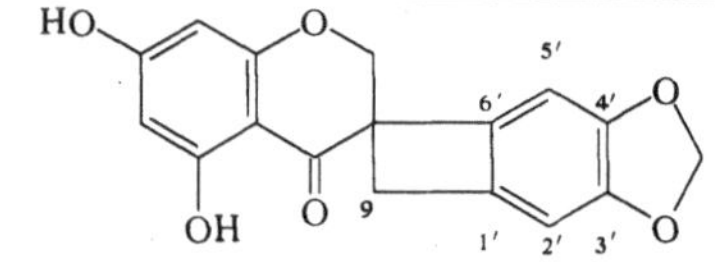	
(20)	2-Hydroxy-7-O-methyl-scillascillin	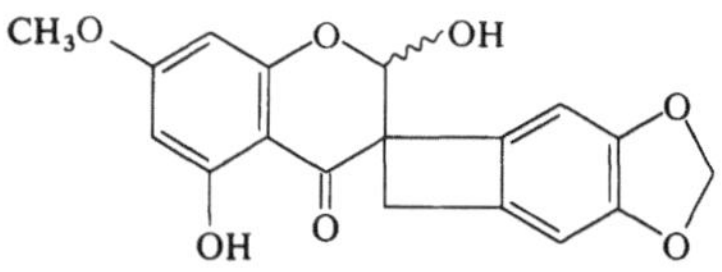	
(21)	Brazilin	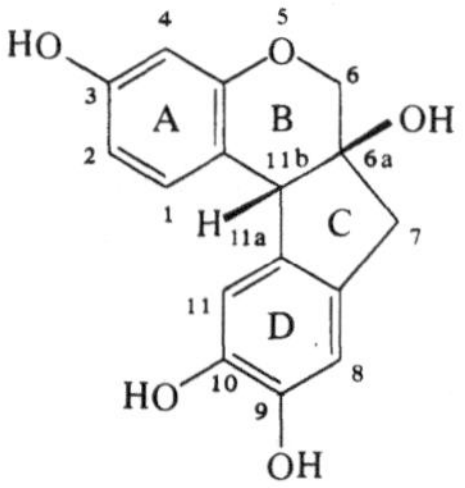	
(22)	Hematoxylin	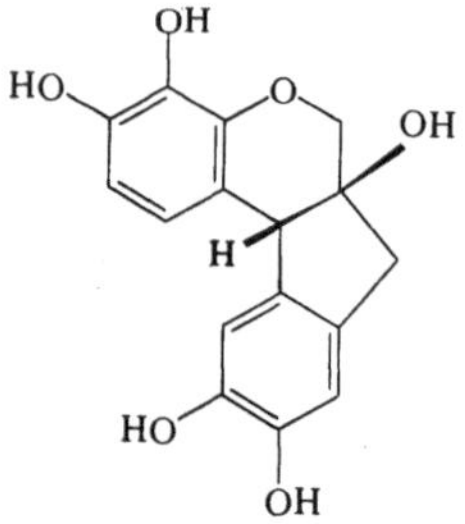	

[1] (a) E. autumnalis, (b) E. bicolor, (c) E. comosa, (d) Scilla scilloides, (e) Haematoxylon spp., (f) Caesalpinia spp.

[2] Double bond.

[3] No mark means hydrogen.

[4] Synthesized from (6).

Melting Point (C)	$[\alpha]_D$	(solvent)	Source[1]	References
210—211°	+40°	(dioxane)	(d)	(44)
188—189°	−11,5 → +79.2° (18 h, dioxane)		(d)	(44)
130°/250°[7]	+121°	(methanol)	(e) (f)	(14, 63, 76)
100°/120°/140°[8]	+99°	(water)	(e)	(14, 16, 63, 76)

[5] Synthesized from (10).

[6] (16).

[7] Colourless needles (1.5 H_2O), prisms (1 H_2O). Values are only approximate due to decomposition.

[8] Values uncertain, probably due to decomposition.

small amount of methanol (81). With chloroform emulsions which separated in three phases were formed. Centrifugation was necessary. The homoisoflavanones are found in the diethyl ether fraction and can be further purified by extraction with 2 N sodium carbonate solution (26). No additional material is obtained after treatment of the crude extract with mineral acids, thus demonstrating the absence of glycosidic components (19).

Later on, morphological studies of fresh bulbs revealed that the homoisoflavanones are located exclusively between the fleshy storage leaves accompanied by waxy material (69). They can be washed off easily with diethyl ether in a soxhlet apparatus. The general validity of this method is discussed by Martin and Batt (50). Washing the extract with water, evaporation of the solvent and partition of the sticky material between petroleum ether (b.p. 50—70°) and methanol-water (8 : 2) for removing the waxy compounds yields the homoisoflavanones in remarkable purity (38).

2.2. Chromatography

Separations are usually performed on silica gel columns using methylene chloride and methanol as solvents. The homoisoflavanones are eluted by methylene chloride containing 0.5—5% methanol. The elution depends upon the number of hydroxyl groups and the degree of methylation. The main metabolites can be purified directly by crystallization of the crude fractions. Further purification, especially of the minor components, is achieved by repeated column or layer chromatography. Mixtures of methylene chloride or chloroform with methanol or benzene with methanol are generally used, the latter showing higher resolution (35). An effective separation of 3,9-saturated compounds from those containing a 3(9)-double bond is achieved by mixtures of petroleum ether (b. p. 50—70°)-diethyl ether with 1% methanol in order to prevent tailing (38) or diethyl ether-dioxane-methanol (80 : 19 : 1) (64).

In some cases pure diethyl ether or mixtures of benzene or cyclohexane with diethyl ether or ethyl acetate have been used (26).

For proving the homogeneity of radioactively labelled eucomin and degradation products, Dewick (19) used five different solvent systems: (A) benzene-ethyl acetate-methanol-petroleum ether (b. p. 60—80°) (6 : 4 : 1 : 8); (B) toluene-ethyl formate-formic acid (5 : 4 : 1); (C) chloroform-isopropanol (10 : 1); (D) benzene-ethyl acetate-methanol-petroleum ether (b.p. 60—80°) (6 : 4 : 1 : 3); (E) benzene-ethanol (92 : 8).

Since (Z), (E)-isomerization of naturally-occurring homoisoflavanones with a 3(9)-double bond takes place very readily, rapid and mild purification procedures are required. Thus (Z)-eucomin (2) has been isolated by preparative layer chromatography of a crude extract of E. *bicolor* on silica gel with benzene-methanol (95 : 5, twice) and elution with chloroform-methanol (95 : 5). Careful evaporation and crystallization from chloroform-hexane gave (Z)-eucomin (2) in fine yellow needles. Mixtures of isomers are only separated by TLC on polyamide (*38*).

For two-dimensional chromatographic analysis of crude extracts, a small set of solvent systems has been selected (*35*) (Table 2). — The combinations of chloroform-methanol (95 : 5) (No. 4) with petroleum ether (b. p. 50—70°)-diethylether-methanol (40 : 60 : 1, twice) (No. 2) for less polar and of chloroform-methanol (9 : 1) (No. 5) and diethyl ether-dioxane-methanol (90 : 10 : 1) (No. 6) for more polar compounds have been used on silica gel. Good separation on commercially available polyamide-11 plates is achieved by petroleum ether (b. p. 50—70°)-benzene-butanone-methanol (45 : 35 : 10 : 10) (No. 7) and chloroform-butanone-methanol (60 : 26 : 14) (*39, 67*).

Polyamide has also been used in column chromatography. A mixture of (Z)-eucomin (2), 3,9-dihydroeucomin (6) and eucomol (10) has been sucessfully separated by petroleum ether (b. p. 50—70°)-benzene (6 : 5) and increasing amounts of butanone-methanol (1 : 1), beginning with 2 per cent (*38*).

The chromatographic data are summarized in Table 2. For the detection of the spots in TLC ammonia vapors can be used as a general reagent. It intensifies the yellow colour of the unsaturated metabolites as well as the originally slight fluorescence of eucomol and its derivatives under long wave ultraviolet light. Subsequent spraying with 0.05% aqueous fast blue B salt produces very specific colors. With 1% ethanolic aluminium chloride all saturated compounds show a white fluorescence which turns yellow on heating.

3. Structure and Nomenclature

3.1. General Aspects and Nomenclature

The spectroscopic behaviour of the homoisoflavanones resembles very closely that of flavonoids (*48*). The data clearly indicate the presence of sixteen carbon atoms in the basic skeleton and are consistent only with a 3-benzyl(idene)chroman-4-one structure. This has been proven by synthesis of (E)-5,7-dimethoxy-3(4-methoxybenzylidene)chroman-4-one (33) which was identical with the di-O-methyl derivative of (E)-eucomin (3) (*9*).

Table 2. Chromatographic Data of Natural Homoisoflavanones and Chromanones

Eucomins

Structural Type[1]	Compound	R[1]	R[2]	R[3]	Solvent Systems[2]								FBB[3]
					1	2	3	4	5	6	7	8	
I (Z)-Δ[3(9)]	(2)	CH₃	H	H	4[4]						36		brv
II (E)-Δ[3(9)]	(1)	H	H	H	15	25	8	19	40	58	6	22	brv
	(3)	CH₃	H	H	29	46	31	49	62	68	30	58	brv
	(4)	CH₃	H	CH₃	47	67	75	76	80	76	77	89	och[5]
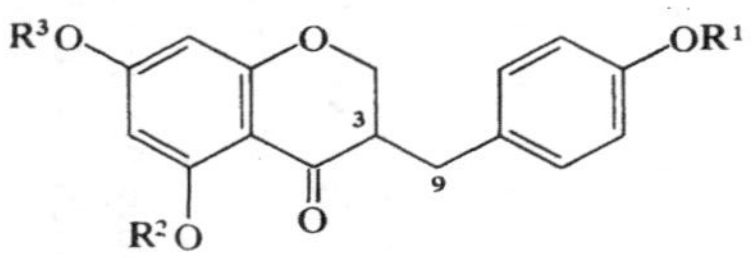 III	(5)	H	H	H	19	32	10	20	41	63	16	42	v
	(6)	CH₃	H	H	36	54	34	52	64	72	44	70	bjv
	(7)	CH₃	CH₃	CH₃	55	75	76	77	82	78	88	93	rv[5]
	(8)	H	CH₃	H		4	4	11	38	21	13	46	r
IV 3-OH	(9)	H	H	H	12	21	3	14	33	55	8	42	rshv
	(10)	CH₃	H	H	25	40	19	40	55	67	27	59	v
	(11)	CH₃	H	CH₃	42	60	58	72	78	72	70	88	rv[5]

Punctatins and Eucomnalins

Structural Type[1]	Compound	R[1]	R[2]	R[3]	Solvent Systems[2]								FBB[3]
					1	2	3	4	5	6	7	8	
I (Z)-Δ[3(9)]	(15)	CH₃	H	OCH₃	4[4]						43		y
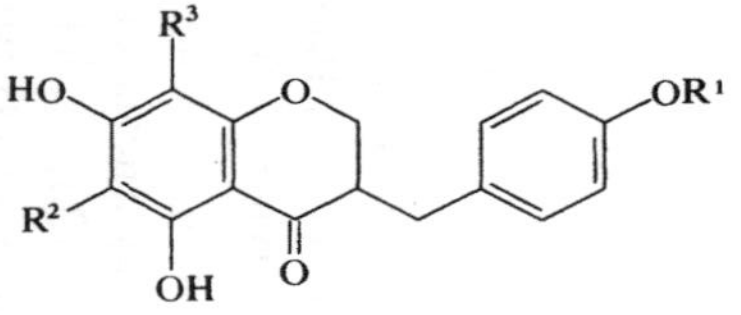 II (E)-Δ[3(9)]	(12)	H	OCH₃	H	14	22	11	24	43	55	9	31	ybr
	(14)	H	H	OCH₃	10	16	10	20	42	50	10	34	ybr
	(16)	CH₃	H	OCH₃	21	35	33	58	68	61	41	71	y
III	(13)	H	OCH₃	H	19	31	12	25	45	60	19	50	orr
	(17)	H	H	OCH₃	14	26	11	22	44	58	21	53	vr
	(18)	CH₃	H	OCH₃	29	46	36	59	69	67	52	78	rv

Scillascillins and Chromanones

	R[1]		1	2	3	4	5	6	7	8	
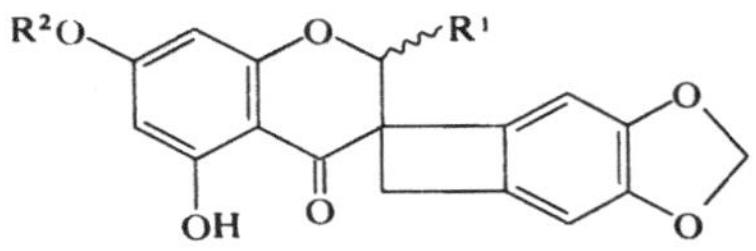 (19)	H	H	35	54	34	47	62	70	38	66	v
(20)	OH	CH$_3$	36	56	35	60	69	69	56	83	vr[5]
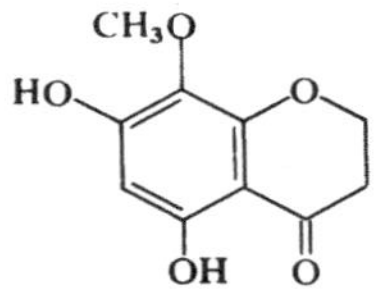 (23)			21	36	26	50	61	60	39	69	rv

[1] For structural types see Table 1.

[2] For chromatography with solvent systems 1—6 HPTLC-silica gel plates *(E. Merck)*, with 7 and 8 precoated polyamide-11 sheets *(E. Merck)* have been used. Developing distance was 5 cm. Data are hRf-values (Rf × 100). Solvent systems: (1) petroleum ether (b. p. 50—70°)-diethyl ether-methanol (40 : 60 : 1); (2) like (1) but twice; (3) toluene-methanol (93 : 7); (4) chloroform-methanol (95 : 5); (5) chloroform-methanol (9 : 1); (6) diethyl ether-dioxane-methanol (90 : 10 : 1); (7) petroleum ether (b. p. 50—70°)- benzene-butanone-methanol (45 : 35 : 10 : 10); (8) chloroform-butanone-methanol (60 : 26 : 14).

[3] FBB: ammonia vapors followed by 0.05% aqueous fast blue B salt. Colours: bl = blue, br = brown, och = ochre, or = orange, r = red, rsh = reddish, v = violet, y = yellow.

[4] Same values as *(E)*-isomer, except with system Nr. 7.

[5] Colour slowly developing.

Dewick (*19*) suggested abandonment of the term "homoisoflavanoid" in favor of the more systematic name 3-benzylchroman-4-one, because biosynthetic studies revealed that the 2,3-rearrangement of the C_6-C_3-C_6 moiety typical of isoflavonoids had not occurred. For practical reasons however we wish to maintain the term homoisoflavanone which allows one to include the structurally more complex scillascillins.

This structural analogy may also be expressed in a common numbering system. The chroman-4-one part is chosen as the basic unit as in the flavonoids, with the benzylic carbon carrying number 9. Ring B is numbered starting from the point of benzylic substitution. For the skeleton of brazilin (**21**) and hematoxylin (**22**) a separate numbering system characteristic of polycycles is used, starting at ring A as shown in Table 1.

3.2. Ultraviolet-Visible Spectroscopy

The naturally occurring homoisoflavanones exhibit a uniform ultraviolet-visible absorption behaviour depending only on the presence or absence of the 3 (9)-double bond. The long wave absorption maxima in ethanol appear between 358 and 367 nm for unsaturated and between 285 and 297 nm for saturated systems. The log ε values range from 4.2 to 4.7[2]. Addition of sodium acetate induces bathochromic shifts of 20—40 nm if a free hydroxyl at C-7 is present. An analogous shift of 20—30 nm is observed with aluminum chloride if the compound contains a free hydroxyl at C-5 (*48*).

3.3. Infrared Spectroscopy

From a large series of infrared spectroscopic data (KBr) the following general conclusions can be drawn (*8*):

1. The unsaturated compounds show a complex pattern of four to five maxima between 1550 and 1700 cm^{-1}, the highest often being set off distinctly. The saturated ones show two or three distinct absorption bands in this region, one near or higher than 1640 cm^{-1}, one around 1600 cm^{-1} and a third near 1590 cm^{-1} or lower. In some cases only two bands are observed; the intermediary band is missing.

2. On methylation at C-7 and/or C-4′ the absorption with the highest wave number still appears around 1640 cm^{-1}. Selective methyla-

---

[2] Values of log ε ca. 3 (*8, 69*) have been recorded erroneously (*19, 38*).

tion at C-5 results in a shift to 1650 cm^{-1} (+10). If both C-5 and C-7 are methylated the band is shifted to 1660 cm^{-1} (+20). Deviations of ±5 cm^{-1} are possible. Further shifts can be observed by gradual acetylation as shown for eucomol in Table 3.

3. Between 3000 and 3500 cm^{-1}, a broad absorption is observed whose intensity depends on the number of hydroxyl groups. Eucomol and its derivatives show a sharp signal at 3450 cm^{-1} which originates from the hydroxyl at C-3.

Table 3. *Infrared Carbonyl Absorption of Eucomol and its Derivatives in KBr (8)*

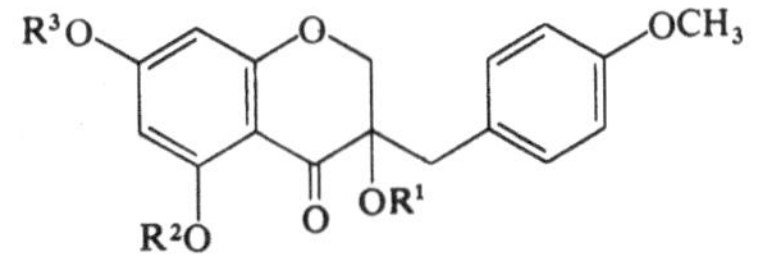

Compound	R^1	R^2	R^3	$\gamma_{C=0}$ (cm^{-1})
(10)	H	H	H	1635
(11)	H	H	Me	1635
(24)	H	Me	Me	1665
(25)	Ac	Me	Me	1680
(26)	Ac	Ac	Ac	1695

3.4. Nuclear Magnetic Resonance Spectroscopy

3.4.1. ^{1}H-NMR Spectra

With regard to ring A and B the proton magnetic resonance spectra of the homoisoflavanones are very similar to those of other flavonoids (48). However, signal assignment of ring A protons on the basis of the rules published by MASSICOT and MARTHE (52) might be misleading in certain cases (38). Selective base-catalyzed deuterium exchange of the high field proton of 3,9-dihydroeucomin (6) allowed unequivocal assignment of the signal to C-8 by means of ^{13}C-NMR spectroscopy (35).

Specific signals arise from the heterocyclic ring C including C-9. The spectra of 3(9)-unsaturated compounds show signals near δ = 5 ppm and in the aromatic region for the protons at C-2 and C-9 respectively. Geometric isomerism at this double bond greatly influences the chemical shift of these protons, including those at C-2'/6'. Due to the anisotropic effects of the carbonyl group and the aromatic ring B adjacent pro-

Table 4. *Assignment of the Hydrogen Atoms in the Nuclear Magnetic Resonance Spectra of Some Natural Homoisoflavanones*[1]

Compound	Solvent	C-2	C-3	C-5	C-7	C-6/8	C-9	C-2',6'/3',5' (C-6'/C-3')	C-4'
(2)	CDCl$_3$	4.91 s	—	12.75 s^2	5.3 br^2	5.93/5.99 AB (2.3)	6.87 s	6.90/7.81 AA' BB' (8.8)	3.85 s
(3)	CDCl$_3$	5.31 d (1.7)	—	12.77 s^2	5.43 br^2	5.90/6.00 AB (2.3)	7.80 t (1.7)	6.96/7.27 AA' BB' (8.8)	3.87 s
(6)	CDCl$_3$	4.12/4.28 ABX (4.2; 6.5; 11.5)	2.8 m	12.13 s^2	5.5 br^2	5.91/5.98 AB (2.3)	2.8/3.2 ABC	6.86/7.14 AA' BB' (8.8)	3.80 s
(8)	(CD$_3$)$_2$CO	4.06/4.26 ABX (11.2)	2.2—3.5 m	3.79 s	8—9 br^2	6.00/6.13 AB (2.3)	2.2—3.5 m	6.79/7.09 AA' BB' (8.8)	8—9 br^2
(9)	(CD$_3$)$_2$CO	4.04/4.11 AB (11.3)	4.17 br^2	11.75 s^2	8.44 br^2	5.97/5.99 AB (2.3)	2.93 s	6.77/7.12 AA' BB' (8.8)	8.44 br^2
(10)	CDCl$_3$	4.06/4.21 AB (11.2)	3.34 s^2	11.26 s^2	5.82 br^2	5.99/6.03 AB	2.95 s	6.87/7.11 AA' BB' (8.8)	3.80 s
(12)	(CD$_3$)$_2$SO	5.34 t (2)	—	12.89 s^2	3.4 br^2	3.72 s/5.96 s	7.69 t (2)	6.90/7.34 AA' BB'	3.4 br^2
(13)	(CD$_3$)$_2$SO	4.10 m	2.90 m	12.14 s^2	10.48 br^{2,3}	3.60 s/5.87 s	2.90 m	6.62/6.96 AA' BB'	9.16 s^{2,3}
(16)	(CD$_3$)$_2$SO	5.34 t (1.8)	—	12.49 s^2	10.55 s^2	5.93 s/3.63 s	7.66 t (1.8)	6.97/7.35 AA' BB' (9)	3.81 s
(17)	CDCl$_3$	4.1—4.5 ABX	2.4—3.5 m	12.00 s^2		6.17 s/3.87 s	2.4—3.5 m	6.82/7.13 AA' BB' (9)	

(19)	(CD$_3$)$_2$SO	4.52/4.62 AB (11)	—	—	12.05 s^2	5.90/5.94 AB (2.3)	3.03/3.32 AB (13.5)	(6.69 d/6.87 d) (<1)	5.93 s
(20a)	CDCl$_3$	3.22 s^2/5.73 s	—	3.84 s	11.84 s^2	6.06/6.11 AB (3)	3.23/3.41 AB (14)	(6.49 s/6.71 s^3)	5.84/5.87 AB
(20b)	CDCl$_3$	3.30 s^2/5.65 s	—	3.84 s	11.94 s^2	<6.06/6.11 AB (2.3)	3.23/3.41 AB (14)	(6.69 s/6.72 s^3)	5.89/5.92 AB

[1] Chemical shifts (δ) are given in ppm downfield from tetramethyl silane (TMS) used as internal standard. Coupling constants (J) are given in Hz. For unequivocally recognized fine structures the following abbreviations are used: s = singlet; d = doublet; t = triplet; m = multiplet; br = broad; AB, ABC, ABX = respective coupling systems, values are given for AB-part.
[2] Protons exchange with D$_2$O.
[3] Values may be interchanged.

Table 5. Assignment of the Hydrogen Atoms in Nuclear Magnetic Resonance Spectra of the Tetraacetate of Brazilin (21) and of Hematoxylin (22) (14)[1]

Com-pound	Solvent	C-1[2]	C-2	C-4	C-6α	C-6β	C-7	C-8	C-11	C-12
(21) (Ac)	CDCl$_3$	7.35 d (8)	6.85 dd (8; 2)	6.69 d (2)	4.75 dd (12.5; 2)	3.73 d (12.5)	3.45 s	7.10 s	7.04 s	4.50 br. s (~2)
(22)	CDCl$_3$/CF$_3$CO$_2$H	6.62/6.82 AB (7)		—	4.17/3.80 AB (12)		2.85/3.25 AB (17)	6.73 s	6.73 s	4.17 s

[1] See Table 4, footnote 1; dd = doublet of doublet.
[2] Numbering of skeleton see Table 1.

tons are shifted downfield (*8, 38*). In (*E*)-isomers an additional long-range coupling (J ~ 2 Hz) between the protons at C-2 and C-9 is observed. The 3,9-dihydro compounds show a complex signal pattern in which only the protons at C-2 can be clearly assigned as the AB-part of an ABX-system. After base-catalyzed deuterium exchange at C-3, the spectrum of 3,9-dihydroeucomin (**6**) shows two AB-systems, one for the C-2 protons at δ = 4.27 and δ = 4.12 ppm (J = 11.4 Hz) and another at δ = 3.17 and δ = 2.71 ppm (J = 14.1 Hz) for the benzylic protons at C-9 (*38*). Metabolites which are fully substituted at C-3 are also characterized by two AB-systems in this region.

The substitution patterns in ring A of the eucomnalins and punctatins have been elucidated by making use of the anisotropic solvent effect of benzene on aromatic methoxy groups. Compared with what is observed in chloroform, the signals are significantly shifted upfield if the methoxy group has at least one free ortho-position (*12, 75*). Two and three signals respectively are shifted in the spectra of the methyl derivatives of the eucomnalins and punctatins indicating the attachment of the substituents to carbon atoms No. 5, 6, 7 and 5, 7, 8 (*26, 65*).

The spectrum of 2-hydroxy-7-O-methylscillascillin (**20**) is quite complex. Due to the hemiacetal structure the compound shows mutarotation (*44*). The mixture of the diastereomeric forms gives rise to a double set of signals for most protons. Data for the 2α- and 2β-hydroxy compounds (**20 a**) and (**20 b**) are presented in Table 4.

The spectra of brazilin (**21**) and hematoxylin (**22**) or their respective derivatives are very similar to each other (*14*). For structural reasons however they differ in certain respects from the spectra of chroman-4-ones. The aromatic protons appear close together near δ = 7 ppm, the protons at C-6 and C-7 show signals at about δ = 4 ppm and δ = 3 ppm respectively, and the proton at C-11b appears as singlet between δ = 4—5 ppm (*14*) (Table 5). The spectra of the acetyl derivatives in chloroform show excellent fine structure which allows complete signal assignment. Long-range coupling of 2 Hz between the α-proton of C-6 and the proton at C-11b is observed. This value does not allow one to decide between cis- (dihedral angle 0°) and trans-fusion (dihedral angle 180°) of rings B and C (*14*), a problem which was solved by chemical transformations (see Chapter 4.8.).

3.4.2. ^{13}C-NMR Spectra

A review of carbon-13 magnetic resonance spectra of flavonoids including isoflavonoids (*57*), chalcones and biflavonoids (*22*) has been published recently by Chari and Wagner (*12*). Signal assignments of the

Table 6. *Assignment of the Carbon Atoms in the Nuclear Magnetic Resonance Spectra of Eucomins and Related Compounds*[1]

Compound	C-2	C-3	C-4	C-4a	C-5	C-6	C-7	C-8	C-8a	C-9	C-1'	C-2'/6'	C-3'/5'	C-4'	CH$_3$O-4'
Homoisoflavanones															
(2)	74.1 t (150)	125.5 s	186.6 s	103.2 s	164.8 s	96.2 d (—)	166.7 s	94.7 d (163)	162.5 s	140.5 d (159)	126.6 s	133.1 d (162)	113.4 d (159)	160.6 s	55.2 q (146)
(3)	67.1 t (150)	127.1 s	184.1 s	101.6 s	164.5 s	96.2 d (163)	166.9 s	94.9 d (163)	161.9 s	136.0 d (155)	126.2 s	132.4 d (160)	114.3 d (162)	160.6 s	55.2 q (144)
(6)	68.9 t (—)	45.5 d (—)	197.5 s	101.2 s	163.7 s	95.9 d (—)	166.5 s	94.7 d (163)	162.8 s	31.1 t (—)	129.9 s	129.9 d (157)	133.8 d (160)	157.9 s	55.0 q (144)
(10)	71.8 t (150)	71.6 s	198.0 s	100.1 s	164.0 s	96.3 d (165)	166.8 s	95.0 d (163)	162.5 s	38.7 t (127)	127.0 s	131.4 d (159)	113.3 d (159)	158.2 s	54.9 q (144)
Chromones, Chromanones															
(27)	157.2 d (199)	110.6 d (171)	181.4 s	105.2 s	161.8 s	99.2 d (160)	164.5 s	94.1 d (165)	157.9 s						
(28)	66.4 t (149)	36.0 t (130)	196.0 s	102.3 s	163.8 s	96.0 d (162)	166.7 s	94.9 d (163)	163.2 s						

[1] Spectra are run in (CD$_3$)$_2$SO as solvent. Chemical shifts (δ) are given in ppm downfield from tetramethyl silane (TMS) used as internal standard. Coupling constants (J) are given in Hz and are taken from spectra recorded under gated decoupling conditions. Only 1.2-coupling constants are given and the following abbreviations are used: s = singlet, d = doublet, t = triplet, q = quartet.

homoisoflavanones (Z)- and (E)-eucomin (2) and (3), 3,9-dihydro-
eucomin (6) and eucomol (10) as well as 5,7-dihydroxy-4-chromone
(27), the corresponding chroman-4-one (28), (E)-4'-methoxycinnamic
(29), 4'-methoxyhydrocinnamic (30) and (±)-4'-methoxyphenyllactic
acids (31) are presented in Table 6 and 7 and are in good agreement
with published data.

Table 7. *Assignment of the Carbon Atoms in Nuclear Magnetic Resonance Spectra of Cinnamic and Related Acids*[1]

Com-pound	C-1	C-2	C-3	C-1'	C-2'/6'	C-3'/5'	C-4'	CH₃O-4'
(29)	167.8s	116.7d (159)	173.7d (159)	127.0s	129.8d (160)	114.4d (162)	161.1s	55.3q (144)
(30)	173.7s	35.6t (129)	29.6t (128)	132.9s	129.1d (157)	113.8d (159)	157.8s	55.0q (144)
(31)	175.0s	71.3d (144)	39.2t (129)	130.0s	130.3d (157)	113.5d (160)	157.9s	55.0q (143)

[1] See footnotes Table 6.

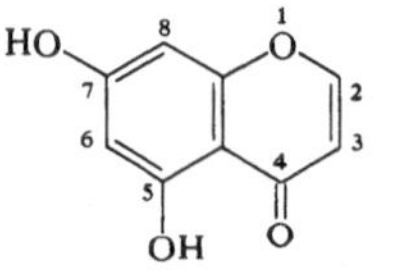

(27) 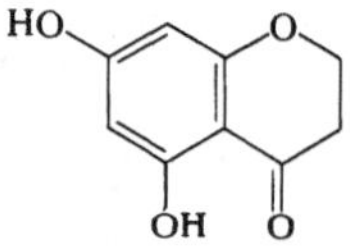   (28)

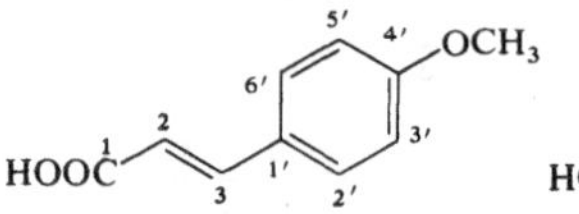

·(29) 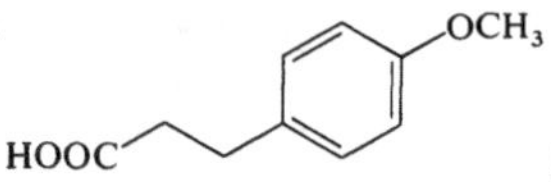   (30) 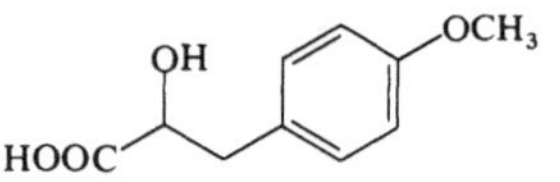   (31)

3.5. Mass Spectrometry

In the mass spectrum of eucomin (3) which serves as representative
example for the 3(9)-unsaturated compounds the molecular ion at
m/z 298 (100)[3] is also the base peak. The main fragmentation is a

---

[3] Values in brackets indicate the intensities in per cent of the base peak.

References, pp. 149—152

retro-Diels Alder reaction (RDA) which yields the fragments m/z 146(50) and m/z 152(5). As the result of an effective hydrogen transfer from C-2 to ring A, a well known behaviour of flavonoids (4, 21, 56), a peak of much higher intensity occurs at m/z 153(84) (*9, 69*). This fragment happens to be the base peak of several related compounds (*64*).

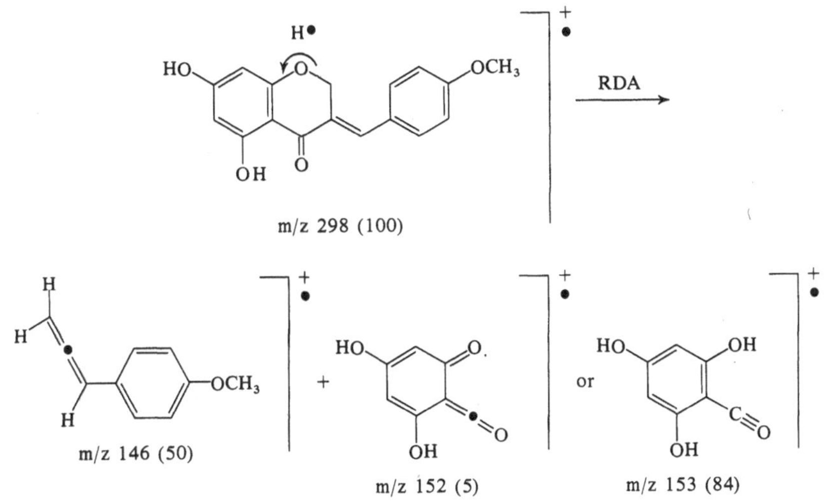

m/z 298 (100)   RDA →

m/z 146 (50)   +   m/z 152 (5)   or   m/z 153 (84)

The common fragmentation behaviour of the saturated metabolites is an A-4 type breakdown of the molecule. In most cases the tropylium ion gives rise to the base peak of the spectrum which for eucomol (**10**) appears at m/z 121 (100). The chromanone fragment at m/z 195 may eliminate water, CO or undergo an RDA cleavage to give a fragment at m/z 152 or, due to a hydrogen shift, at m/z 153. A minor pathway is subsequent loss of water, CO and methyl from the molecular ion (*9, 69*). In 5-O-methylated derivatives the chromanone moiety becomes the most abundant fragment. This reaction is accompanied by a nearly exclusive hydrogen transfer (*26, 64*).

3.6. Optical Activity and Absolute Configuration

The natural homoisoflavanones which are saturated at C-3 are optically active (Table 1). The 3-benzylchroman-4-ones show negative rotation values. The only exception so far is 3,9-dihydroeucomin (**6**) from *E. bicolor* ($[\alpha]_D = +38°$) whose dimethyl derivative ($[\alpha]_D = +12°$) is enantiomeric to the analogous compound prepared from (−)-4-demethyl-5-O-methyl-3,9-dihydroeucomin (**8**) isolated from *E. comosa* (*26, 27, 38*).

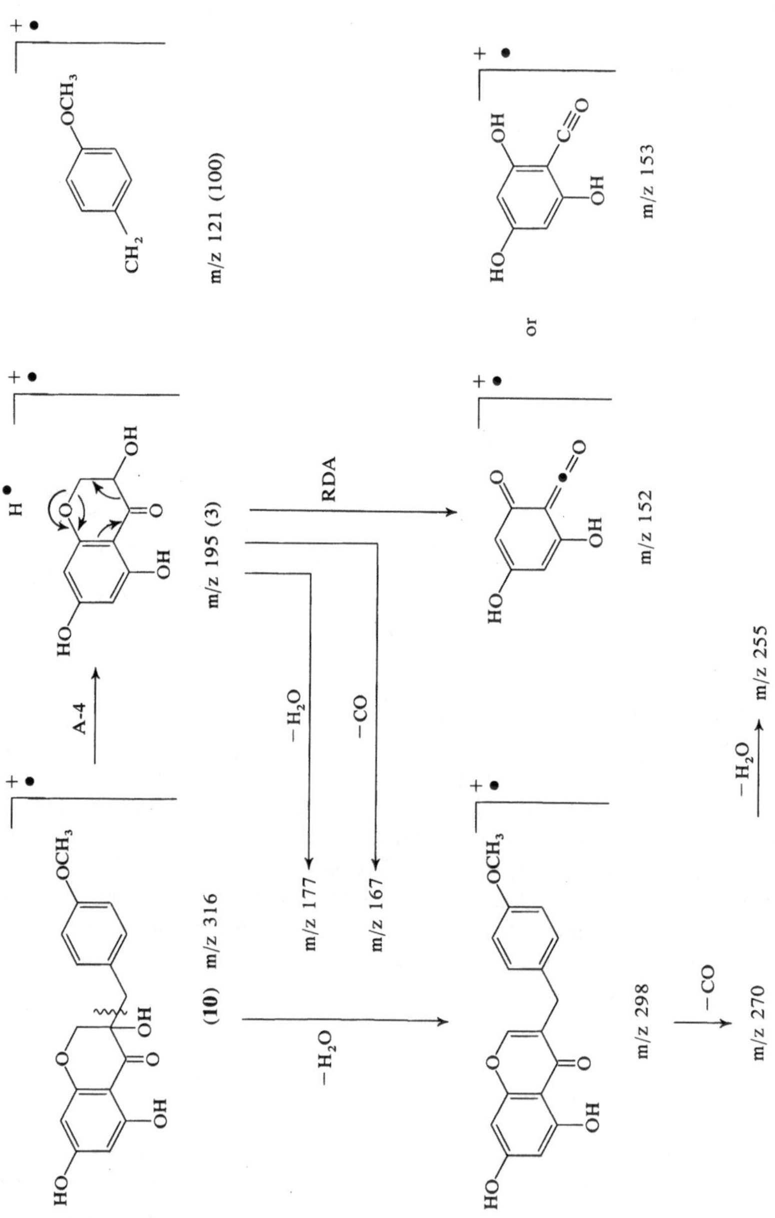
(10) m/z 316
A-4
m/z 195 (3)
m/z 121 (100)
RDA
−H2O
−CO
m/z 177
m/z 167
m/z 152
or
m/z 153
m/z 298
m/z 270
m/z 255

From CD/ORD measurements it was tentatively concluded that (−)-eucomol (**10**) possesses the (R)-configuration at C-3 (9, 69). However, an X-ray analysis of (−)-7-O-(p-bromophenacyl)eucomol (**32**) in which use was made of the anomalous dispersion effect of bromine, unequivocally revealed the (S)-configuration (74).

(**32**)

Scillascillin (**19**) exhibits an $[\alpha]_D$-value of +40° whereas the 2-hydroxy-7-O-methyl derivative (**20**) shows mutarotation, the $[\alpha]_D$-value in dioxane changing from −11,5° to +79,2° within 18 hours (44).

The structures of scillascillin (**19**) and 2-hydroxy-7-O-methylscillascillin (**20**) have been independently determined by X-ray analyses (44). However, no heavy atoms were introduced into the molecules. Thus the absolute configuration could not be deduced. Natural brazilin (**21**) and hematoxylin (**22**) are both optically active (Table 1). The similarity of the ORD-curves of their acetyl derivatives proves their common configuration at C-6a and C-11b (14).

4. Chemical Transformations and Syntheses

4.1. Synthesis of the Skeleton

In general 3-benzylidenechroman-4-ones are prepared by condensation of chroman-4-ones with aromatic aldehydes (72). Strong alkaline (58) or acidic (15) conditions can be applied only if eventual sensitive substituents are suitably protected. Using dry hydrogen chloride in acetic acid BOEHLER and TAMM synthesized 5,7-di-O-methyleucomin (**33**) from 5,7-dimethoxychroman-4-one and anisaldehyde (9). Later KRISHNA-MURTY et al. (45) observed the simultaneous formation of a C-6 or C-8 α-chlorobenzylated product (**34**) in this reaction. Starting from O-benzoylated compounds 5,7-di-O-benzoyleucomin (**35**) and 4′,5,7-tri-O-benzoyl-4′-demethyleucomin (**36**) are easily prepared by this method using diethyl ether and ethyl acetate as solvents respectively (45). The unprotected phenolic compounds can be condensed successfully in boiling acetic anhydride (23, 24). The relatively low yield of 5,7-di-O-acetyleucomin from the reaction of 5,7-dihydroxychroman-4-one (27) with anisaldehyde and its tedious purification were shown to be the consequence of the formation of large amounts of di-O-acetyl-anisyl-acetat (45).

Ravisé and Kirkiacharian (60) reported on a new and effective condensation procedure which employs a mixture of acetic acid and piperidine as solvents. This combination is known to possess mild acid as well as base-catalytic properties.

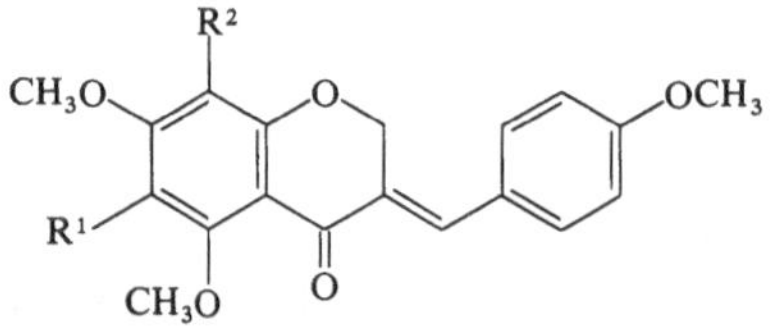

(33) $R^1 = R^2 = H$

(34) $R^1 = H$; $R^2 = C_6H_5CHCl$ or

$R^1 = C_6H_5CHCl$; $R^3 = H$

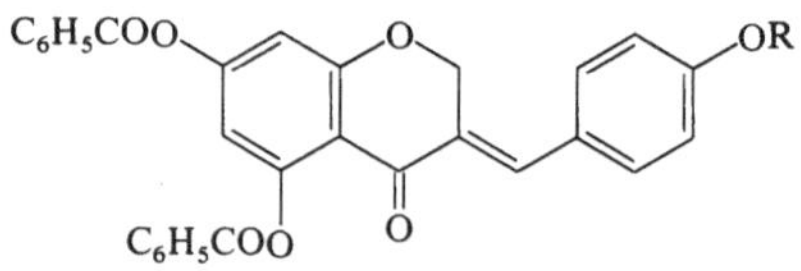

(35) $R = CH_3$

(36) $R = C_6H_5CO$

In a different approach dihydrochalcones have been transformed to 3-benzyl-4-chromones. Treatment of (37) with ethyl formate and sodium led to (38) which on hydrogenation and selective demethylation yielded racemic 4'-O-methyl-3,9-dihydropunctatin (18) (25). A modified method

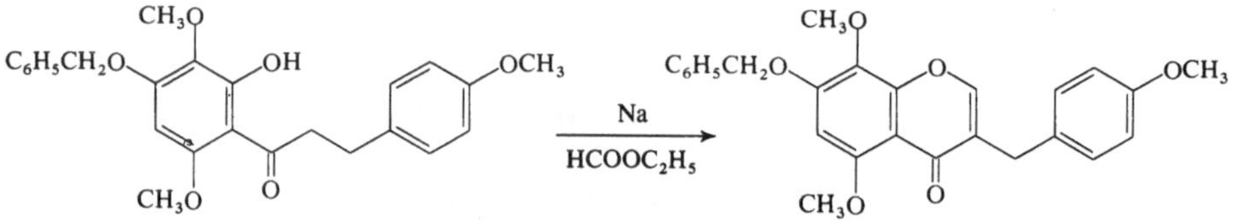

(37)

(38)

for the ring closure was described by Bass (7). Starting from 2',4',6'-trihydroxy-dihydrochalcone (39), reaction with boron trifluoride-diethyl ether complex in dimethylformamide followed by methanesulfonyl chloride gave the chromone (40) in excellent yield. Subsequent hydrogenation led to the corresponding homoisoflavanones.

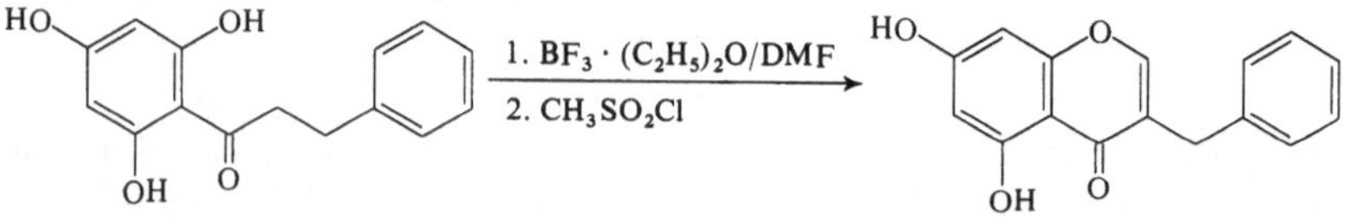

(39)

1. $BF_3 \cdot (C_2H_5)_2O/DMF$

2. CH_3SO_2Cl

(40)

Other possible intermediates are 3-benzyl-4-hydroxycoumarins. Starting from the parent compound (41) the reaction to 3-benzyl-4-chromone (42) is achieved in moderate yield by reduction with diborane, followed by oxidation with sodium bichromate in aqueous sulfuric acid (42).

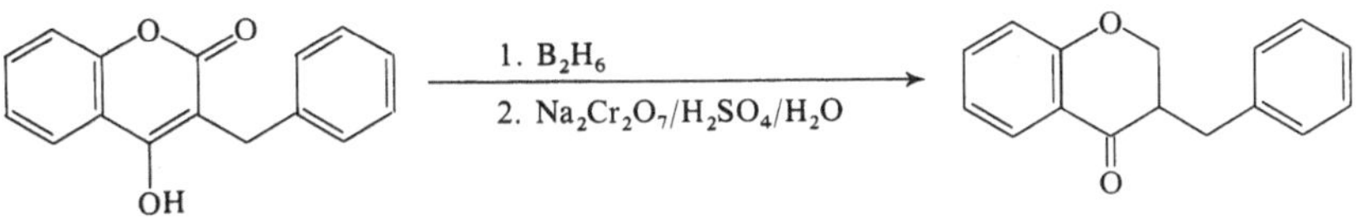

4.2. Routes to Eucomol

A two step synthesis of (±)-eucomol (10) has been published by FARKAS et al. (23, 24). 5,7-Di-O-benzyleucomin (43) was transformed with alkaline hydrogen peroxide to the 3,9-epoxy compound (44). Hydrogenation with a special palladium catalyst under carefully controlled conditions led to (10) in moderate yield. Proof for structure (44) was adduced by boiling the compound with p-toluenesulfonic acid in methanol which gave the 9-methoxy derivative (45).

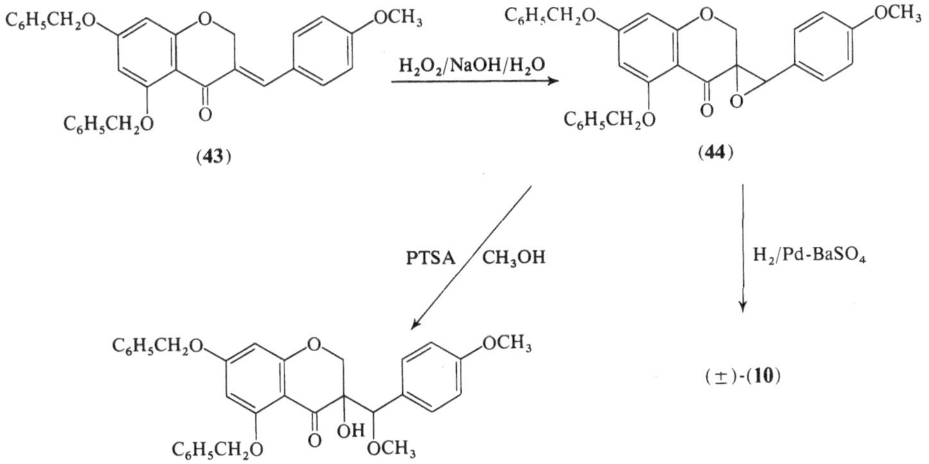

KRISHNAMURTY et al. (45) opened a more direct way to derivatives of eucomol (10). By subjecting 5,7-dimethoxy-3(4-methoxybenzyl)-chroman-4-one (46) to lead tetraacetate oxidation the 3-acetoxy derivative (25) was formed in good yield.

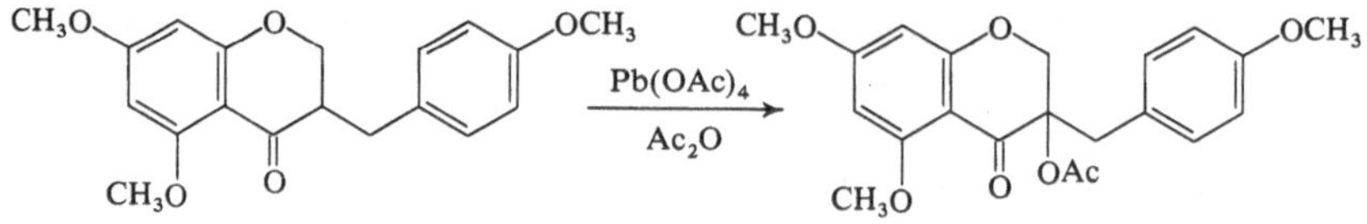

(46)   (±)-(25)

4.3. Isomerization and Hydrogenation Reactions of the 3(9)-Double Bond

As mentioned above the genuine naturally-occurring homoisoflavanones eucomin and 4'-O-methylpunctatin possess the (Z)-configurations
as shown by the formulae (2) and (15). Gradual isomerization to the
thermodynamically more stable (E)-forms (3) and (16) is observed under
several conditions. Quantitative transformation is successfully carried
out with hydrogen chloride in methanol at reflux temperature (35). In
synthesis the (E)-isomers usually predominate and are easily purified
by crystallization. In the case of 5-O-methyleucomin an equilibrium
between the two isomers is observed. After chromatographic separation
only the (E)-isomer can be obtained in homogeneous form by crystallization from tetrahydrofuran (64). In permethylated systems the (Z)-
isomers can be prepared by acid catalysis or by irradiation. Treatment
of (E)-4',5,7-tri-O-methylpunctatin with dry hydrogen chloride in acetic
acid at room temperature yields 20% of the corresponding (Z)-isomer,
but by chromatography (silica; benzene-ethyl acetate 1 : 1) the latter is
readily converted to the (E)-form (26). On the other hand irradiation
of 5,7-di-O-methyleucomin (33) at 300—400 nm gives the (Z)-isomer
in up to 54% yield (9).

Isomerization of an exo to an endo double bond requires more drastic
conditions and is irreversible. Transformation of 5,7-di-O-methyleuco-

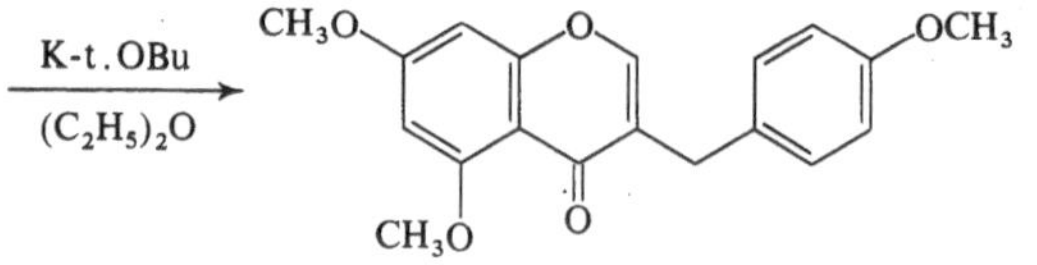

(47)

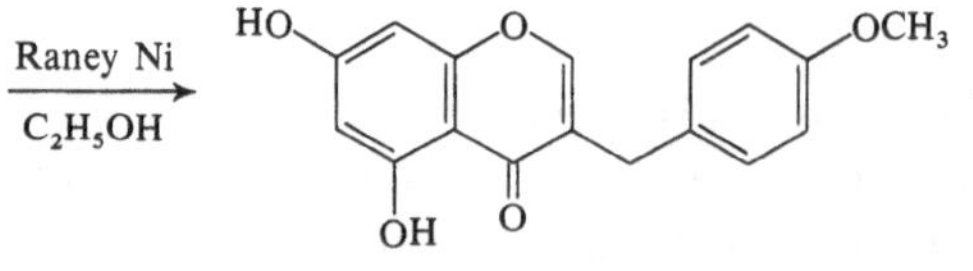

(48)

min (33) to (47) is only achieved by potassium t-butoxide in diethyl ether (9). The product is easily identified by its ultraviolet (several maxima between 230 and 280 nm) and ^{1}H-NMR spectra (H-2 at 7.34 ppm, H-9 at 3.62 ppm).

CHATTERJEA et al. (13) have described the possible use of Raney nickel to catalyze this isomerization. However, the reaction was incomplete and substantial reduction of the double bond took place. As can be shown with eucomin (3) large quantities of 3,9-dihydroeucomin (6) are formed along with small amounts of (48) (35). This isomerization was also observed in hydrogenation reactions using palladium as catalyst. From the reaction with 4′-demethyleucomin (1) 6% endo-isomer have been isolated (64). Chromatographic detection and purification of endo-isomers is generally performed on silica using solvent systems No. 1, 2 or 6 or suitable modifications (Table 2). Under these conditions they migrate somewhat slower than the 3,9-dihydro compounds. More recently ANDRIEUX et al. (2) have found that migration of the exocyclic double bond to the endocyclic position can be achieved satisfactorily by rhodium trichloride trihydrate using a suitable solvent.

4.4. Deuterium Exchange Reactions

Deuteration experiments have been carried out in order to assign signals and coupling constants in the ^{1}H-NMR spectra of 3,9-dihydro-eucomin (6) (38). Not only the aliphatic proton at C-3 is exchanged by sodium methoxide in monodeutero methanol and even by potassium carbonate in hexadeutero acetone, but also the aromatic proton at C-8 undergoes selective exchange (see Chapter 3.4.1.). In 5,7-dihydroxy-chroman-4-one both aromatic protons are exchanged equally well together with those at C-3. Similar observations have been made with the flavonol kielcorin (55) and other aromatic systems (40).

4.5. Acylation and Deacylation Reactions

Acetylation of eucomin (3) with boiling acetic anhydride leads to 7-O-acetyleucomin (49) (9). In later experiments additional formation of di-O-acetyleucomin (50) was observed (64). This is exclusively formed in presence of pyridine (9). In the case of eucomol (10) the aliphatic hydroxyl group is also acetylated under both conditions (9).

Benzoylation of eucomin (3) with benzoyl chloride and potassium carbonate in acetone at room temperature occurs selectively at the 7-hydroxyl group (64). This is probably due to the low solubility of

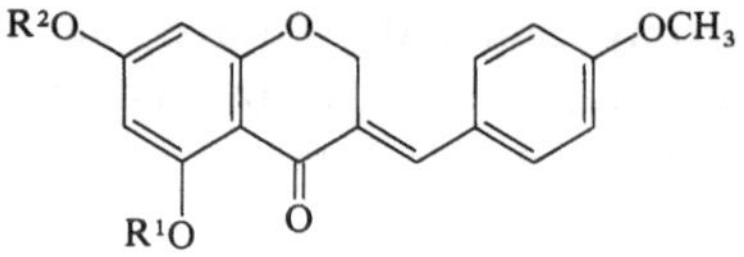

W. HELLER and CH. TAMM:

(49) R¹ = H; R² = Ac

(50) R¹ = R² = Ac

the product in this solvent. Using 5,7-dihydroxychroman-4-one as substrate selective reaction to form (51) is only achieved with sodium pivalate in acetone (35). Demethyleucomol (9) reacts under these conditions in a similar manner and yields compound (52). However, during

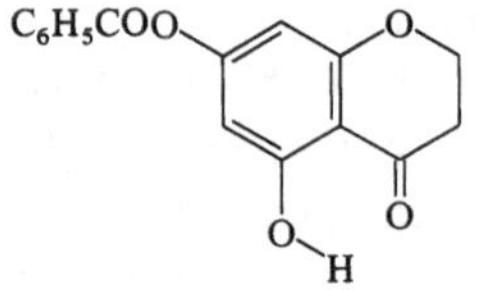

(51)

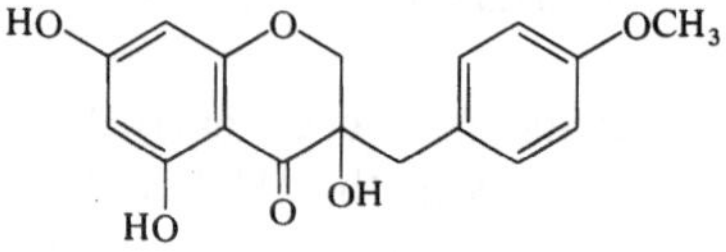

(9)

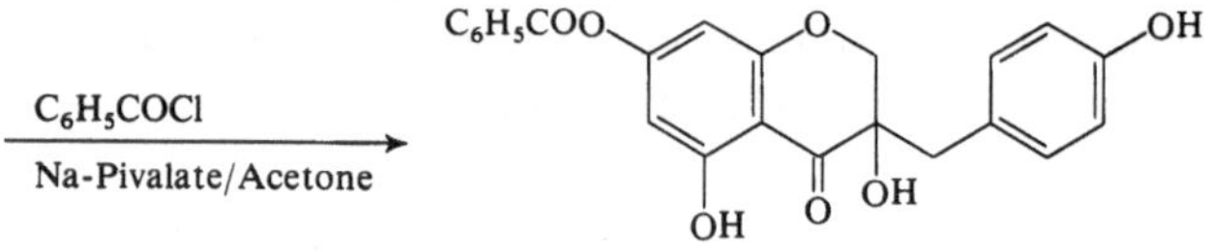

(52)

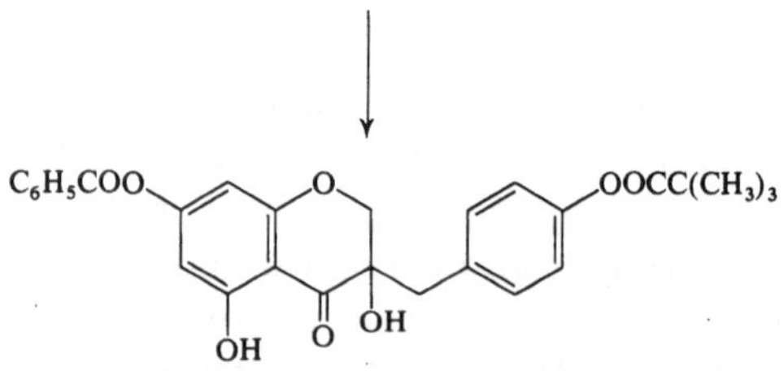

(53)

the usual work-up a substantial amount of pivaloylation takes place at C-4' leading to the derivative (53) (64). Cleavage of both acetoxy and benzoyloxy groups is successfully achieved by alcoholic alkali. With 5% potassium hydroxide in ethanol 5,7-di-O-benzoyleucomin (35) is hydrolysed within 2 hours at room temperature (45). The milder reagent potassium carbonate in methanol-water (3 : 2) gives only 25% of demethyleucomin (1) from the tri-O-benzoyl compound (36) within 90 minutes (64).

4.6. Alkylation and Dealkylation Reactions

In saturated homoisoflavanones such as eucomol (10) diazomethane in diethyl ether leads selectively to the 7-O-methyl derivative (11) (9). On addition of a more polar solvent like methanol, complete methylation to (24) is observed (9, 38).

With dimethyl sulfate and potassium carbonate in acetone complete methylation is usually observed to give 5,7-di-O-methyleucomin (33) from eucomin (3) even at room temperature (9, 38). Using proton acceptors of appropriate basicity like sodium acetate or pivalate, only 7-O-methyleucomin (4) is formed (38). Similar results are obtained with methyl iodide and silver oxide in dimethyl formamide (9). Methylation of eucomin (3) has also been performed in benzene and 2 N sodium hydroxide under controlled conditions, yielding 7-O-methyleucomin (4) together with the C-methylated derivative (54) (9).

By treatment of 5,7-diacetoxychroman-4-one (55) with boiling dimethyl sulfate and potassium carbonate in acetone also selective methylation at C-7 takes place giving compound (56) (45). This behaviour is also observed with 7-O-acetyleucomin (49) where even room tem-

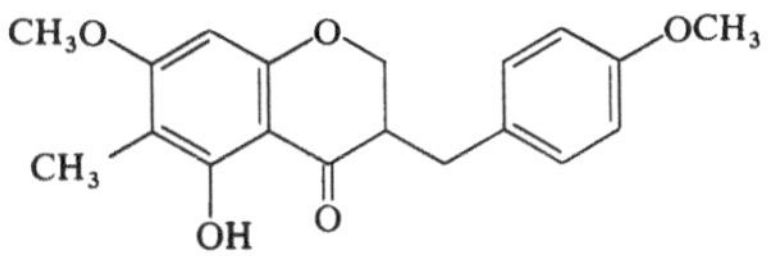

(54)

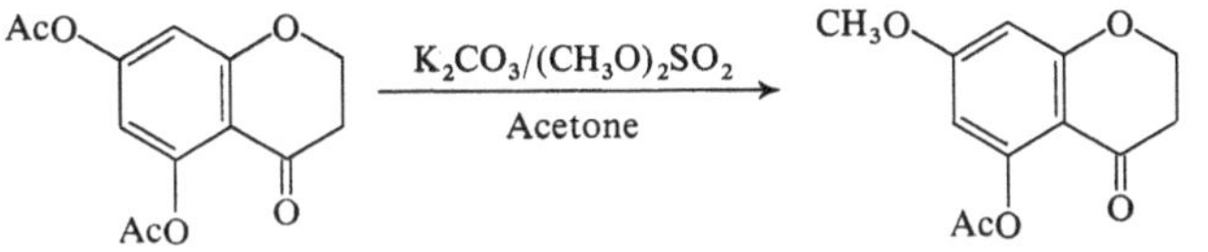

$K_2CO_3/(CH_3O)_2SO_2$

Acetone

(55)   (56)

perature reaction leads to substantial amounts of 5,7-di-O-methyl-eucomin (33) together with 7-O-acetyl-5-O-methyleucomin (64). In contrast, with 7-benzoyloxy-5-hydroxychroman-4-one (51) only mono-methylation to (57) is found under mild conditions (35).

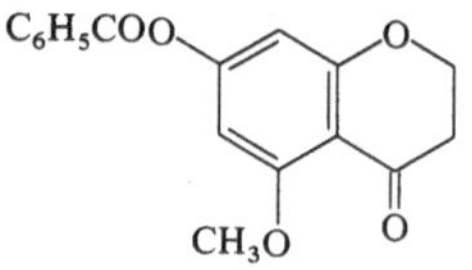

(57)

Benzylation of eucomin (3) with benzyl bromide and potassium carbonate in dimethyl formamide gives 5,7-di-O-benzyleucomin (43). If acetone is used as solvent additional C-benzylation at C-6 or C-8 takes place (23, 24). Again sodium acetate shows complete selectivity towards the activated hydroxyl at C-7 (64). Similarly 7-O(p-bromophenacyl)euco-mol (32) is isolated from the reaction of eucomol (10) with p-bromo-phenacyl bromide and sodium acetate or pivalate in acetone (74).

Complete dealkylation is achieved by boron tribromide in chloroform at 0° (46). Eucomin (3) and eucomol (10) are transformed to their respective demethyl derivatives (1) and (9) in good yields (64). Hydrogen

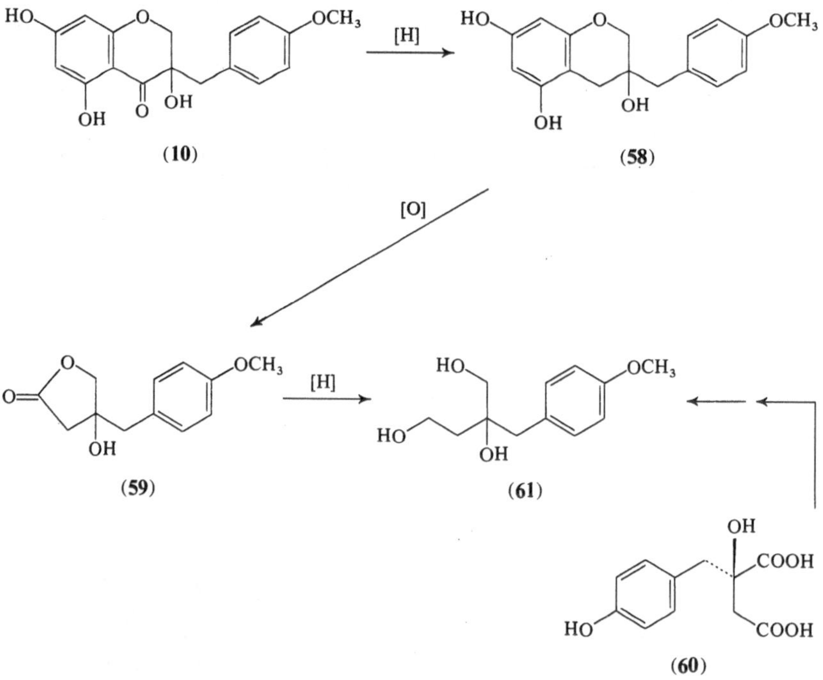

(10)　　　　　　　　　　(58)

(59)　　　　　　(61)

(60)

bromide in glacial acetic acid completely decomposes eucomin (3). With hydrogen iodide and red phosphorous in acetic anhydride (17) additional reduction to 4'-demethyl-3,9-dihydroeucomin (5) is observed (64).

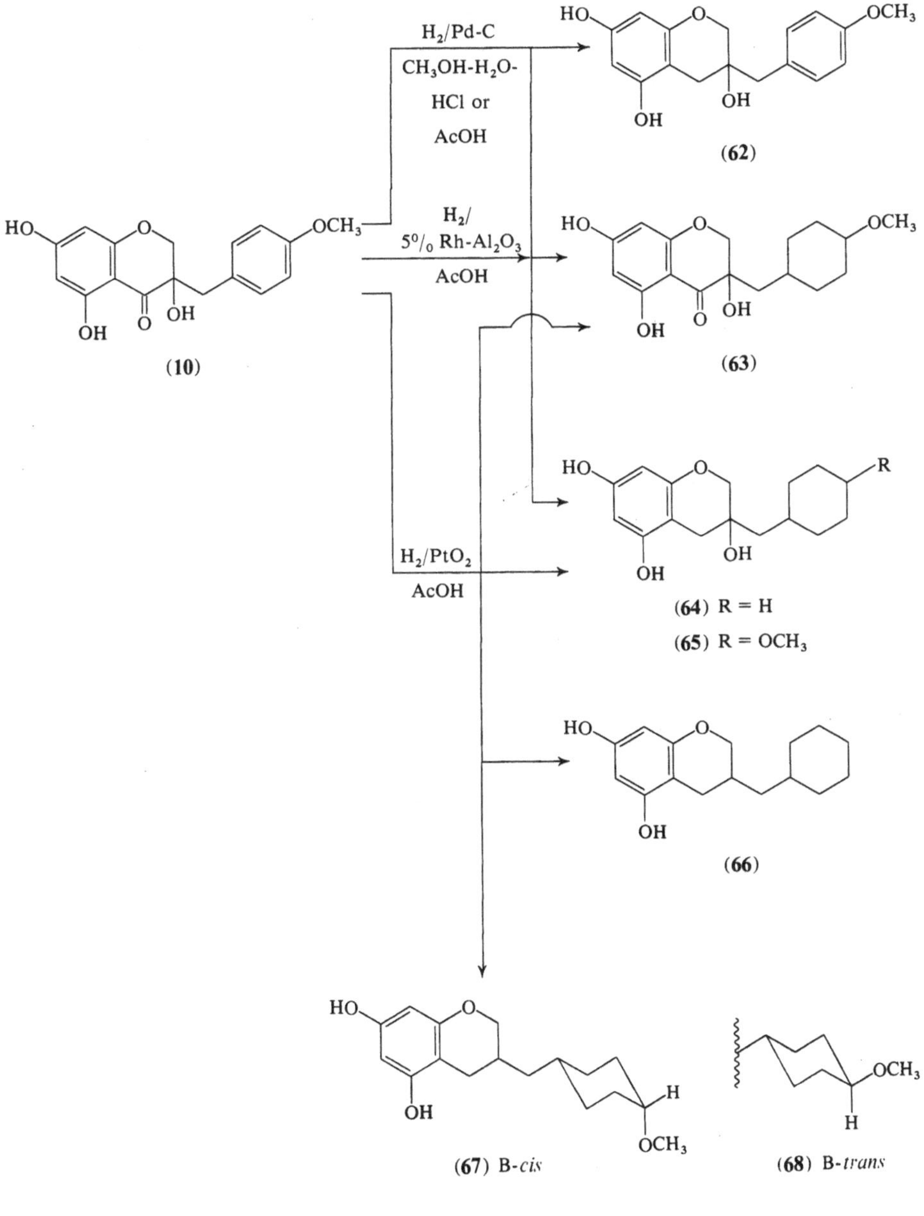

(10)

H$_2$/Pd-C
CH$_3$OH-H$_2$O-
HCl or
AcOH

(62)

H$_2$/
5% Rh-Al$_2$O$_3$
AcOH

(63)

(64) R = H
(65) R = OCH$_3$

H$_2$/PtO$_2$
AcOH

(66)

(67) B-cis

(68) B-trans

4.7. Further Reactions of Eucomol

The crucial step in the chemical interrelation of eucomin (3) and eucomol (10) was elimination of the tertiary hydroxyl group. This reaction can be achieved with 5,7-di-O-methyleucomol (24) only by using a mixture of phosphorous pentoxide and alumina in boiling pyridine and gives 45% of crude (47) (9).

An attempt to determine the absolute stereochemistry of natural eucomol (10) by chemical degradation to a product of known chirality led to a series of new transformations (64). It was anticipated that reduction of (10) to the deoxo derivative (58) followed by oxidative cleavage of ring A would lead to the lactone (59). From this intermediate the key compound (61) should be obtained by subsequent reduction. The derivative (61) is easily prepared from eucomic acid (60) whose absolute stereochemistry has been elucidated recently (34, 36).

Hydrogenation experiments were carried out using different catalysts. However, the yields were very small or complex mixtures of products were obtained. With 10% palladium-on-charcoal in methanol and a trace of conc. hydrochloric acid or in acetic acid, small amounts of 4-deoxoeucomol (62) as well as the hexahydro derivatives (63) and (64) were formed.

Platinum oxide in acetic acid gave six products with intact ring A which showed a violet colour with fast blue B salt reagent in TLC. The main product hexahydroeucomol (63), which is also found in the reaction with 5% rhodium-on-alumina in acetic acid, is easily identified by its ^{1}H-NMR spectrum. Further hydrogenation led to deoxohexahydroeucomol (65) from which the demethoxy derivative (64) is conceivably formed by elimination of methanol and subsequent reduction. The 3-deoxy-4-deoxo derivatives (66), (67) and (68) were also formed, but only in trace amounts. Detailed analyses of the mass (10) and ^{1}H-NMR spectra (3, 6) allowed unequivocal assignment of the cis- and trans-configuration to the 3-deoxy-deoxohexahydroeucomols (67) and (68), respectively.

With zinc in methanol-conc. hydrochloric acid (1 : 1) eucomol (10) was reduced to the racemic 7-O-methylchromane derivative (69). This compound is also obtained by catalytic hydrogenation of 7-O-methyleucomin (4) with 10% palladium-on-charcoal in ethanol (64).

(69)

A more promising approach to 4-deoxo derivatives of eucomol appeared to be reduction of a suitable thioketal derivative. Eucomol (10) itself did not react with ethanedithiol and boron-trifluoride-diethyl ether complex as catalyst. The negative result is probably due to chelate formation. 5,7-Di-O-methyleucomol (24), however, was smoothly transformed to the thioketal (70) which gave 4-deoxo-5,7-di-O-methyleucomol (71) and the racemic chromane (73) on treatment with Raney nickel in ethanol. An attempt to interrelate these compounds via an elimination of water from (71) in boiling acetic anhydride yielded only the 3-O-acetyl derivative (72) (64).

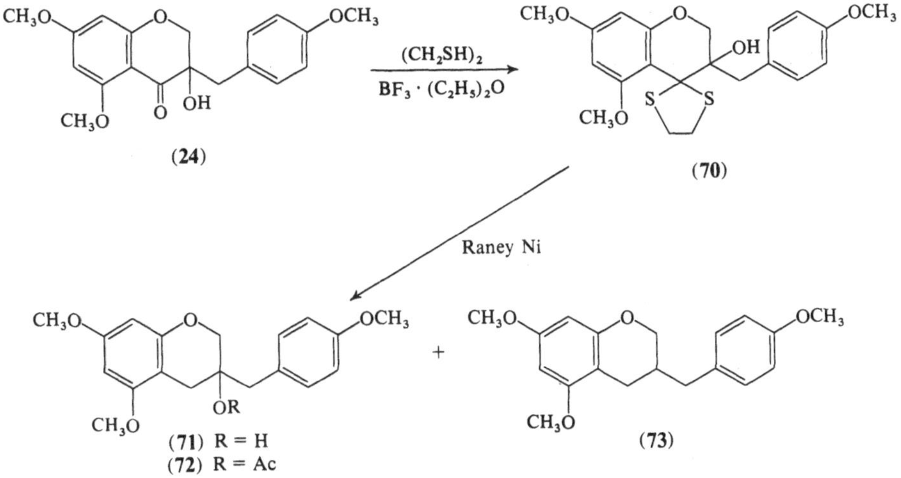

In order to achieve better selectivity in a future oxidative cleavage of ring A, e. g. by ozone after transformation to a quinone system (5), 7-O-methyleucomol (11) has been transformed to its 8-hydroxy derivative (74) by potassium peroxodisulfate and potassium hydroxide in pyridine (47). Treatment of (74) with benzyl bromide and potassium carbonate in aceton gave the dibenzyl product (76) which on thioketalization yielded

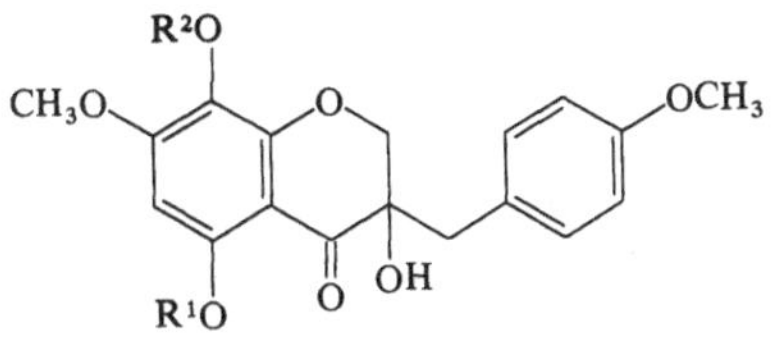

(74) R^1 = R^2 = H
(75) R^1 = H; R^2 = C$_6$H$_5$CH$_2$
(76) R^1 = R^2 = C$_6$H$_5$CH$_2$

8-benzyloxy-7-O-methyleucomol (75) as the only product. Obviously benzylic ethers in the, peri-position are cleaved by boron trifluoride under these conditions.

4.8. Chemistry of Brazilin and Hematoxylin

The chemistry of brazilin (21) and hematoxylin (22) has been reviewed by ROBINSON (62, 63) and DONNELLY (20). In the first successful approach to a synthesis of brazilin the trimethyl ether (79) of anhydrobrazilin (deoxybrazilone) (78) was prepared by treating 7-methoxy-3(3,4-dimethoxybenzyl)chroman-4-one (77) with phosphoric anhydride (16). The hydroxyl group at C-6a was introduced by a series of transformations (63). Later KIRKIACHARIAN (41) found that hydration of the double bond can be achieved more directly by sequential treatment with diborane and alkaline hydrogen peroxide. The brazilin and hematoxylin derivatives ·obtained in this fashion were identical with products of earlier experiments. The course of the hydroboration reaction established the cis-fusion of rings B and C.

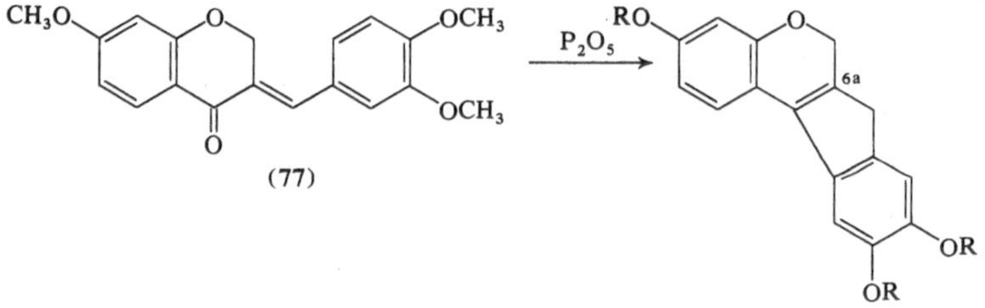

(77)

(78) R = H
(79) R = CH₃

A slightly different route was chosen by DANN and HOFFMANN (16). Starting from 7,8-dibenzyloxy-3(3,4-dibenzyloxybenzylidene)chroman-4-one (80) epoxidation of the double bond with alkaline hydrogen peroxide to derivative (81) and reduction of the latter with sodium borohydride gave the epoxy alcohol (82). Subsequent reductive cleavage of the epoxide ring with lithium aluminium hydride to the diol (83) and treatment with 1% perchloric acid in acetic acid led to racemic hematoxylin tetrabenzyl ether (84) which gave racemic hematoxylin (22) on catalytic hydrogenation. Resolution of both racemic brazilin (21) and hematoxylin (22) was achieved by fractional crystallization of their l-menthoxyacetyl derivatives. It was demonstrated that the synthetic (+)-enantiomers of both compounds were identical with the natural products (54).

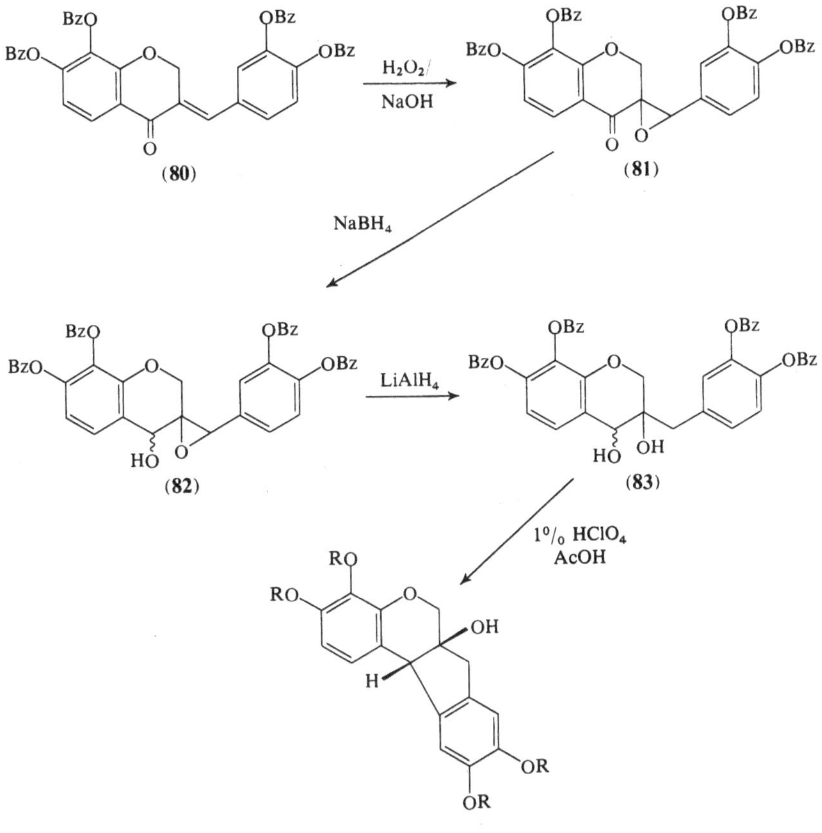

$(\pm)$-(22) R = H

$(\pm)$-(84) R = Bz = $C_6H_5CH_2$

5. Biosynthesis

Preliminary experiments to guarantee suitable feeding conditions showed that eucomin content reaches a maximum in the dormant bulb and it decreases as growth progresses and the flower spike is produced. The concentration increases again during the fruiting period and again reaches its maximum in the dormant bulb (19). It was also observed that at comparable growth stages incorporation rates are best in autumn (35).

Different administration methods have been tested (Table 8). In one case feeding through the roots proved to be more satisfactory than an injection procedure (19). However, other experiments indicated equivalent results provided the precursor solution is efficiently distributed

Table 8. *Incorporation Experiments with Eucomis bicolor*

Compound administered / Application method[2]	Feeding period (d)	Specific activity (Ci/mol)	Products[1] Eucomin (3)	3,9-Dihydro-eucomin (6)	Eucomol (10)	7-O-Methyl-eucomol (11)	References
			Specific activity (× 10^6) (Ci/mol); Incorporation (%) ((Dilution value (× 10^{-6})))				
[U-14C]-L-Phenylalanine							
Injection	1—5[3]		90 / 0.05	190 / 0.01	180 / 0.56	340 / 0.08	
Injection	14	10	220 / 0.25	400 / 0.04	450 / 2.54	590 / 0.40	
Root	3	10	12 / 0.001				(19)
[1-14C]-D,L-Phenylalanine							
Root	3	59	31 / 0.002				(19)
Injection	5	44	(0.87) / 0.033		(1.1) / 0.026		(29)
Sodium[2-14C]-acetate							
Injection	14	60	200 / 0.013	890 / 0.006	660 / 0.211		(19)
Root	3	56	4.2 / 0.0002				(29)
Root	5		(0.09) / 0.016		(3.8) / 0.005		(29)
Bulb stem	0.7		(0.49) / 0.003		(1.0) / 0.009		
[Methyl-14C]-L-Methionine							
Injection	14	56	260 / 0.076	770 / 0.019	770 / 0.91	2050 / 0.27	
Injection	21		280 / 0.035	540 / 0.006	500 / 0.33	1490 / 0.12	

Root	2	59		21	0.0014	(19)
Root	3	59		10	0.0005	(19)
Bulb slice	3.5		(0.24)		0.004	(29)
[Methyl-^{14}C]-2',4',4'-Tri-hydroxy-6'-methoxy-chalcone (90) Root	3	0.093		2.1	0.0009	(29)
[6'-Methyl-^{14}C]-2',4'-Di-hydroxy-4,6'-dimethoxy-chalcone (91) Root	3	0.093		9.2	0.005	(29)

[1] In experiments without reference cited products have been isolated by TLC on silica gel using (a) benzene-methanol (9 : 1) and (b) petroleum ether (b.p. 50—70°) diethyl ether-methanol (40 : 60 : 1, twice) for the subsequent separation of (3) and (6). On analytical scale the metabolites were quantitized at 360 or 290 nm before radioactivity measurements. On a preparative scale the metabolites were recrystallized to constant specific activity. Both methods gave comparable results.

[2] See text.

[3] Combined samples of a time dependent incorporation experiment (see text).

in the bulb (*1*). Uptake through the bulb stem after cutting off the lower part including the roots, or incubation of bulb slices in a citrate buffer solution at pH 4.75 also gave reasonable results (*29*).

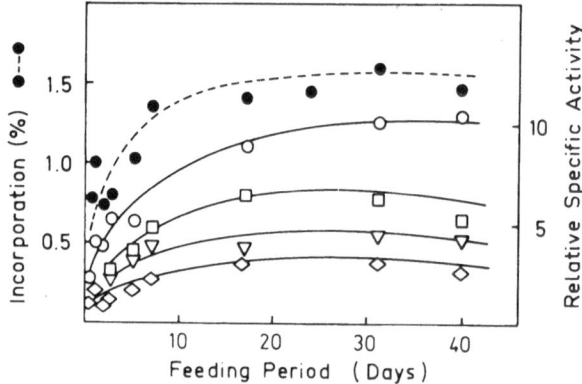

Fig. 1. Time-course of radioactivity incorporation from [U-¹⁴C]-L-phenylalanine into total bulb wax phenolics (●) given in %, and eucomin (**3**) (◇), 3,9-dihydroeucomin (**6**) (□), eucomol (**10**) (▽), and 7-O-methyleucomol (**11**) (○) given in arbitrary values for relative specific activities

Semiquantitative measurements indicated maximum incorporation rates at about two weeks (*1*). A more detailed study with *E. bicolor* var. *alba* and [U-¹⁴C]-L-phenylalanine revealed rapid accumulation of radioactivity in homoisoflavanones within the first few days followed by a slower but steady increase over several weeks (*35*) (Fig. 1).

Possible mechanisms of the biosynthetic formation of homoisoflavanones were first discussed by TAMM in 1972 (*69*). Feeding experiments carried out in different laboratories have demonstrated the specific incorporation of L-phenylalanine, sodium acetate and the methyl group of L-methionine (*19, 29, 69*) (Table 8). Preliminary indications for the distribution of the radioactivity have been obtained by DEWICK (*19*).

Eucomin (**3**) was purified as its dimethyl ether (**33**) and degraded in several steps to the products (**85**)—(**89**). Their relative specific activities are summarized in Table 9. The results indicate that a C_6—C_3 unit derived from phenylalanine is incorporated intact into the eucomin molecule and gives rise to carbon atoms 4, 3, 9 and the aromatic ring B. The O-methyl group of ring B as well as carbon atom 2 are supplied by the methyl group of methionine. The role of 2'-methoxychalcones as biosynthetic intermediates for homoisoflavanones has been demonstrated by the successful incorporation of [methyl-¹⁴C]-2',4',4-trihydroxy-6'-methoxychalcone (**90**) and [6'-methyl-¹⁴C]-2',4'-dihydroxy-4,6-dimethoxychalcone (**91**) into eucomin (**3**), the latter compound being the better precursor.

Table 9. *Relative Specific Activities of Degradation Products Obtained from Labelled Eucomin (19)*

Compound administered	Relative specific activities			
	5,7-Di-O-methyl-eucomin (33)	Anisic acid (85)	MeEt₃N⁺I⁻ (4'-OMe)	4,6-Dimethoxy-salicyclic acid (88)
[1-¹⁴C]-D,L-Phenyl-alanine	1.00	0.02	—	1.00[1]
[U-¹⁴C]-L-Phenylalanine	1.00	0.73	—	0.14
Sodium[2-¹⁴C]-acetate	1.00	0.08	—	0.91
[Methyl-¹⁴C]-L-Methionine	1.00	0.46	—	0.06
[Methyl-¹⁴C]-L-Methionine	1.00[2]	0.40	0.38	—

[1] Relative specific activity of $BaCO_3$ (C-4) was 0.92.
[2] Relative specific activity of 2'-hydroxy-2,4',6'-trimethoxyacetophenone (89) was 0.07.

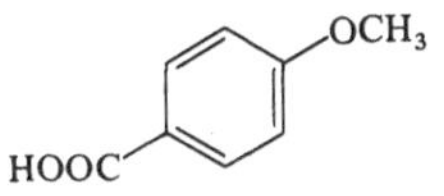

(85)

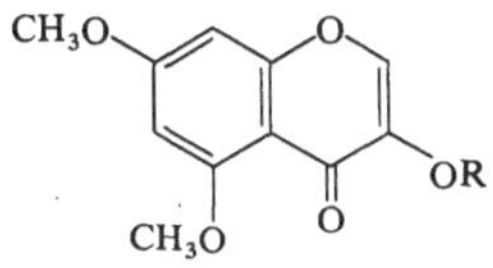

(86) R = H
(87) R = CH₃

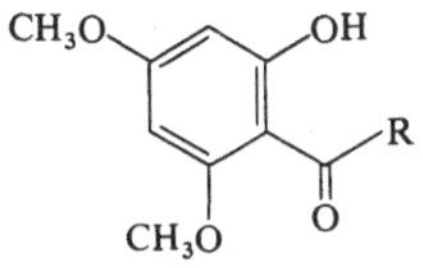

(88) R = OH
(89) R = CH₂OCH₃

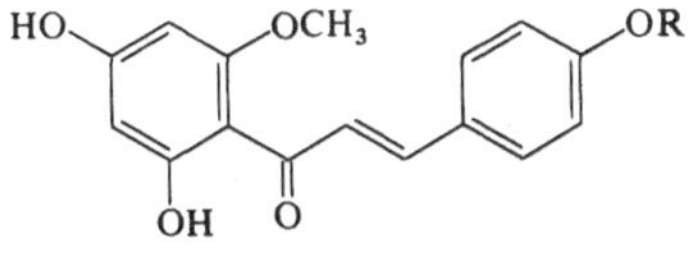

(90) R = H
(91) R = CH₃

Mechanisms for the formation of homoisoflavanones have been thoroughly discussed by DEWICK (*19*). In analogy to flavonoid biosynthesis (*32*) an intermediate (93) produced from a chalcone (92) could be formed. The latter would cyclize to the chromanone (95) giving the homoisoflavanones (94) or (97) either by elimination of a proton or by hydride transfer. Addition of water to (94) or hydroxylation of (97) would finally lead to the 3-hydroxy derivative (96).

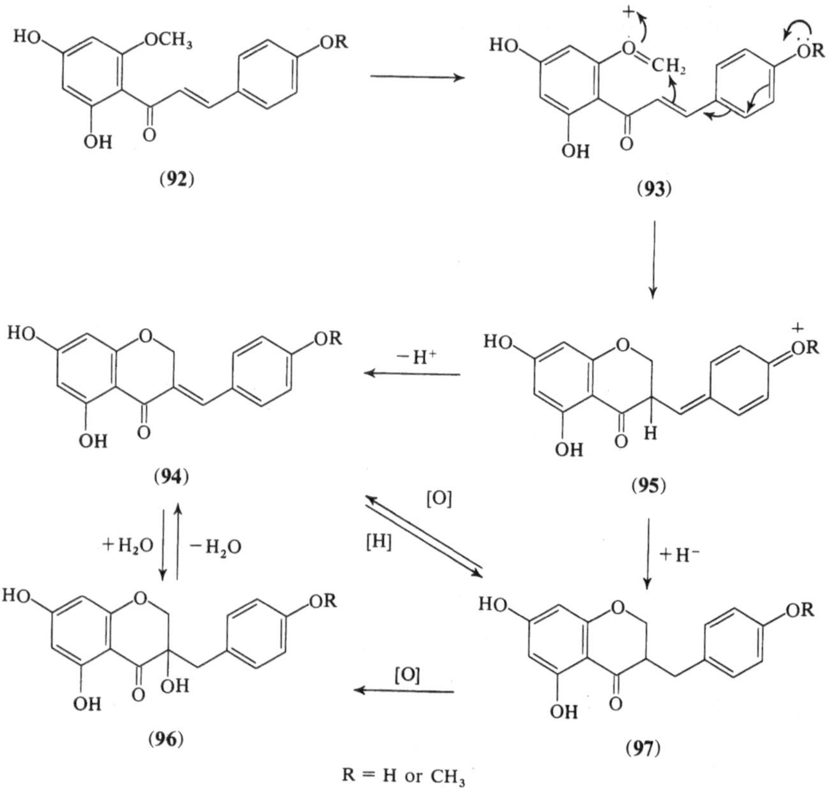

(92)   (93)

(94)   (95)

(96)   (97)

R = H or CH₃

Comparison of the specific activities of eucomin (3), 3,9-dihydro-eucomin (6), and eucomol (10) (see Table 8, Fig. 1) resulting from incorporation experiments with *E. bicolor* lead to the conclusion that the dihydro compound (6) is formed first. It then may be oxidized to substance (10) or desaturated to eucomin (3). Eucomin could also be derived from eucomol (10) by elimination of water. However, in *E. comosa* and *E. pallidiflora* ssp. *pallidiflora* not even traces of eucomol (10) have been detected besides a reasonable quantity of eucomin (3) and some amounts of its 3,9-dihydro derivative (6) (35). These results indicate that formation of eucomin (3) and eucomol (10) is not correlated although such a correlation has been proposed earlier (69).

The co-occurrence of the 3-benzylchroman-4-one derivative 3,9-dihydroeucomnalin (13), scillascillin (19) and 2-hydroxy-7-O-methyl-scillascillin (20) in *Scilla scilloides* opened a discussion on their possible biogenetic relationship (19). Starting from the 3,3′-dioxy compound (98), cyclization to the spirocyclic intermediate (99) and formation of the scillascillin skeleton (101) may easily take place. In a similar manner

brazilin (21) and hematoxylin (22) could be derived from the intermediate (98). Reduction of the latter to compound (100) and ring
closure would lead to the tetracyclic compound (102) which would give
the brazilin skeleton (103). Preference should be given to this hypothesis
over the neoflavonoid pathway proposed earlier by GRISEBACH and OLLIS
(31). It parallels the synthetic routes leading to the natural products
(16, 62).

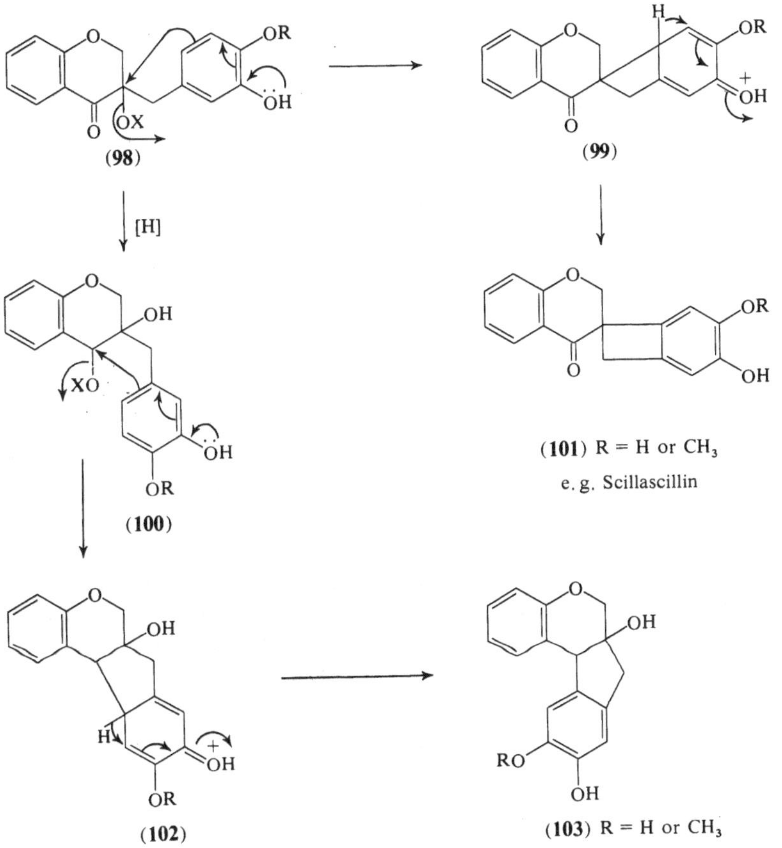

(98)

(99)

(100)

(101) R = H or CH$_3$

e. g. Scillascillin

(102)

(103) R = H or CH$_3$

e. g. Brasilin

6. Biological Activity

The highly specific morphological distribution of the homoisoflavanones in *Liliaceae* bulbs raised the question of their biological significance. Phenolic compounds have been shown to be widely distributed in
waxy surface material of buds (77, 78, 80), leaves (11, 49, 50, 51, 79),

fruits (*11, 70*) and many other plant organs (*71*). Their possible role in the protective action against microorganisms has been extensively discussed by several authors (*11, 28, 68*). Since polyphenols are produced by plants after infection by microorganisms it has been suggested that the homoisoflavanones are not constitutive plant metabolites but rather phytoalexins[4]. However, their high concentration especially at the base of all storage leaves, their presence even at the vegetative apex as well as in sterile-grown seedlings of *E. autumnalis* (*29*) and the simultaneous occurrence of large quantities of lipid material make this hypothesis unprobable.

Recently Ravisé and Kirkiacharian (*60*) have analyzed the effect of a series of natural and synthetic homoisoflavanones on *Phytophthora parasitica* and on the activity of some of its enzymes. They have found inhibition of *in vitro* growth and sporogenesis of the microorganism. Enzymatic studies revealed no remarkable inhibition of a polyphenol oxidase or of α- and β-amylases, whereas β-glucosidase and five pectinolytic enzymes (endo PTE/PATE and endo PMG/PG) which are directly involved in the infection mechanism are inhibited to a varying extent. No structure-dependent effects were readily perceptible except that fully methylated compounds seemed to be relatively ineffective.

Aqueous extracts of the heartwood of *Haematoxylon braziletto* and *H. campechianum* were shown to possess antibacterial activity (*59*). They are bactericidal for *Salmonella typhosa* and *Micrococcus pyrogenes* var. *aureus* and bacteriostatic for *Escherichia coli*. The active principle has not been identified unequivocally. However it has been suggested that the antibacterial activity is due to brazilin (**21**) or its oxidation product brazilein (**104**) which is the actual pigment. Neither the unpigmented sapwood nor the bark of the plant contain the active compound. The comparable hematoxylin derivatives are somewhat less active.

(**104**)

---

[4] Phytoalexins are plant metabolites whose production is induced by the presence of pathogenic microorganisms. For reviews see (*30, 31 a*).

7. Chemotaxonomy

So far reports of simple homoisoflavanones are restricted to the two genera *Eucomis* and *Scilla*, both belonging to the family *Liliaceae*, subfamily *Scilloideae*. A preliminary investigation of 28 species from 10 genera of this subfamily revealed that phenolic compounds occur quite commonly in the bulb waxes (Table 10).

One hundred micrograms of partially purified bulb extracts have been analyzed by two-dimensional chromatography on precoated silica gel plates using solvent systems No. 4 and No. 2 in sequence (see Chapter 2.). The phenolic compounds were detected by (a) 0.05% fast blue B salt (FBB) as described and (b) vanillin-sulfuric acid reagent (*53*) at room temperature, and (c) after careful heating. In this manner the known homoisoflavanones were unequivocally identified in the genera *Eucomis, Chionodoxa, Scilla*, and *Muscari*. The members of the genus *Eucomis* so far investigated may be divided into two groups: (a) Groups containing only metabolites with 4′,5,7-substitution (eucomins), large amounts being hydroxylated at C-3 (*E. bicolor*-type, 1 species), and (b) Groups containing also metabolites bearing an additional methoxy group in ring A either at C-6 (eucomnalins) or at C-8 (punctatins) and no detectable amount of C-3 hydroxylated compounds (*E. comosa*-type: 4 species) (Fig. 2).

It is interesting to note that the inclusion of *Chionodoxa* in *Scilla* based on a close morphological relationship with *Scilla bifolia* (*66*) is

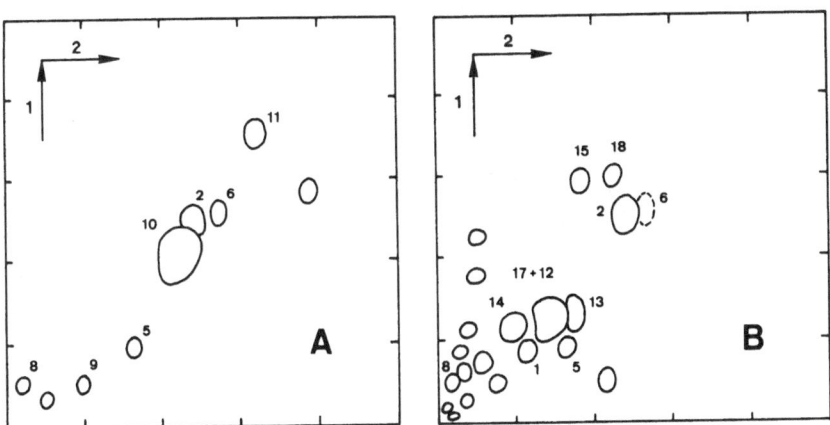

Fig. 2. Two-dimensional chromatograms of bulb wax phenolics of (A) *Eucomis bicolor* and (B) *E. comosa* (see text). The individual numbers refer to the compounds listed in Table 1

Table 10. *Distribution of Homoisoflavanones in the Subfamily Scilloideae*[1]

	Phenolics[2], FBB[3] positive	Eucomins I, II[4]		III				IV			
		(1)	(2),(3)	(4)	(5)	(6)	(7)	(8)	(9)	(10)	(11)
Ornithogalum (3 species)	×										
Hyacinthus (2 species)	×										
Veltheimia capensis	•										
Lachenalia tricolor	+										
Drimopsis maculata	+							○		○	
Camassia esculenta	+										
Muscari M. racemosum	•										
M. paradoxum	+				○						
M. latifolium	+										
M. comosum	+			○	+						
M. armeniacum	•				+						
Scilla S. violacea	•				○						
S. tubergeniana	+			○	○						
S. sibirica	•										
S. scilloides	++				+						
S. bifolia	++			○	+						
S. amethystina	+										
Chionodoxa Ch. tmoli	+				+						
Ch. sardensis	++										
Ch. luciliae	++				+						
Ch. gigantea	++				+						
Eucomis E. zambesiaca	++	•	++			•	•	+			
E. pallidi- ssp. pole-evansii	++	+	+				•	+			
flora ssp. pallidiflora	+	+	+			+	•	+			
E. comosa	++	•	+			+	•	+			
E. autumnalis	++	•	+			•	•	+			
E. bicolor var. alba	++	•	+	•	+	+	•	+	+	++	•
E. bicolor	++	•	+	•	+	+	•	+	+	++	•

References, pp. 149—152

Eucomnalins													
II	**(12)**	●	+	+	●	+	●	+					
III	**(13)**	+	+	+	+	+	●	●			+		
Punctatins													
I, II	**(14)**	+	+		+	++	+	+	+	+			
	(15), (16)	+	+		+	+	++	+	++	++			○	○
III	**(17)**	++	++	++	++	++	++	+	++				
	(18)	+	+	●	+	●	+	●	+				
Scillascillins													
	(19)	●	●	●	●	●	●	+	+	+	●		
	(20)	●	●	●	●	●	●	●	+	○	+		○ ○
Chromanone													
	(23)	●	●										
Blue metabolite[5]						●	●			+ ● ●	●		

[1] Bulbs were either purchased from C. G. van Tubergen, Haarlem, The Netherlands, or obtained from the Botanical Garden in Basel, Switzerland.

[2] (±) very abundant, (+) abundant, (●) rare, (○) tentative, (×) bulb wax not visible.

[3] FBB: fast blue B salt reagent (see text).

[4] See Table 1.

[5] See text.

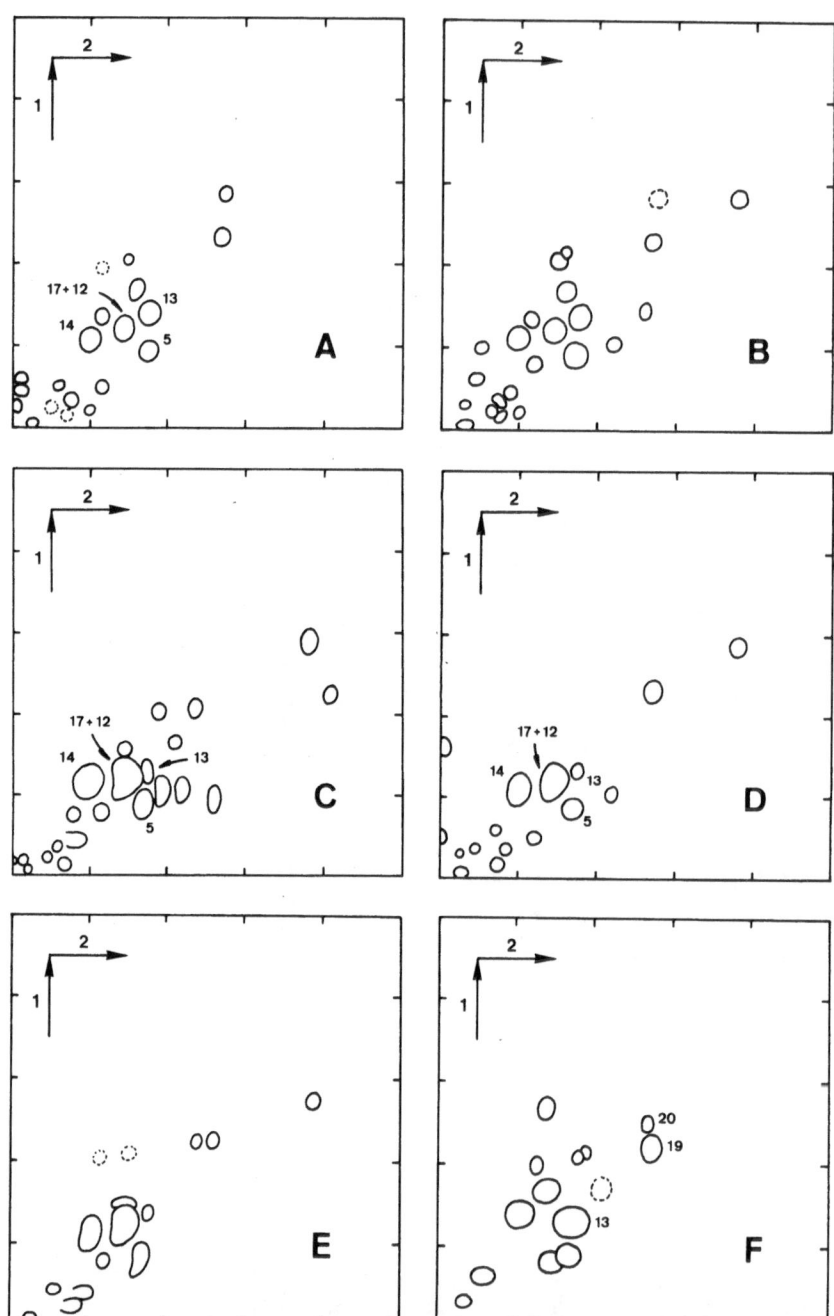

Fig. 3. Two-dimensional chromatograms of bulb wax phenolics of (A) *Chionodoxa gigantea*, (B) *Ch. tmoli*, (C) *Scilla bifolia*, (D) *Ch. luciliae*, (E) *Ch. sardensis*, and (F) *S. scilloides* (see text)

supported by the presence of several common metabolites (Fig. 3). However, the infrageneric order of the species might be subject possible to revision. Metabolite patterns indicate that *Chionodoxa gigantea* and *Ch. tmoli* constitute one group which contains 4'-demethyl-3,9-dihydroeucomin, eucomnalins and punctatins in comparable amounts, and *Ch. luciliae, Ch. sardensis* and *Scilla bifolia* a second one which contains punctatins as main metabolites.

Scillascillins could be detected in the genera *Chionodoxa, Scilla* and *Muscari* (Table 10). A metabolite of unknown structure at hRf 21 and 44 in solvent systems No. 4 and No. 2 respectively showing a brilliantly blue color with spray methods (b) and (c) (see above), was also observed in some species of these genera.

More detailed studies will be necessary to evaluate clearly the taxonomic significance of the bulb wax phenolics; in particular more information has to be collected on their intraspecific variability.

References

1. ANDERMATT, P.: Beitrag zur Kenntnis der Biogenese der Spirostanole und Untersuchungen über die Biogenese der Homo-Isoflavone. Ph. D. Thesis, Basel 1971.

2. ANDRIEUX, J., D. H. R. BARTON, and H. PATIN: Rhodium-catalyzed Isomerization of Some Unsaturated Organic Substrates. J. C. S. Perkin I, **1977,** 359.

3. ANTENUIS, M., and D. TAVERNIER: New Models for Conformational Analysis by Nuclear Magnetic Resonance. Tetrahedron Letters **1964,** 3949.

4. AUDIER, H.: Etude des Composés Flavoniques par Spectrométrie de masse. Bull. Soc. Chim. France **1966,** 2892.

5. BALDWIN, R. M., C. D. SNYDER, and H. RAPOPORT: Biosynthesis of Menaquinones. Dissymmetry in the Napththalenic Intermediate. J. Amer. Chem. Soc. **95,** 276 (1973).

6. BARKER, S. A., J. HOMER, M. C. KEITH, and L. F. THOMAS: Proton Resonance Studies of Methoxy and Acetoxy Derivatives of Pyranose Molecules Applied to the Conformation of Methyl 3-O-Carbamoyl-α- and -β-L-novioside. J. Chem. Soc. **1963,** 1538.

7. BASS, R. J.: Synthesis of Chromones by Cyclization of 2-Hydroxyphenyl Ketones with Boron Trifluoride-Diethyl Ether and Methanesulphonyl Chloride. J. C. S. Chem. Commun. **1976,** 78.

8. BOEHLER, P.: Isolierung und Strukturaufklärung der Homo-Isoflavone Eucomol und Eucomin. Ph. D. Thesis, Basel 1967.

9. BOEHLER, P., and CH. TAMM: The Homo-Isoflavones: A New Class of Natural Products. Isolation and Structure of Eucomin and Eucomol. Tetrahedron Letters **1967,** 3479.

10. BUDZIKIEWICZ, H., C. DJERASSI, and D. H. WILLIAMS: "Mass Spectrometry of Organic Compounds", p. 470. San Francisco, Cambridge, London, Amsterdam: Holden-Day. 1967.

11. CALDICOTT, A. B., and G. EGLINTON: Surface Waxes in "Phytochemistry" (ed. L. P. MILLER), Vol. 3, p. 162. New York, Cincinnati, Toronto, London, Melbourne: Van Nostrand Reinhold. 1973.

12. CHARI, V. M., and H. WAGNER: Advances in the Spectroscopy of Plant Phenolics. Recent Adv. Phytochemistry **12,** 29 (1979).

13. CHATTERJEA, J. N., S. C. SHAW, and J. N. SINGH: A Synthesis of Anhydrobrazilic Acid. Isomerization of Arylidenechroman-4-ones to Homoisoflavones. J. Indian Chem. Soc. **51**, 281 (1974).

14. CRAIG, J. C., A. R. NAIK, R. PRATT, E. JOHNSON, and N. S. BHACCA: Nuclear Magnetic Resonance Spectra and Stereochemistry of the Antibacterial Principle from *Haematoxylon braziletto*. J. Org. Chemistry **30**, 1573 (1965).

15. DANN, O., and H. HOFMANN: Die Epoxydation von 3-Benzylidenchromanonen-(4) zu 3,α-Oxido-3-benzyl-chromanonen-(4). Chem. Ber. **95**, 1446 (1962).

16. — — Die Synthese von (±)-Haematoxylin. Chem. Ber. **98**, 1498 (1965).

17. DEULOFEU, V., and T. J. GUERRERO: N-Methyl-3.4-Dihydroxyphenylalanine. Org. Synth. Coll. Vol. 3, 586 (1955).

18. DEWICK, P. M.: Biosynthesis of the 3-Benzylchroman-4-one Eucomin. J. C. S. Chem. Commun. **1973**, 438.

19. — Biosynthesis of the 3-Benzylchroman-4-one Eucomin in *Eucomis bicolor*. Phytochemistry **14**, 983 (1975).

20. DONNELLY, D. M. X.: Neoflavonoids in "The Flavonoids" (eds. J. B. HARBORNE, T. J. MABRY, and H. MABRY), p. 801. London: Chapman and Hall. 1975.

21. DREWES, S. E.: Chroman and Related Compounds. Vol. 2 of Progress in Mass Spectroscopy (ed. H. BUDZIKIEWICZ). Weinheim/Bergstrasse: Verlag Chemie. 1974.

22. DUDDECK, H., G. SNATZKE, and S. S. YEMUL: ^{13}C NMR and CD of Some 3,8″-Biflavonoids from *Garcinia* Species and of Related Flavanones. Phytochemistry **17**, 1369 (1978).

23. FARKAS, L., A. GOTTSEGEN, and M. NOGRADI: Synthesis of Eucomin and (±)-Eucomol. Tetrahedron Letters **1968**, 4099.

24. — — — The Synthesis of Eucomin and (±)-Eucomol. Tetrahedron **26**, 2787 (1970).

25. FARKAS, L., A. GOTTSEGEN, M. NOGRADI, and J. STRELISKY: Synthesis of Homoisoflavanones-II. Constituents of *Eucomis autumnalis* and *E. punctata*. Tetrahedron **27**, 5049 (1971).

26. FINCKH, R. E.: Isolierung und Strukturaufklärung homoisoflavonoider Verbindungen aus *Eucomis punctata* L'Hérit. Ph. D. Thesis, Basel 1970.

27. FINCKH, R. E., and CH. TAMM: The Homoisoflavones III. Isolation and Structure of Punctatin, 3,9-Dihydropunctatin, 4′-O-Methyl-3,9-dihydropunctatin, 4′-Demethyleucomin, and 4′-Demethyl-5-O-methyl-3,9-dihydroeucomin. Experientia **26**, 472 (1970).

28. FRIEND, J.: Phenolic Substances and Plant Disease. Recent Adv. Phytochemistry **12**, 557 (1979).

29. GRISEBACH, H.: Personal communication.

30. GRISEBACH, H., and J. EBEL: Phytoalexine, chemische Abwehrstoffe höherer Pflanzen? Angew. Chem. **90**, 668 (1978).

31. GRISEBACH, H., and W. D. OLLIS: Biogenetic Relationship Between Coumarins, Flavonoids, Isoflavonoids, and Rotenoids. Experientia **17**, 4 (1961).

31a. GROSS, D.: Phytoalexine und verwandte Pflanzenstoffe. Fortschr. Chem. Org. Naturst. **34**, 187 (1977). Wien, New York: Springer.

32. HAHLBROCK, K., and H. GRISEBACH: Biosynthesis of Flavonoids in "The Flavonoids" (eds. J. B. HARBORNE, T. J. MABRY, and H. MABRY), p. 905. London: Chapman and Hall. 1975.

33. HEGNAUER, R.: Chemotaxonomie der Pflanzen, Vol. 2, p. 329. Basel und Stuttgart: Birkhäuser. 1963.

34. HELLER, W.: Über einige Inhaltsstoffe von *Eucomis punctata* L'Hérit. Isolierung, Konstitution und Synthese von (−)-*R*-Eucominsäure. Ph. D. Thesis, Basel 1973.

35. HELLER, W., and CH. TAMM: Unpublished results.

36. — — Isolierung, Konstitution und Synthese der (*R*)-(−)-Eucominsäure. Helv. Chim. Acta **57**, 1766 (1974).

37. — — 5,7-Dihydroxy-8-methoxychroman-4-on aus dem Zwiebelwachs von *Eucomis comosa.* Helv. Chim. Acta **61,** 1257 (1978).

38. HELLER, W., P. ANDERMATT, W. A. SCHAAD, and CH. TAMM: Homoisoflavanone IV. Neue Inhaltsstoffe der Eucomin-Reihe von *Eucomis bicolor.* Helv. Chim. Acta **59,** 2048 (1976).

39. JAY, M., J.-F. GONNET, E. WOLLENWEBER, and B. VOIRIN: Sur L'Analyse Qualitative des Aglycones Flavoniques dans une Optique Chimitaxonomique. Phytochemistry **14,** 1605 (1975).

40. KIRBY, G. W., and L. OGUNKOYA: Deuterium and Tritium Exchange Reactions of Phenols and the Synthesis of Labelled 3,4-Dihydrophenylalanines. J. Chem. Soc. **1965,** 6914.

41. KIRKIACHARIAN, B. S.: Sur une Nouvelle Synthèse de la (±)-Triméthylbraziline. C. R. Acad. Sci. Ser. C **274,** 2096 (1972).

42. — Hydroborations: New Routes to Isoflavanones and Homoisoflavanones. J. C. S. Chem. Commun. **1975,** 162.

43. KIRKIACHARIAN, B. S., and M. GARNIER: Hydroborations. Sur une Nouvelle Synthèse de la (±)-Tétraméthylhématoxyline. C. R. H. Acad. Sci. Ser. C **277,** 1037 (1973).

44. KOUNO, I., T. KOMORI, and T. KAWASAKI: Zur Struktur der neuen Typen Homo-Isoflavanone aus Bulben von *Scilla scilloides* Druce. Tetrahedron Letters **1973,** 4569.

45. KRISHNAMURTY, H. G., B. PARKASH, and I. R. SESHADRI: Synthesis of Eucomin, 4′-Demethyleucomin and 5,7-Di-O-methyleucomol. Indian J. Chemistry **12,** 554 (1974).

46. McOMIE, J. F. W., M. L. WATTS, and D. D. WEST: Demethylation of Aryl Methyl Ethers by Boron Tribromide. Tetrahedron **24,** 2289 (1968).

47. MARKHAM, K. R.: Gentian Pigments-II. Xanthones from *Gentiana bellidifolia.* Tetrahedron **21,** 1449 (1965).

48. MARKHAM, K. R., and T. J. MABRY: Ultraviolet-Visible and Proton Magnetic Resonance Spectroscopy of Flavonoids in "The Flavonoids" (eds. J. B. HARBORNE, T. J. MABRY, and H. MABRY), p. 45. London: Chapman and Hall. 1975.

49. MARTIN, J. T.: Studies on the Natural Protective Covering of Plants: I. Plant Wax in Relation to Resistance to Infection by Fungi. Ann. Rep. Long Ashton Res. Sta. for 1956. **1957,** 94.

50. MARTIN, J. T., and R. F. BATT: Studies on Plant Cuticle. I. The Waxy Coverings of Leaves. Ann. Appl. Biol. **46,** 375 (1958).

51. MASAYUKI, S., D. DIFEO JR., N. NAKATANI, B. TIMMERMANN, and T. J. MABRY: Flavonoid Methyl Ethers on the External Leaf Surface of *Larrea tridentata* and *L. divaricata.* Phytochemistry **15,** 727 (1976).

52. MASSICOT, J., and J.-P. MARTHE: Résonance magnétique nucléaire de produits naturels. III. — Etude de quelques dérivés flavoniques et substances apparentées. Bull. Soc. Chim. France **1962,** 1962.

53. MATTHEWS, J. S.: Steroids-CCXXIII. Color Reagent for Steroids in Thin-Layer Chromatography. Biochim. Biophys. Acta **69,** 163 (1963).

54. MORSINGH, F., and R. ROBINSON: The Synthesis of Brazilin and Haematoxylin. Tetrahedron **26,** 281 (1970).

55. NIELSEN, H., and P. ARENDS: Structure of the Xantholignoid Kielcorin. Phytochemistry **17,** 2040 (1978).

56. PELTER, A., P. STAINTON, and M. BARBER: The Mass Spectra of Oxygen Heterocycles. II. The Mass Spectra of Some Flavonoids. J. Heterocyclic Chemistry **2,** 262 (1965).

57. PELTER, A., R. S. WARD, and R. J. BASS: The Carbon-13 Nuclear Magnetic Resonance Spectra of Isoflavones. J. C. S. Perkin I **1978,** 666.

58. PFEIFFER, P., E. BREITH, and H. HOYER: Oxy-benzyl-chromanone. 11. Mitteilung zur Brasilin- und Haematoxylinfrage. J. prakt. Chemie **237,** 31 (1931).

59. PRATT, R., and Y. YUZURIHA: Antibacterial Activity of the Heartwood of *Haematoxylon braziletto.* J. Amer. Pharm. Assoc. **48,** 69 (1959).

60. RAVISÉ, A., and B. S. KIRKIACHARIAN: Influence de la structure de composés phénoliques sur l'inhibition du *Phytophthora parasitica* et d'enzymes participant aux processus parasitaires. III. Homoisoflavanones. Phythopath. Z. **92**, 36 (1978), and references therein.
61. REYNEKE, W. F.: N Monografiese Studie va die Genus *Eucomis* L'Hérit. in Suid-Afrika. M. Sc. Thesis, Pretoria 1972.
62. ROBINSON, R.: Chemistry of Brazilin and Haematoxylin. Bull. Soc. Chim. Fr. **1958**, 125.
63. — Brazilin and Haematoxylin in "Chemistry of Carbon Compounds" (ed. E. H. RODD), Vol. IV, part B, p. 1005. New York: Elsevier. 1959.
64. SCHAAD, W.: Isolierung und Ermittlung der Konstitution neuer Homoisoflavanone aus *Eucomis bicolor* Bak., Ph. D. Thesis, Basel 1977.
65. SIDWELL, W. T. L., and CH. TAMM: The Homo-Isoflavones II: Isolation and Structure of 4'-O-Methylpunctatin, Autumnalin and 3,9-Dihydroautumnalin. Tetrahedron Letters **1970**, 475.
66. SPETA, F.: Über *Chionodoxa* Boiss., ihre Gliederung und Zugehörigkeit zu *Scilla* L., Naturk. Jahrb. Stadt Linz 1975, **21**, 9 (1976).
67. STAHL, E.: Dünnschicht-Chromatographie, 2nd ed. p. 671. Berlin Heidelberg New York: Springer. 1967.
68. STOESSL, A.: Antifungal Compounds Produced by Higher Plants. Recent Adv. Phytochemistry **3**, 143 (1970).
69. TAMM, CH.: Die Homo-isoflavone, eine neue Klasse von Naturstoffen. Arzneim. Forsch. (Drug Res.) **22**, 1776 (1972).
70. TRONCHET, J.: Flavone Derivatives of Fruits. Localization, Distribution and Evolution. Bull. Soc. Bot. Fr. **119**, 25 (1972).
71. — Role of Protection and Control of Surface Flavonic Derivatives. First Experimental Results. Bull. Liaison, Groupe Polyphénols **48**, 18 (1973).
72. WAGNER, H., and L. FARKAS: Synthesis of Flavonoids in "The Flavonoids" (eds. J. B. HARBORNE, T. J. MABRY, and H. MABRY), p. 202. London: Chapman and Hall. 1975.
73. WATT, J., and M. G. BREYER-BRANDWIJK: Medicinal and Poisonous Plants of Southern and Eastern Africa, 2nd edn., p. 298. Edinbourgh, London: E. S. Livingstone. 1962.
74. WEBER, H. P., W. HELLER, and CH. TAMM: Homoisoflavanones. V. Crystal and Molecular Structure of (−)-7-O-(p-Bromophenacyl)eucomol. The Absolute Configuration of (−)-Eucomol. Helv. Chim. Acta **60**, 1388 (1977).
75. WILSON, R. G., J. H. BOWIE, and D. H. WILLIAMS: Solvent Effects in NMR Spectroscopy. Solvent Shifts of Methoxyl Resonances in Flavones Induced by Benzene; an Aid in Structure Elucidation. Tetrahedron **24**, 1407 (1968).
76. WINDHOLZ, M. (ed.): The Merck Index, 9th edn. Rathway: Merck. 1976.
77. WOLLENWEBER, E.: Flavonoidmuster als systematisches Merkmal in der Gattung *Populus*. Biochem. Syst. Ecol. **3**, 35 (1975).
78. — Flavonoidexkret der Betulaceen. Biochem. Syst. Ecol. **3**, 47 (1975).
79. — Einige Neufunde externer Flavonoide bei amerikanischen Farnen. Flora **168**, 138 (1979).
80. WOLLENWEBER, E., P. LEBRETON, and M. CHANDESON: Flavonoids in Bud Excretions of *Prunus* and *Rhamnus* Species. Z. Naturforsch. **27b**, 567 (1972).
81. ZIEGLER, R., and CH. TAMM: Isolation and Structure of Eucosterol and 16β-Hydroxyeucosterol, Two Novel Spirocyclic Nortriterpenes and of a New 24-Nor-5α-chola-8,16-diene-23-oic Acid from Bulbs of Several Eucomis Species. Helv. Chim. Acta **59**, 1997 (1976).
82. — Recent Studies on Homoisoflavanones in "Flavonoids and Bioflavonoids". Proc. 5th Hung. Biflavonoid Symposium, Matrafüred, Hungary, May 25—27, 1977, p. 95. Amsterdam, Oxford, New York: Elsevier Scientific Publ. Company. 1977.

(Received March 3, 1980)

Naturally Occurring Phenalenones
and Related Compounds

By R. G. Cooke, Chemistry Department, University of Melbourne, Parkville, Australia, and J. M. Edwards, Pharmacy School, University of Connecticut, Storrs, Connecticut, U.S.A.

Contents

I. Introduction

Secondary metabolites containing the phenalenone nucleus, or having structures which can be reasonably presumed to have been derived from an intact phenalenone, are rarely found in nature. The first compounds were isolated in 1955 and their occurrence seems to be restricted to one family of higher plants (Haemodoraceae), four genera of Hyphomycetes (Fungi Imperfecti), and one genus *(Roesleria)* within the class Discomycetes (Ascomycotina). The plant and fungal phenalenones are structurally quite different and are derived from unrelated bio-synthetic pathways. Different numbering systems have traditionally been used for the two classes of compounds and we have retained this dichotomy in the present review.

II. Phenalenones and Related Metabolites from Fungi

A. Occurrence

The first phenalenones isolated from fungal sources were the yellow pigment atrovenetin (**1**) and the red-coloured norherqueinone (**2**), herqueinone (**3**), and isoherqueinone (**4**), all elaborated by *Penicillium herquei* and *P. atrovenetum* (*1—6*); subsequent investigations have revealed deoxyherqueinone (**5**) (*7*), isonorherqueinone (**6**) (*2, 8*) and herqueichrysin (**7**) (*9, 10*) as additional metabolites of *P. herquei*. The related naphthalic anhydride (**8**) was also isolated from the same source (*6*), and has been found as a metabolite of *Fusicoccum putrefaciens* (*11*) and of *Roesleria pallida* (*12*).

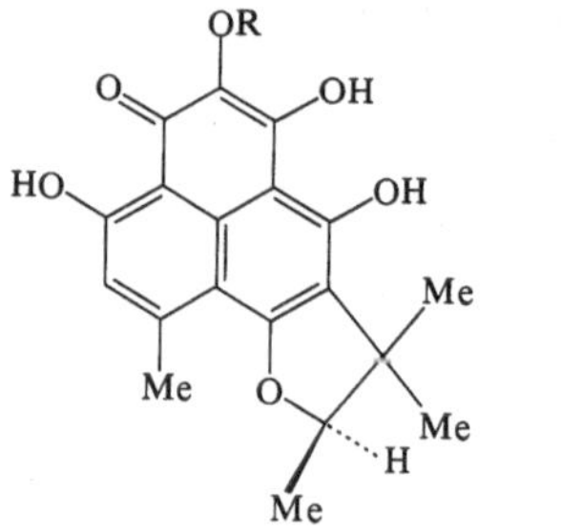

Atrovenetin (**1**) R = H
Deoxyherqueinone (**5**) R = Me

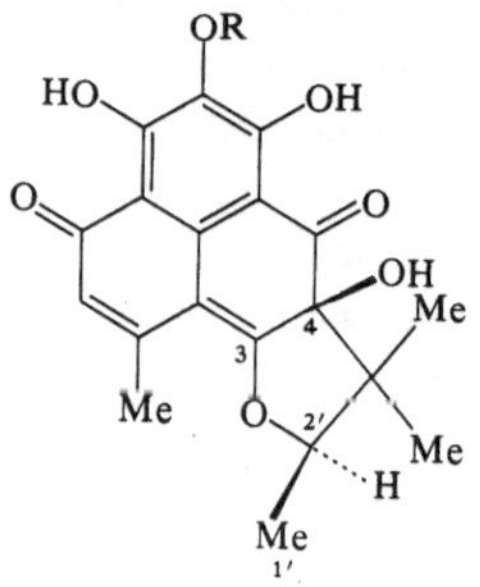

Norherqueinone (**2**) R = H
Herqueinone (**3**) R = Me
Iso-herqueinone (**4**) R = Me,
   C-2′ epimer of (**3**)
Iso-norherqueinone (**6**) R = H
   C-2′ epimer of (**2**)

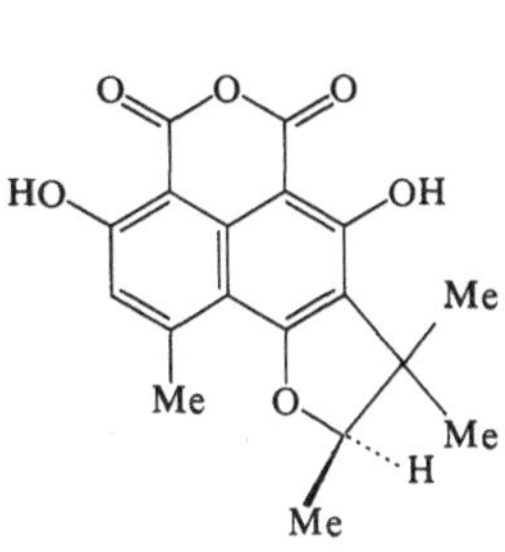

Phenalenones of *P. herquei* and *P. atrovenetum*

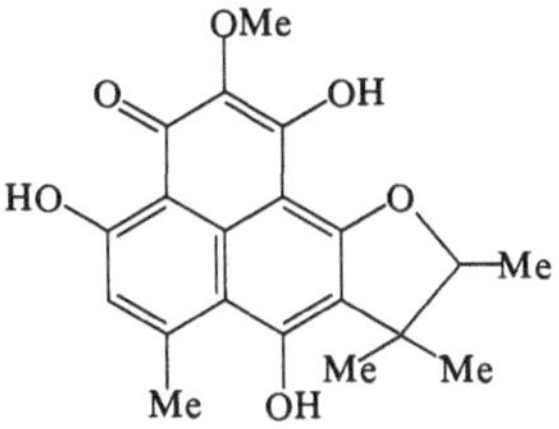

The remaining fungal phenalenone derivatives either contain modified phenalenone rings or, as in the case of the naphthalic anhydride (**8**), are presumed to be products of oxidative metabolism of intact phenalenones. Thus, the structures of the dimeric compounds duclauxin (**9**), xenoclauxin (**10**), and cryptoclauxin (**11**), three of thirteen pigments isolated from *P. duclauxi* (*13*), bear an obvious relationship to those of the intact phenalenones, although no monomeric or unoxidized compounds seem to have been isolated from the producing organism. Duclauxin has also been found in the culture filtrates of *P. stipitatum* (*14*), and recently two similar pigments gilmaniellin and dechlorogilmaniellin have been isolated from *Gilmaniella humicola* (*16*). Similarly, the chlorinated naphthalic ester (**12**), lamellicolic anhydride (**13**), and its O-carbomethoxy analog (**14**), all isolated from *Verticillium lamellicola*, are presumed to be derived from an intact phenalenone by oxidative degradation (*15*).

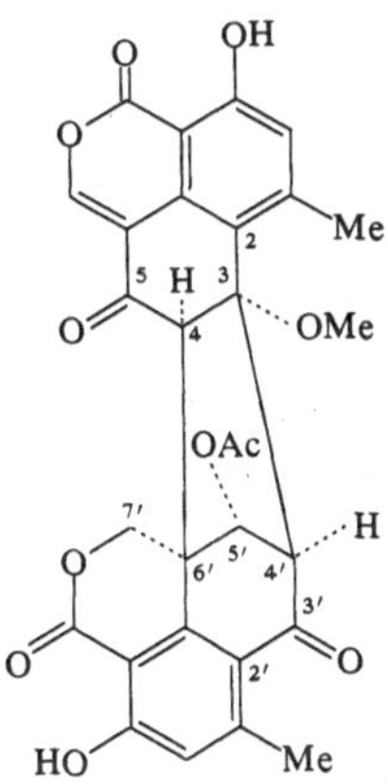

Duclauxin (**9**) 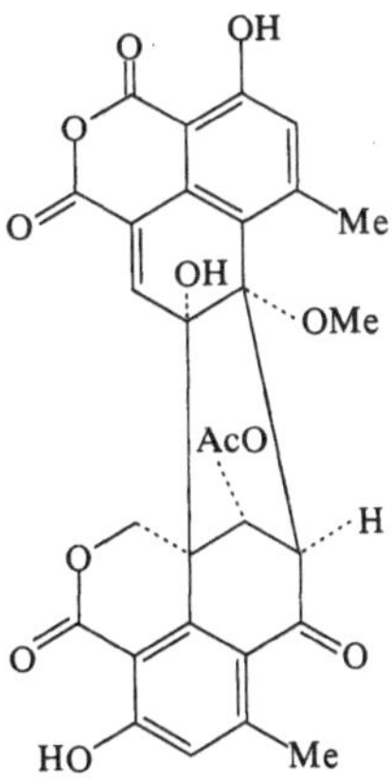   Cryptoclauxin (**11**)

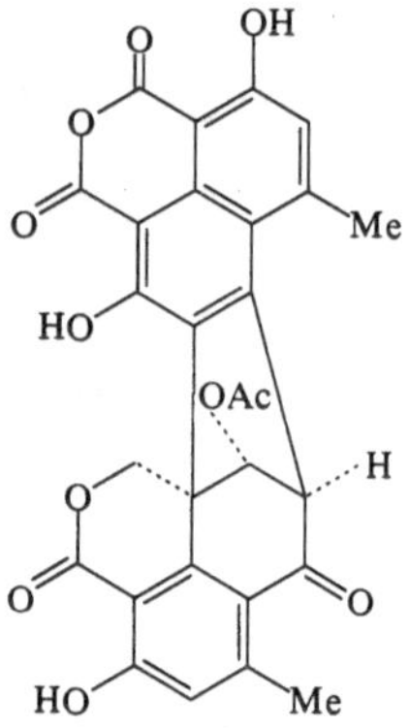

Xenoclauxin (10)

Phenalenones of *P. duclauxi*

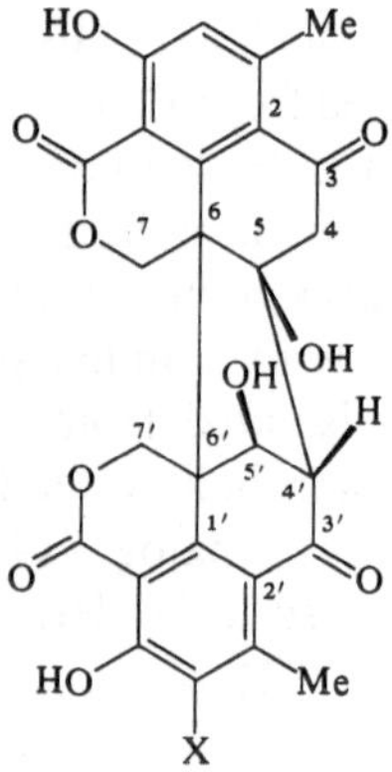

Gilmaniellin X = Cl
Dechlorogilmaniellin X = H

Phenalenones of *Gilmaniella humicola*

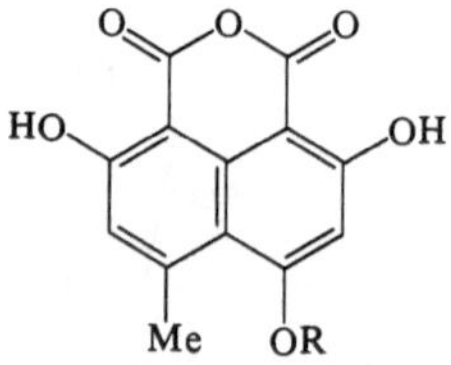

(13) R = H
(14) R = COOMe

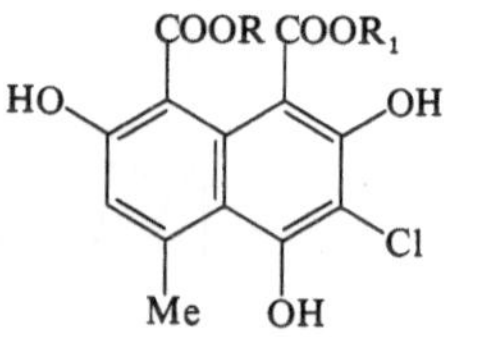

(12) R = H, R₁ = Me or R = Me, R₁ = H

Phenalenone-derived metabolites of *V. lamellicola*

B. Structure and Chemistry

The interrelationships between atrovenetin (1), norherqueinone (2), herqueinone (3), deoxyherqueinone (5), and the naphthalic anhydride (8) are shown in the scheme. X-Ray crystallographic analysis (17) of the ferrichloride salt of the orange trimethyl ether of atrovenetin

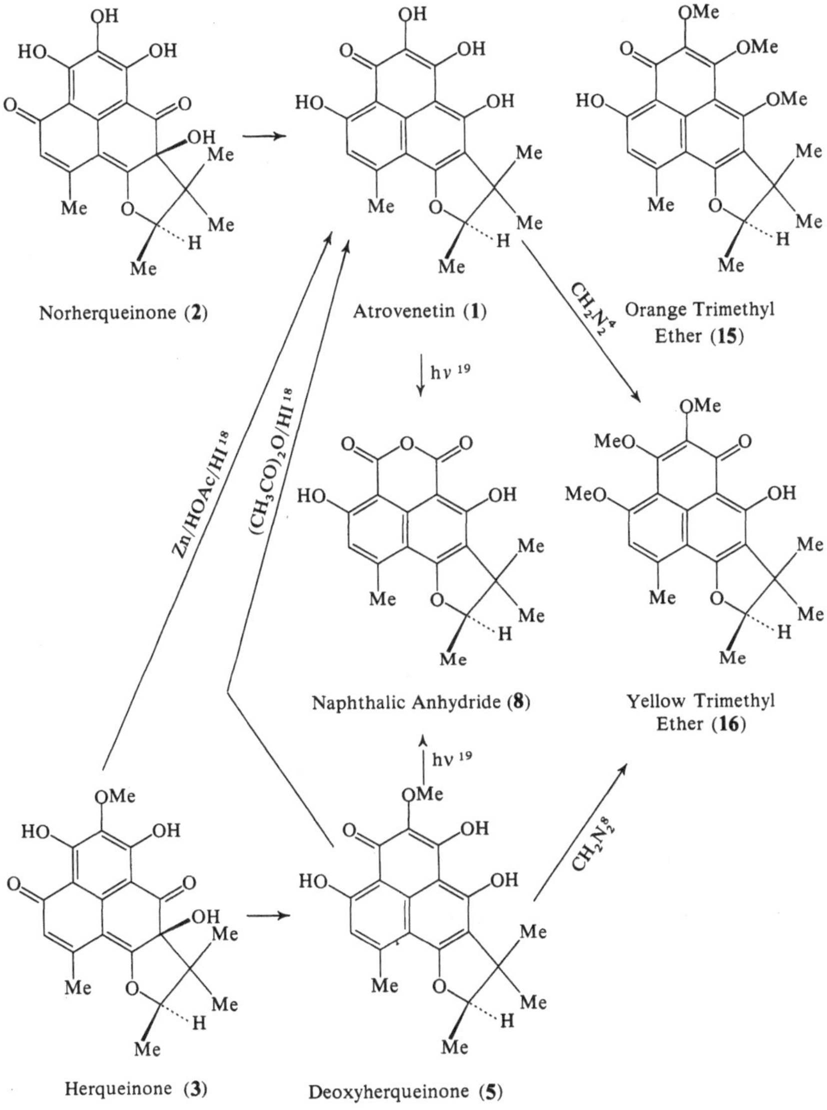

Scheme I. Chemical relationships between the phenalenones of *P. herquei*

established structure (15); this ether, and the isomeric yellow trimethyl
ether (16), are also available from deoxyherqueinone (5) on treatment
with diazomethane (8). Herqueinone, which is a methyl ether of norherqueinone, is converted into (1) by reductive demethylation (18).
Similarly (5) is a monomethyl ether of (1) and yields (1) on treatment
with acetic anhydride and HI (18). Both (1) and (5) are easily converted
into the anhydride (8) by photochemical oxidation (19) or by treatment
with alkaline hydrogen peroxide (20).

The extensive work which led to the structures of atrovenetin, norherqueinone, and the anhydride (8), as well as much of the chemistry
of (3), has been reviewed previously in this series (21); since 1966,
the structures of atrovenetin and anhydride (8) have been confirmed
by synthesis (vide infra), and the configuration (R) at C-2' shown in
the structures of the scheme, has been established in the case of (1)
and (3) by the correlation of (1) with (−)-S-ethyl lactate (22). The
controversy over the structure of herqueinone (8, 23, 24) (orientation
of the C_5 unit, and position of the OMe group) has been resolved in favor
of structure (3)*.

The position of the methoxyl group present in (3) follows from its
reduction to deoxyherqueinone (5) (2) and the methylation of (5) to the
isomeric ethers (15) and (16) (8). Furthermore, since it has been shown
that the ethers derived from (3) via (5) have the same optical rotation
as those obtained from (1), it follows that (1), (3), and (5) have the same
absolute configuration (R) at C-2' (8, 25).

Herqueinone and isoherqueinone (4) and norherqueinone and isonorherqueinone (6) are two epimeric pairs which differ only in the
configuration of the asymmetric centre C-2'; the paired compounds
are not enantiomeric, having the same configuration at C-4. The
absolute configurations of (3) and (4) (R and S respectively) have been
established (22). Pure (±)-isoherqueinone can be obtained by base
treatment of a mixture of (3) and (4), and its formation can be explained
by assuming epimerization (at C-4) of the herqueinone present in the mixture to yield the enantiomer of (4). The compound formed when impure
samples of herqueinone (i.e. containing isoherqueinone) are treated
with potassium carbonate in acetone, and which has been referred to
in the literature as isoherqueinone, is thus racemic isoherqueinone,
made up of isoherqueinone originally present and the enantiomeric
compound formed by epimerization of (3). A possible mechanism for
the epimerization suggested by Brooks and Morrison (8) is shown in
Scheme II.

---

* The absolute configuration of herqueinone shown in the figures, has recently been
established by X-ray analysis (41).

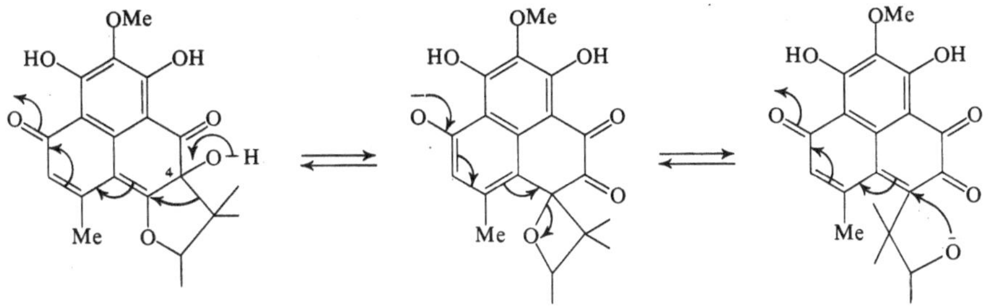

Scheme II. Proposed mechanism for epimerization at C-4 of herqueinone (**8**)

The structure of demethylherqueichrysin (**17**), which has not yet been established as a natural product, has been established by synthesis (*10*) of the racemic compound *(vide infra)*. Treatment of (**17**) with pyridine hydrochloride at 220° converts it partially into atrovenetin (**1**) (*26*). The recovered demethyl herqueichrysin is found to be of high optical purity, implying that the reclosure of the dihydrofuran ring occurs to form (**1**) preferentially. If (**17**) should prove to be a metabolite of the fungus, it may be that it isomerises *in vivo* to (±)-atrovenetin, providing the S-enantiomer of (**1**) which could be involved in the biosynthesis of isoherqueinone (**4**) and other compounds having the S-configuration at C-2′ (*10*).

The structure of herqueichrysin, a methyl ether of (**17**) which is isomeric with deoxyherqueinone, was originally proposed (*9*) to be (**18**), but is probably better represented by (**7**). SIMPSON (*27*) has reported

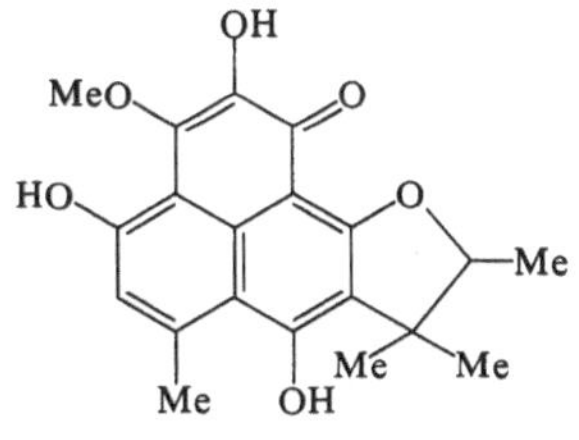

(**18**)

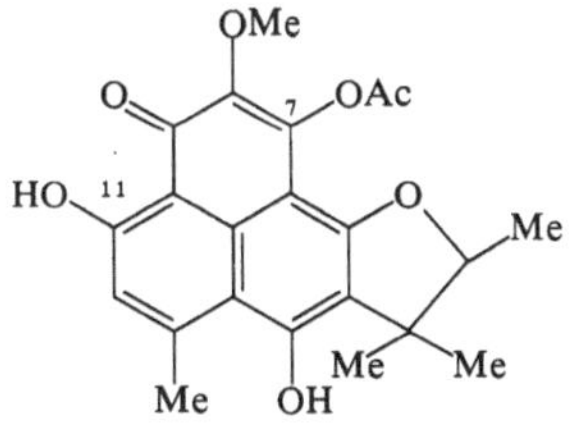

(**19**)

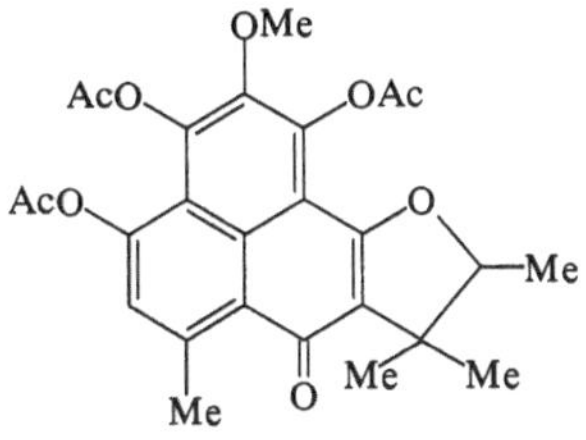

(**20**)

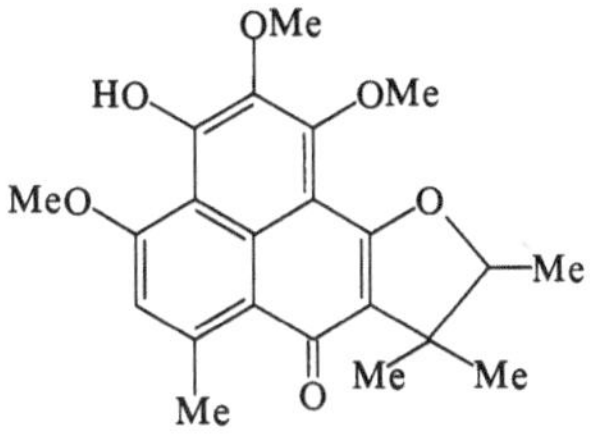

(**21**)

that herqueichrysin forms a single mono acetate to which structure
(19) has been assigned by spectroscopic analysis. The compound has a
resonance in its NMR spectrum at 16.8 ppm presumably due to the *peri*
OH at C-11 (the OH at C-3 is seen at 5.9 ppm); (7) also forms a
triacetate (20) (*9, 10*) and di- and trimethyl ethers (*27*). There are
apparently no isomeric derivatives formed in these reactions, implying
that the tautomerism of (7) is restricted, presumably by the hydrogen
bonding possibilities. Structure (7) is certainly compatible with the
available data; to accommodate the reported (*10*) resonance at 10.6 ppm
in the NMR spectrum of the dimethyl ether of (7) one would have to
postulate structure (21) rather than the tautomeric form with the
carbonyl group at C-9; the structure of the trimethyl ether would
correspond to that of the acetate (20) in which the carbonyl function
is now at the most sterically hindered site C-3.

The structure and absolute configuration of duclauxin (9) were
established by X-ray crystallographic analysis of its monobromo derivative
(*28, 29*). The compound forms a diacetate, a dimethyl ether, and a
dihydro derivative. Treatment of (9) with ammonia leads to proto-
duclauxamide (22), which lacks the acetyl function, and in which the
oxygen of the α-pyrone ring has been replaced by nitrogen; (22) on
treatment with dilute acid yields duclauxamide (23) by loss of the elements
of methanol. A series of orange-coloured compounds homologous with
(23) is obtained when (9) is treated with primary amines (*13*). Xenoclauxin (10) unlike (9), gives no colour with ammonia, exhibits strong
fluorescence in UV light, and has prominent IR absorptions at 1723
and 1676 cm^{-1} very similar to those seen in the spectra of anhydride
(8) and other naphthalic anhydrides. Oxidation of (9) with chromic

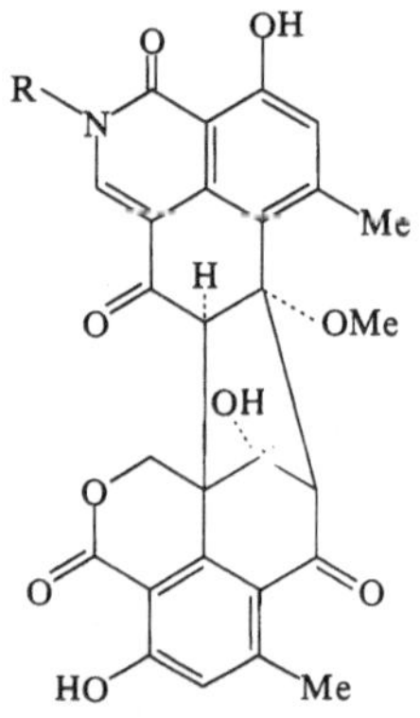

(22) R = H

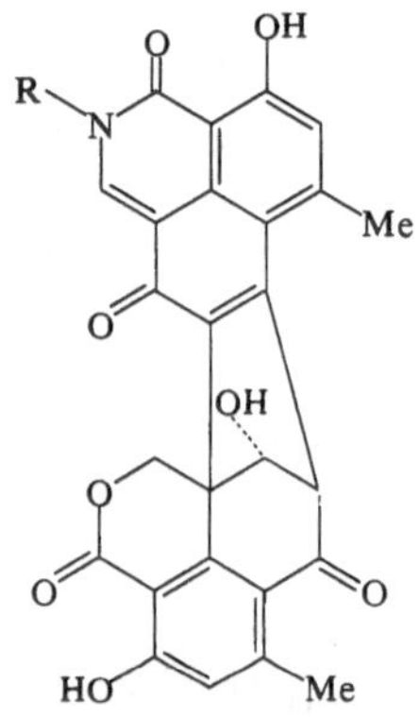

(23) R = H
(24) R = Me

acid in pyridine yields (10); similar oxidation of N-methylduclauxamide (24) yields N-methylxenoclauximide, available from (10) by reaction with methylamine (13, 30).

The structure of cryptoclauxin (11) rests mainly on correlation of its NMR spectrum with those of (9) and (10) (30); the compound contains an anhydride lacking some of the H-bonding present in (10) (1800 and 1748 cm^{-1}), and a tertiary OH group.

The structures of gilmaniellin and its dechloro analog rest on an X-ray study of gilmaniellin and spectroscopic comparisons between the two natural products and their acetates (16). The dimers from *Gilmaniella*, although apparently derived from similar monomeric units which combine to form the duclauxin pigments, are coupled through carbons 6,6′ and 5,4′ rather than through carbons 4,6′ and 3,4′.

C. Synthesis

Exploratory work by BYCROFT and EGLINGTON (31) let to the synthesis of the bisacetylnaphthalene (25) which on treatment with base gave the phenalenone (26). The methoxylated compound could be demethylated by treatment with HBr in acetic acid to yield (27), which is a tautomer of the presumed cyclization product from the C$_{14}$-

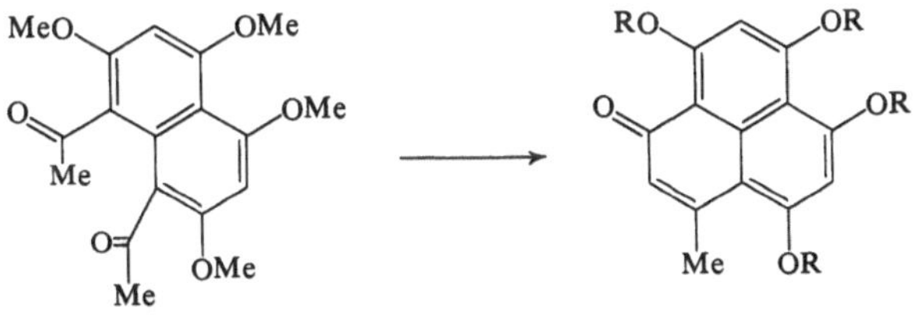

(25)

(26) R = Me
(27) R = H

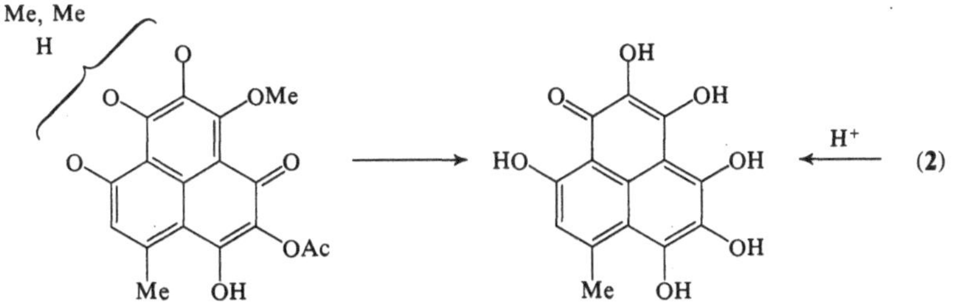

(32)

Norxanthoherquein (33)

(2)

heptaketide precursor of atrovenetin *(vide infra)*. The total synthesis of atrovenetin from the tetramethoxynaphthalene (**28**) was described in 1973 by Frost and Morrison *(32)*.

The substituted naphthalene (available in four steps from 3,4,5-trimethoxybenzaldehyde) reacted with malonic acid and polyphosphoric acid (100°, 15 min) to give the phenalenone (**29**), which was selectively demethylated to (**30**) with hydrochloric acid. The demethyl compound formed a dimethylallyl ether (**31**), which on heating in DMF at 100° gave the yellow trimethyl ether of atrovenetin (**16**). Demethylation of (**16**) with pyridine hydrochloride led to the racemic natural product. A more prolonged reaction between (**28**) and malonic acid in poly-

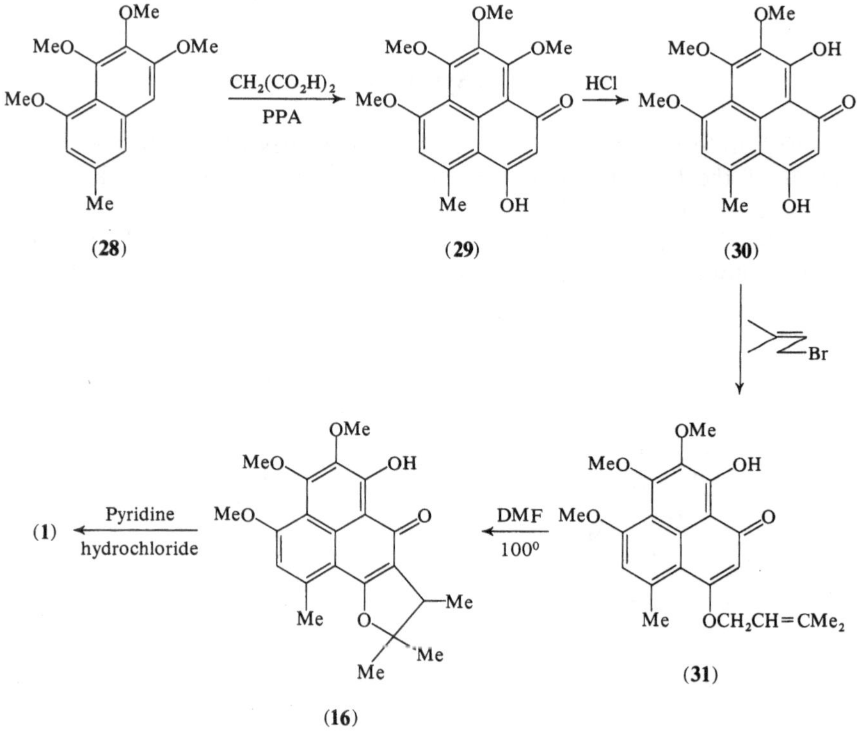

Scheme III. Synthesis of atrovenetin *(32)*

phosphoric acid (100°, 16 hrs) resulted in the formation of a trimethoxy-2-acetylphenalenone which on Baeyer-Villiger oxidation gave the acetoxy compound (**32**) of indeterminate structure. Demethylation and

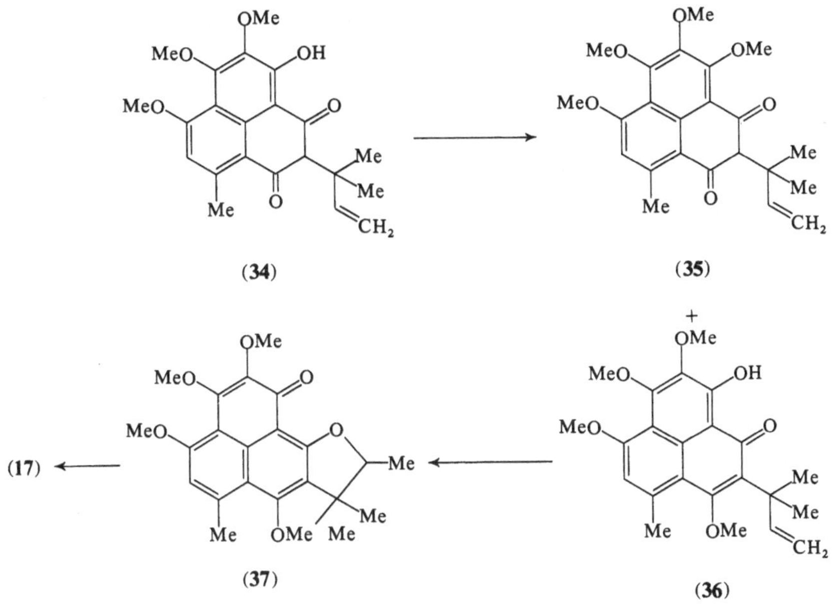

Scheme IV. Synthesis of demethylherqueichrysin (*32*)

hydrolysis of (**32**) with pyridine hydrochloride afforded norxantho-herquein (**33**) identical with material obtained from (**2**) by acid treatment.

The foregoing synthesis also comprises a synthesis of the naphthalic anhydride (**8**) which co-occurs with (**1**) in *P. herquei* and can be obtained from it by oxidation *in vitro* (*4*). A synthesis of (**8**) from lamellicolic anhydride (**13**) has also been reported (*15*) in a preliminary communication.

The allyl ether has also been converted into demethylherqueichrysin (**17**). Rearrangement of (**31**) to the diketone (**34**) was achieved in DMF at 100° in the presence of K_2CO_3, and (**34**) on treatment with diazomethane gave two isomeric methyl ethers (**35**) and (**36**). Treatment of (**36**) with *p*-toluenesulphonic acid effected ring closure to (**37**), which on partial demethylation with aqueous HI gave the racemic product (**17**) (*10*).

D. Biosynthesis

In 1956, consideration of the structure of atrovenetin (**1**) resulted in a proposal by BARTON and co-workers (*4, 20*) that the main ring system of the pigment was probably derived from acetate units, with a C-5 mevalonate unit providing the dihydrofuran ring. In 1961 through the incorporation of radio-labelled precursors into norherqueinone, THOMAS (33) was able to provide verification of this hypothesis. Thus $1\text{-}^{14}C$-acetate and $2\text{-}^{14}C$-mevalonic acid were incorporated (1.8% and 0.75% respectively) into (2) by *P. herquei;* the isolated norherqueinone was subjected to acid hydrolysis yielding norxanthoherquein (33) and 3-methyl-2-butanone (38). Oxidation of (33) under Kuhn-Roth conditions gave acetic acid from the C-methyl group and C-7, and CO_2 from the remaining 12 carbon atoms; the ketone (38) was degraded to yield CHI_3. The results of these experiments showed that $1\text{-}^{14}C$-acetate derived (38) contained two-ninths of its total activity in the C_5 side chain, and one-ninth in the methyl substituted carbon (C-13); the methyl group is inactive. One half of the activity of (38) is present at C-1. The 3-methyl-2-butanone derived from $2\text{-}^{14}C$-mevalonate labelled (2) was found to contain all the activity of the metabolite. Although the possible

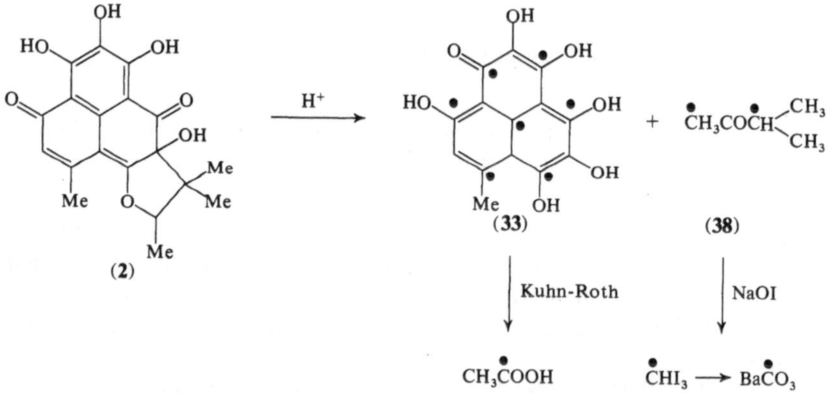

Scheme V. Degradation of norherqueinone. Label from $1\text{-}^{14}C$-acetate *(33)*

acyclic intermediates between the presumed linear C_{14}-polyketide precursor of the fungal phenalenones and the isolated pigments are unknown, the way in which the polyketide is folded has been established by ^{13}C-studies. Three possible modes of folding (39—41) could in theory lead to the methylphenalenone system (34); each would result in the same hydroxylated phenalenone (27). Analysis of the ^{13}C-NMR spectrum of the diacetate of deoxyherqueinone (42) enabled SIMPSON

References, pp. 187—190

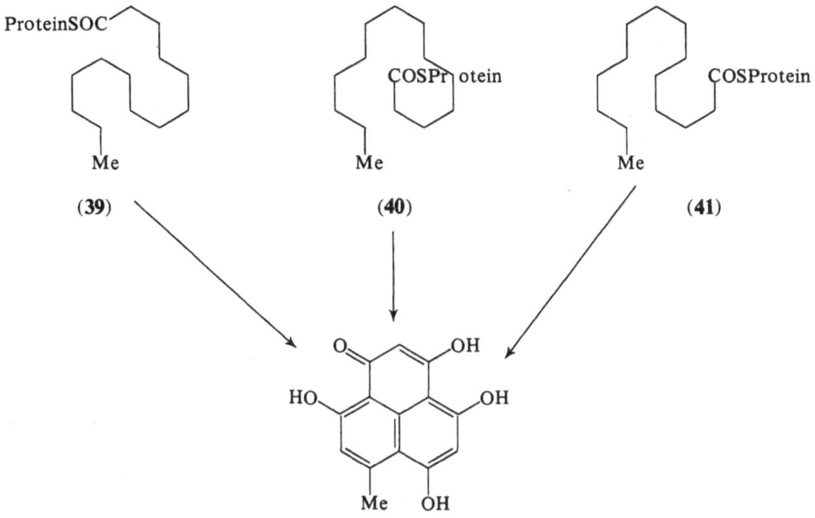

(39)   (40)   (41)

(27)

to determine the sites of labelling of the molecule by 1,2-^{13}C-acetate, 2-^{13}C-malonate, and singly labelled 1- and 2-^{13}C-acetate (27). As indicated in the figure below 2-^{13}C-malonate gave rise to specific enrichments at C-2, C-4, C-6, C-8, C-10, and C-12; 2-^{13}C-acetate gave rise to the same pattern, but C-14 (of the starter unit) was also labelled. The coupled carbons derived from intact acetate units showed the folding of the polyketide chain to have been as in (39). The oxygen function at C-8 is presumably introduced at the stage of the non-cyclized polyketide; the formation of the dimethylallyl ether and its subsequent cyclization would resonably take place later in the biosynthesis after the aromatic system had been formed.

The ability of *P. herquei* to interconvert atrovenetin (1), deoxyherqueinone (5), herqueinone (3) and norherqueinone (2) has been studied

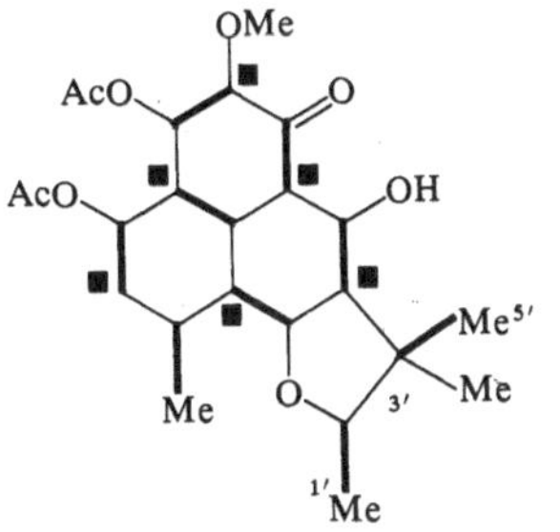

—— Label from 1,2-^{13}C-Acetate.

■ Label from 2-^{13}C-Malonate.

by Kriegler and Thomas (35) and a sequential biosynthesis has been established for the first three metabolites *via* successive methylation and oxydation. Norherqueinone is apparently formed by oxidation of (1) but is not methylated to (3). It has been suggested (10) that the biosynthesis of isoherqueinone (4) and isonorherqueinone (6) could reasonably proceed through a similar sequence from the C-2'-(−)-S-enantiomer of atrovenetin. That (S)-atrovenetin may itself prove to be a metabolite of the fungus, formed directly, by epimerization, or *via* demethylherqueichrysin (17), is supported by the isolation of partially racemic atrovenetin from *P. herquei* (10). The co-occurrence of (7) with

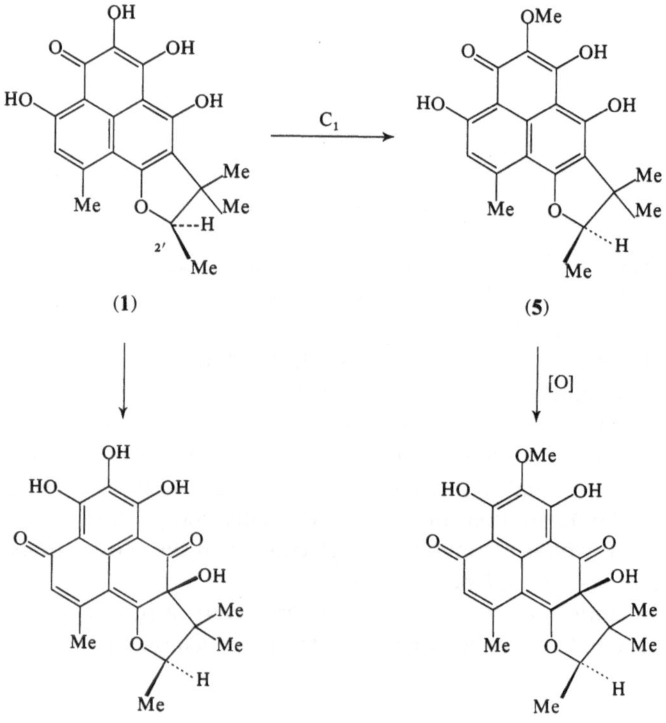

Scheme VI. Biosynthetic relationships between phenalenones of *P. herquei* (35)

the dihydrofuran ring between C-4 and C-5 together with compounds in the atrovenetin series implies the possible existence of a common precursor bearing a C_5-substituent at C-4 [e.g. (43), a compound very similar to (34) used in the synthesis of (17)] (32), and this is supported to some extent by Simpson's observation (27) that in both (7) and (5)

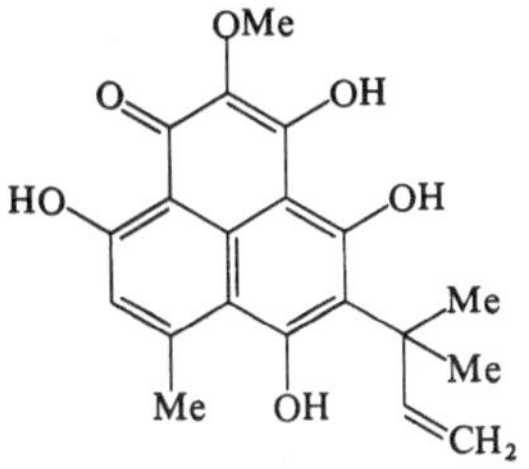

(**43**)

derived from 1,2-^{13}C-acetate C-5′ (*trans* to C-1′) is coupled to C-3′, i. e. that both series of compounds have the same configuration at C-2′.

The problem of the biosynthesis of the dimeric metabolites (**9—11**) of *P. duclauxi* has been addressed by SHIBATA and his co-workers (*37*), who have shown that these pigments, like the monomers, are also of polyketide origin. 1-^{14}C-Acetate, 2-^{14}C-acetate, and ^{14}C-formate are all well incorporated into duclauxin (**9**) by the fungus (2.5%, 2.1%, and 0,65% respectively). Analysis of the products of degradation of the iso-lated phenalenone showed good agreement between the calculated and experimental radioactivities as shown in Scheme VII. There is as yet no evidence as to the nature of the monomeric C$_{13}$ units which combine to form duclauxin and the other dimeric phenalenones; nor is it known whether the oxidative modifications take place before or after the joining of the two units. No C$_{13}$ phenalenones have so far been identified in *P. duclauxi*. One last biosynthetic investigation (*36*) concerns the origin of the 4-O-carbomethoxy group present in the anhydride (**14**) isolated from *V. lamellicola*. Neither 1-^{14}C-acetate nor ^{14}C-bicarbonate

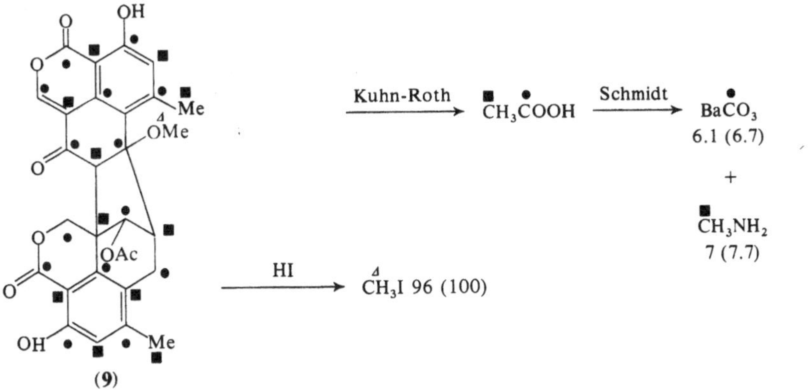

Scheme VII. Degradation of duclauxin (*37*). Figures show the observed and (calculated) % of the total activity

were incorporated into the carbomethoxy group. In view of the ^{13}C studies (28) the carboxyl carbon may be derived from C-2 of acetate, or possibly from methionine.

E. Biological Activity

Atrovenetin (1) (38) and to a lesser extent deoxyherqueinone (5) (39) have antibacterial activity against some Gram-positive organisms. Herqueinone (3) and norherqueinone (2) have no activity.

Duclauxin (9) was identified (14) as one of the agents responsible for the antitumor activity of culture filtrates of *P. stipitatum*. It was shown that the phenalenone decreased the proliferation of cells of Ehrlich's ascites carcinoma *in vitro*, and that the compound exerts its cytotoxic effect on cancer cells by inhibiting the synthesis of nucleic acids without affecting protein synthesis (40).

III. Phenalenones and Related Metabolites in Higher Plants

A. Occurrence

In a general revision of the Monocotyledons the family Haemodoraceae has been defined by HUTCHINSON (42) as two tribes, Haemodoreae and Conostyleae which are restricted almost entirely to the Southern Hemisphere as shown in Table 1.

Table 1. *Family: Haemodoraceae*

Tribe I: Haemodoreae		Tribe II: Conostyleae	
Genera	Habitats	Genera	Habitats
Barberetta	South Africa	*Anigozanthos*	Australia
Dilatris	South Africa	*Blancoa*	Australia
Haemodorum	Australia	*Conostylis*	Australia
Hagenbachia	Brazil	*Macropidia*	Australia
Lachnanthes	North America	*Tribonanthes*	Australia
Lanaria	South Africa	*Lophiola*	North America
Phlebocarya	Australia		
Pyrrorhiza	Venezuela		
Schiekia	Tropical America		
Wachendorfia	South Africa		
Xiphidium	Tropical America		

Table 2. *Arylphenalenones and Related Metabolites from Higher Plants*

Structure	Trivial Name Description Melting Point	Source References
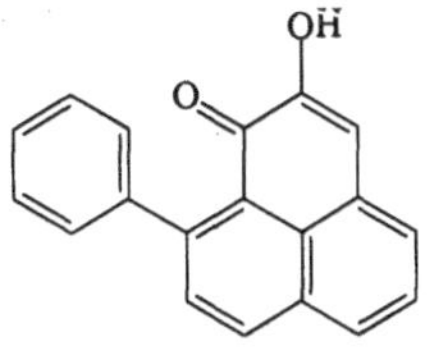	Anigorufone orange needles 123—125°	*Anigozanthos rufus* (*45*) Synthesis (*45*)
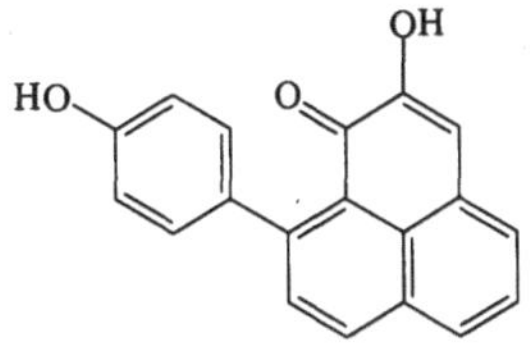	Hydroxyanigorufone red rosettes 247—248° (238—242°)	*Anigozanthos rufus* (*45*) Synthesis (*46*)
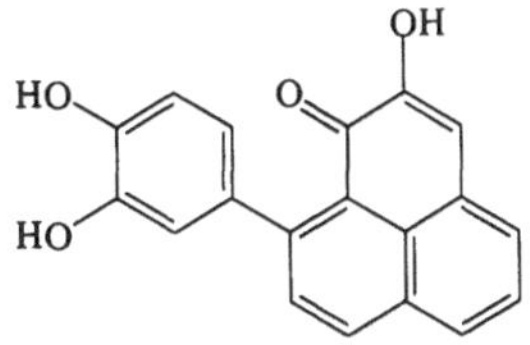	Dihydroxyanigorufone red needles 215—220° decomp.	*Anigozanthos rufus* (*45*) *Conostylis setosa* (*47*) Synthesis of methyl ether (*47*)
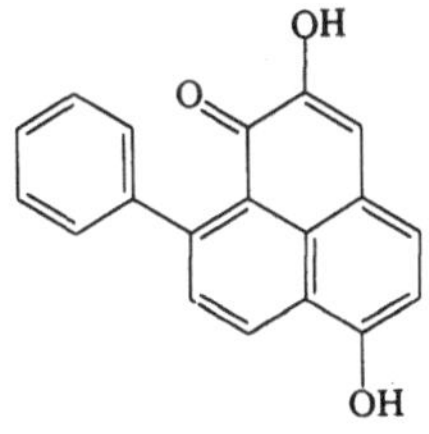	Lachnanthocarpone red needles 218—220°	*Lachnanthes tinctoria* (*48*) *Wachendorfia paniculata* (*49*) Synthesis (*50, 51, 52*)
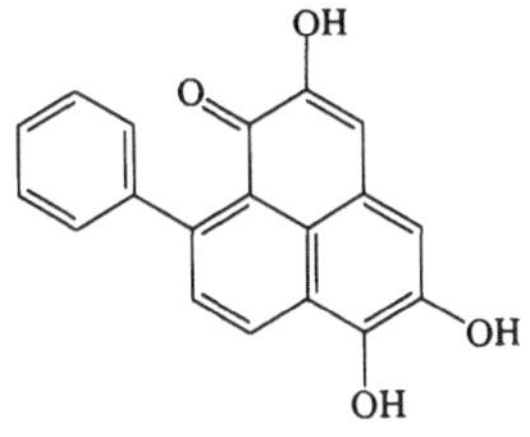	Lachnanthoside aglycone purple crystals 220—223°	*Wachendorfia paniculata* *W. thyrsiflora* (*49*)
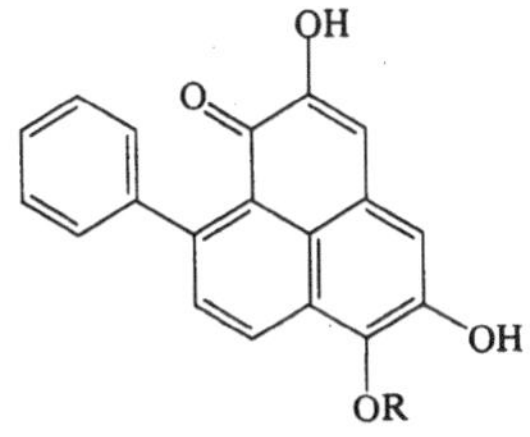	Lachnanthoside (R = unknown biose) red-brown 265—268°	*Lachnanthes tinctoria* (*53*)

Table 2 (continued)

Structure	Trivial Name Description Melting Point	Source References
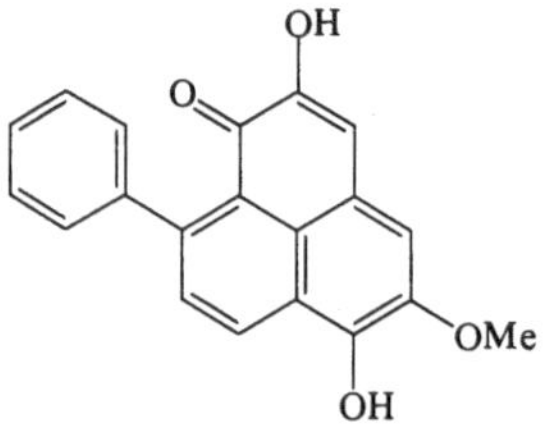	Haemocorin aglycone purple-red plates or almost black needles 277—278° (230—232°)	*Haemodorum corymbosum (54, 55, 56)* *Lachnanthes tinctoria (57)* *Wachendorfia thyrsiflora (49)* Synthesis (58, 59)
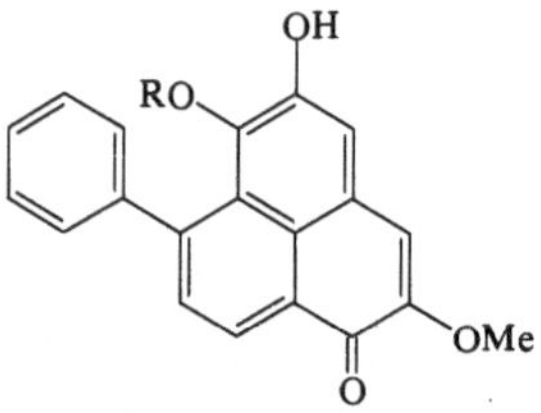 R = cellobiose (position uncertain)	Haemocorin red spangles 263—264° decomp.	*Haemodorum corymbosum (54, 55, 56)*
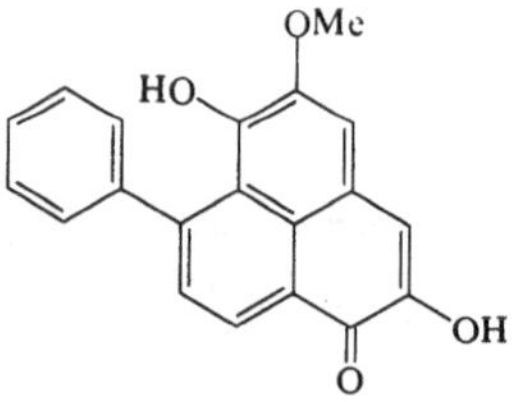	Haemodorin dark red crystals 211—214°	*Haemodorum distichophyllum (60)*
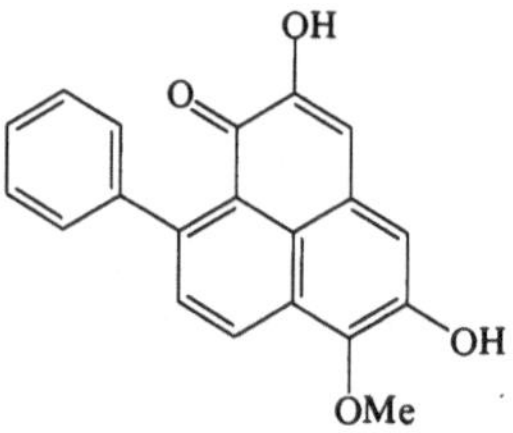	Anigozanthin crimson needles 149—151°	*Anigozanthos rufus (45)*
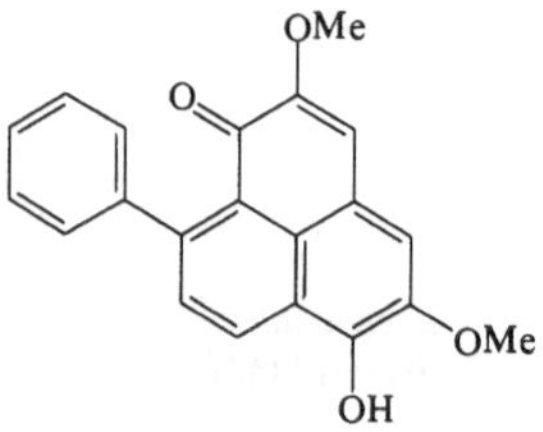	Xiphidone purple crystals 170—172° brown-red rods 184.5—186.5°	*Xiphidium caeruleum (61)* Synthesis (58, 62)

References, pp. 187—190

Table 2 (continued)

Structure	Trivial Name Description Melting Point	Source References
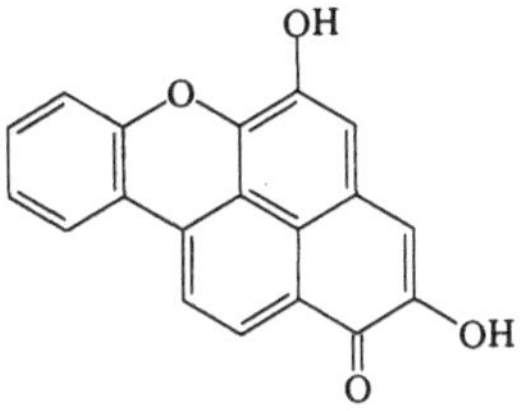	Lachnanthofluorone mauve crystals > 300°	*Lachnanthes tinctoria (53)* Synthesis of methyl ether *(62)*
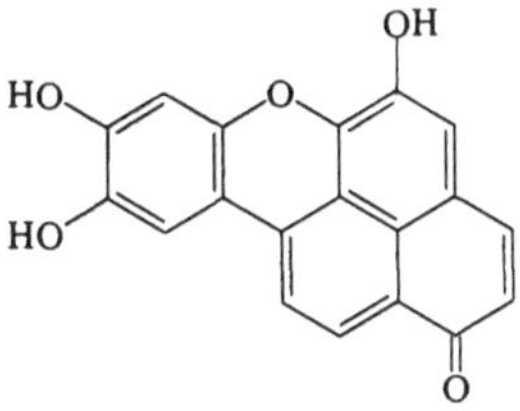	Haemofluorone A black needles > 370°	*Macropidia fuliginosa (62)* *Phlebocarya ciliata (62)* *Anigozanthos rufus (47, 62)* *Conostylis setosa (47, 62)* Synthesis of trimethyl ether *(62)*
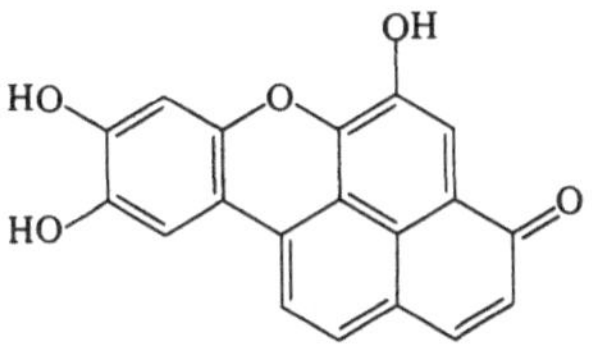	Haemofluorone B black needles > 370°	*Macropidia fuliginosa (62)* *Anigozanthos rufus (47, 62)* *Conostylis setosa (47, 62)* *Phlebocarya ciliata (62)*
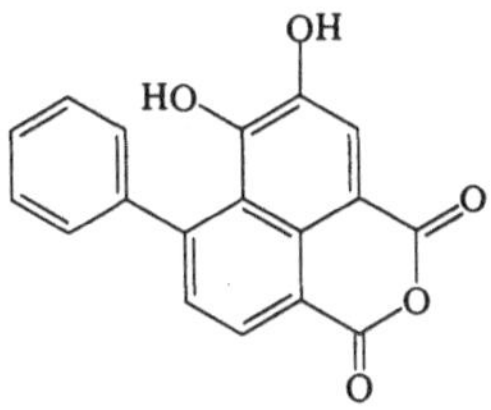	yellow needles 244—245°	*Lachnanthes tinctoria (63)*
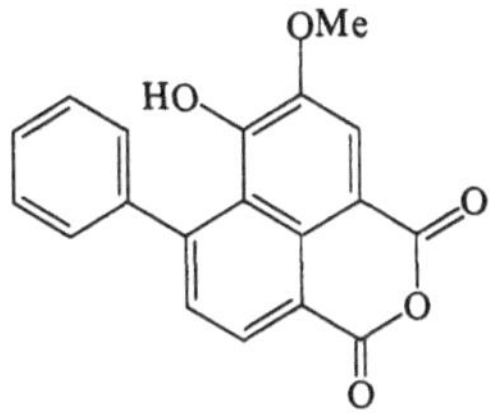	orange-yellow plates 261—262°	*Lachnanthes tinctoria (64)*

Table 2 (continued)

Structure	Trivial Name Description Melting Point	Source References
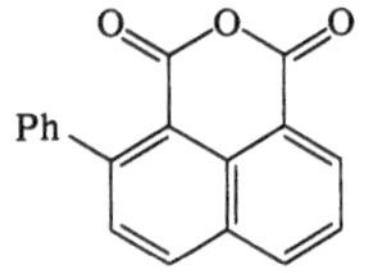	fine needles 241—243°	*Anigozanthos rufus* (*45*) Synthesis (*45*)
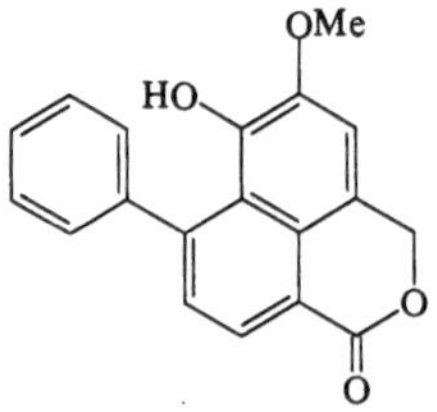	yellow crystals 170—173°	*Lachnanthes tinctoria* (*53*) Synthesis of methyl ether (*65*)
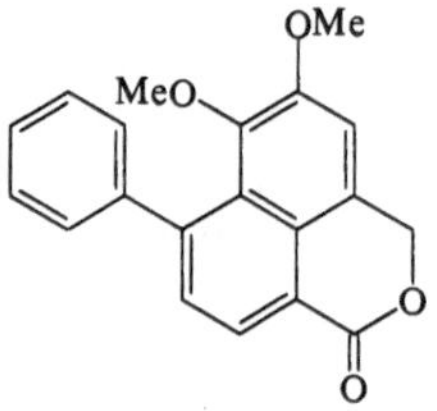	pale yellow plates 172—173°	Methylated fraction of *Lachnanthes tinctoria* extract (*64*) Synthesis (*65*)
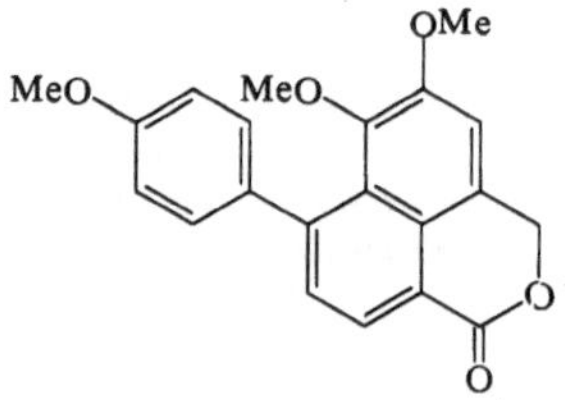	pale yellow needles 182—183°	Methylated fraction of *Lachnanthes tinctoria* extract (*64*)
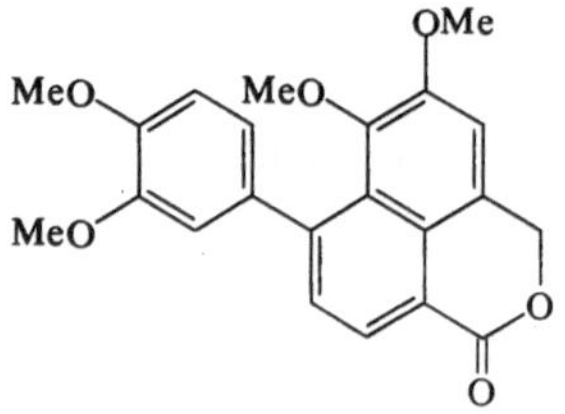	yellow plates 194—196°	Methylated fraction of *Lachnanthes tinctoria* extract (*64*) Synthesis (*65*)
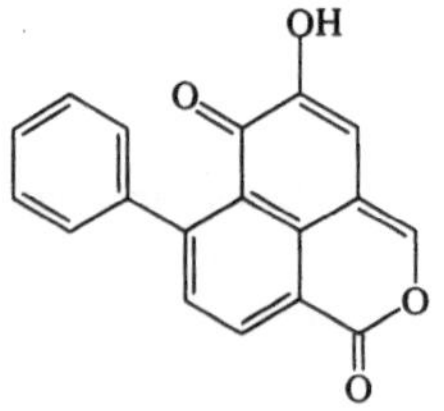	Lachnanthopyrone yellow crystals 222—225°	*Lachnanthes tinctoria* (*66*)

References, pp. 187—190

Table 2 (continued)

Structure	Trivial Name Description Melting Point	Source References
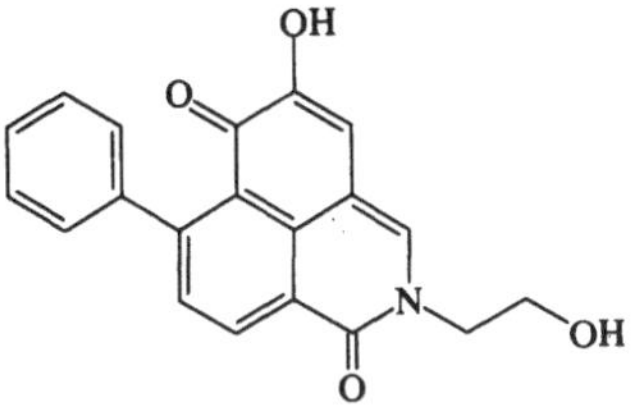	N-(2-hydroxy ethyl) lachnanthopyridone orange crystals 155°	*Lachnanthes tinctoria (66)*
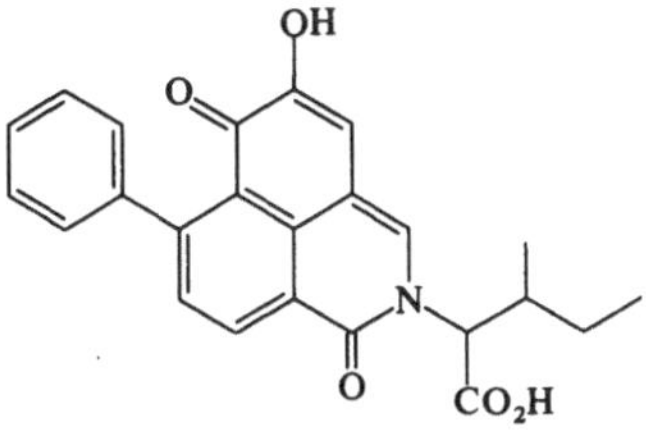	N-(1-carboxy-2-methyl-butyl) lachnanthopyridone orange rosettes 218—219°	*Lachnanthes tinctoria (63)*
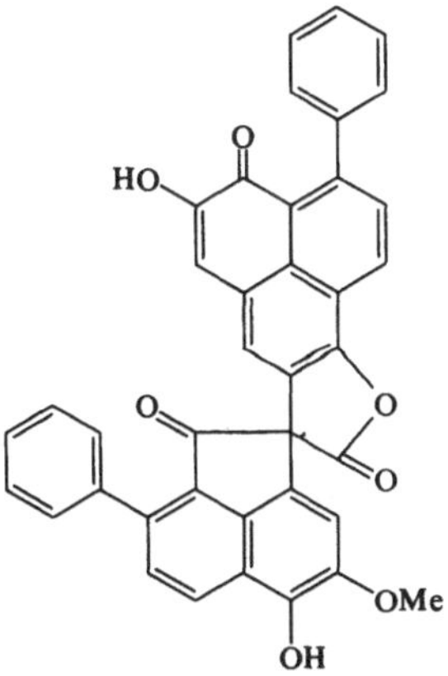	Lachnanthospirone deep orange crystals > 310°	*Lachnanthes tinctoria (57)*

One or more species of nine of these genera have been examined and all but one contained arylphenalenones and/or closely related compounds. Phenalenones have not been found so far in any other family of higher plants so their presence may be a useful chemotaxonomic character. The single exception so far reported among the genera of

Table 1 was *Lophiola*. A careful examination of *L. americana* showed no evidence of any phenalenone pigment (*43*). However this plant has uncertain status and has been included by some authorities in the family Amaryllidaceae (*44*).

Structures, trivial names, physical properties and sources of the pigments isolated from other genera are shown, with relevant references, in Table 2.

B. Structure Determination

All the phenalenones so far isolated from plants are derivatives of 9-phenyl-1H-phenalen-1-one (**44**). All the related compounds occurring with them may be regarded as metabolic oxidation products: phenylnaphthalides, phenylnaphthalic anhydrides, quinonemethides derived from oxa- and aza-phenalenones, derivatives of 1H-naphtho[2,1,8-*mna*]-xanthen-1-one (**45**) and a phenalenone dimer.

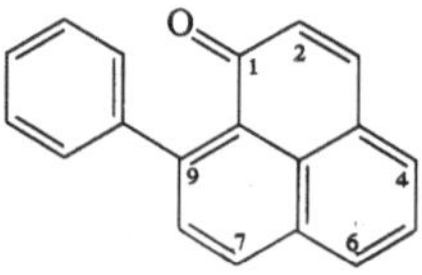

(**44**)

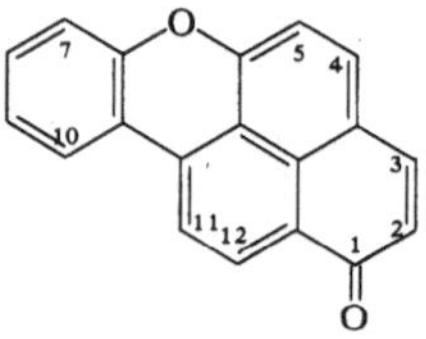

(**45**)

The first phenalenone pigment isolated from a higher plant was the glycoside haemocorin which was first described in 1955 (*54*). On mild acid hydrolysis the pigment gave cellobiose and a purple-red aglycone which was identified as 2,6-dihydroxy-5-methoxy-9-phenyl-1H-phenalen-1-one (**46**) on the basis of the chemical transformations shown in Chart 1. Methylation of (**46**) gave two monomethyl ethers and two corresponding dimethyl ethers (**47**) and (**48**). These derivatives drew attention to the potential tautomerism of hydroxyphenalenones.

Oxidation of the dimethyl ethers gave the two dimethoxyphenylnaphthalic anhydrides (**49**) and (**50**) which were then decarboxylated to give respectively 1,2-dimethoxy-6-phenylnaphthalene (**51**) and 1,2-dimethoxy-8-phenylnaphthalene (**52**). These were identified by unequivocal total synthesis. Further oxidation of both anhydrides (**49**) and (**50**) gave biphenyl-2,3,4-tricarboxylic acid (**53**) which was decarboxylated to biphenyl.

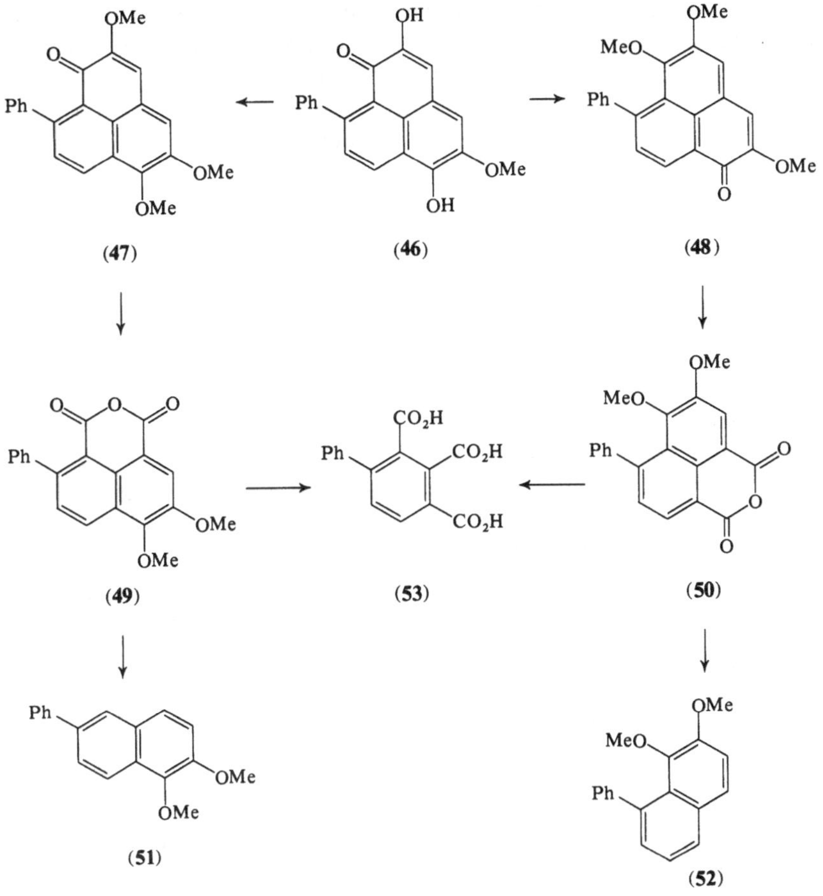

Chart 1

C. Spectroscopic Methods

Subsequent to the investigation of haemocorin other phenalenones and phenalenone derivatives were isolated from other plants. Their structures were determined partly by conversion to known derivatives but largely by spectroscopic methods, using as models the many phenalenone derivatives prepared during the study of haemocorin.

All the natural phenalenones, as derivatives of 2-hydroxy-1H-phenalenone, have distinctive ultraviolet and visible absorption spectra. The naphthoxanthenones also show characteristic spectra which are more complex and more intense, with a pronounced bathochromic shift of the visible absorption maxima in comparison with those of the phenalenones. Some examples of the electronic absorption spectra for pigments with different oxidation levels are shown in Table 3.

Table 3

	2-Hydroxy-9-phenyl-1H-phenalen-1-one (Anigorufone)
λ_{max} (EtOH):	228 sh, 243, 259 sh, 269 sh, 289 sh, 340 sh, 352, 368, 424 nm
log ε:	4.18, 4.28, 4.23, 4.19, 3.98, 3.79, 3.82, 3.88, 3.68
	2,6-Dihydroxy-9-phenyl-1H-phenalen-1-one (Lachnanthocarpone)
λ_{max} (EtOH):	243, 270, 302, 368, 492 nm
log ε:	4.18, 4.19, 4.08, 3.54, 3.72
	2,6-Dihydroxy-5-methoxy-9-phenyl-1H-phenalen-1-one (Haemocorin aglycone)
λ_{max} (Dioxan):	250, 277.5, 300 sh, 355, 372.5, 505 nm
log ε:	4.20, 4.36, 4.23, 3.72, 3.72, 3.71
	5,8,9-Trihydroxy-1H-naphtho[2,1,8-*mna*]xanthen-1-one (Haemofluorone A)
λ_{max} (MeOH):	242, 259 sh, 284, 331, 362, 395, 545 sh, 578 nm
log ε:	4.50, 4.43, 4.24, 3.78, 3.75, 3.76, 4.30, 4.39

In the mass spectra of 9-arylphenalenones the occurrence of the $(M-1)^+$ ion is a useful diagnostic feature. When the structure of the phenalenone is fixed the ion gives the base peak and presumably has a structure of type (**53**). However when tautomerism is possible, as in 6-hydroxy-9-arylphenalenones, this peak is of lower intensity.

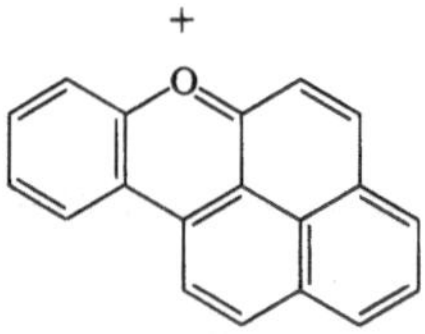

(**53**)

An investigation of carbon-13 nmr spectra of phenalenones has been reported (*67*) and the results have been applied in labelling experiments to study the biosynthesis of the natural arylphenalenones (*68*).

The most generally useful data for structure elucidation have been provided by very comprehensive studies of the proton magnetic resonance spectra of the natural arylphenalenones and the related metabolites, together with a large number of synthetic model compounds. As shown for structures (**54**) and (**55**) the ^{1}H-NMR spectra (δ values in CDCl$_3$) easily distinguish between the dimethyl ethers obtained from 2,6-di-hydroxy-5-methoxy-9-phenyl-1H-phenalen-1-one (haemocorin aglycone).

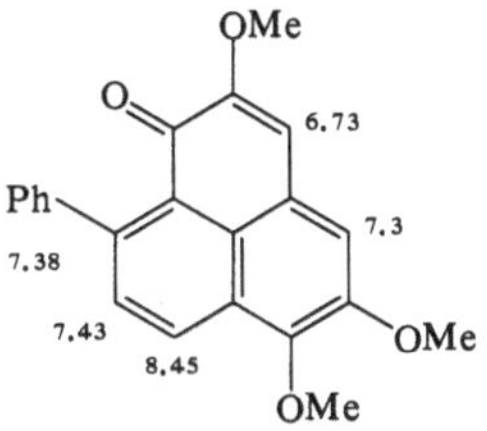

OMe
O
6.73
Ph
7.3
7.38
7.43
8.45   OMe
OMe

(**54**)

OMe
3.22
MeO
7.3
Ph
6.84
7.38
7.50
8.57   OMe
O

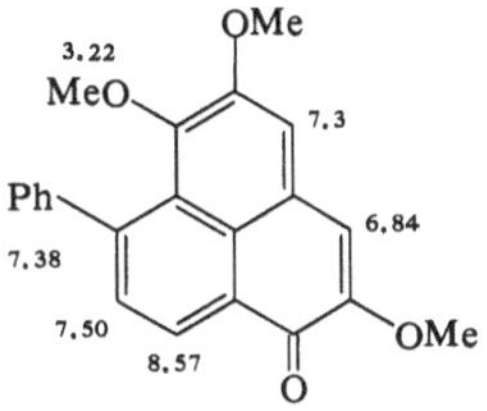

(**55**)

In general the proton *peri* to a carbonyl or hydroxy group gives the signal at lowest field, followed by a proton *peri* to a methoxyl about δ 0.1 upfield. A methoxyl *peri* to a phenyl group is shielded and gives a signal at about δ 3.2 while other methoxy groups give peaks near δ 4.0. The effects of deuterobenzene on the methoxyl resonance positions are also useful indications of molecular structure (*60*).

Similar considerations apply to the spectra of the arylnaphthalic anhydride (**56**) and the arylnaphthalide (**57**).

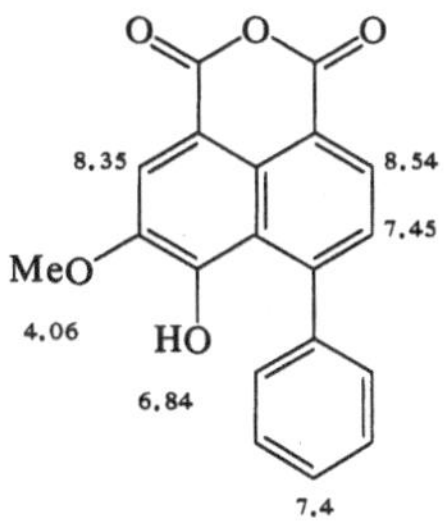

O   O   O
8.35   8.54
7.45
MeO
4.06   HO
6.84
7.4

(**56**)

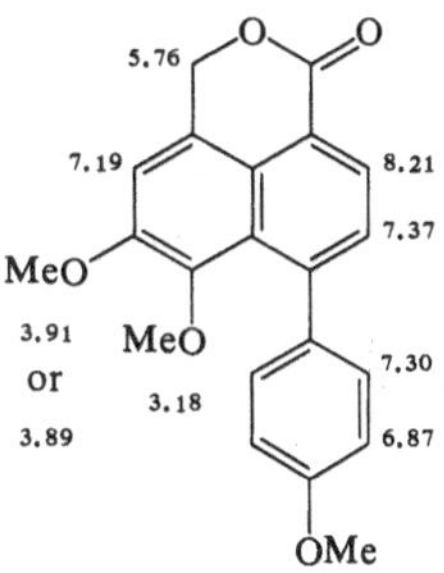

O   O
5.76
7.19   8.21
7.37
MeO
3.91   MeO
or   3.18   7.30
3.89
6.87
6.84
OMe

(**57**)

The published structures of the three natural naphthoxanthenones were also deduced principally by comparison of the ^{1}H-NMR spectra with those of model compounds prepared for the purpose (*62, 69, 70*).

There is also some similarity to the spectra of closely related phenalenones. It has been reported that lachnanthofluorone has been prepared by photolysis of the natural arylphenalenone lachnanthoside aglycone but only the structure of haemofluorone A has been confirmed by un-equivocal total synthesis of the trimethyl ether (62), for which the nmr data are shown on structure (58).

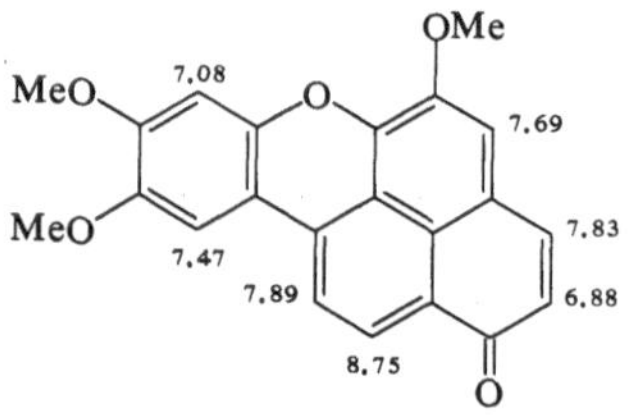

(58)

The structure 8,9,11-trihydroxy-1H-naphtho[2,1,8-*mna*]xanthen-1-one was proposed for the isomeric pigment haemofluorone B (62) on the basis of the nmr data obtained for the trimethyl ether as shown on structure (61). However further consideration of model compounds (71) suggests that the spectrum is more consistent with structure (62). This also seems more in accord with biosynthetic concepts.

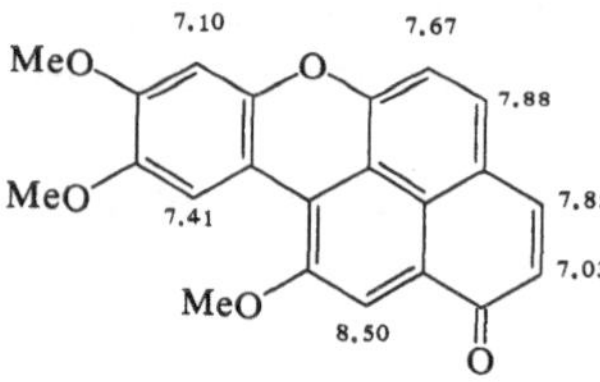

(61)

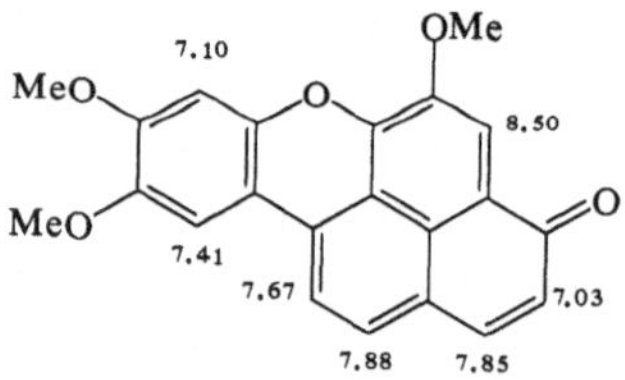

(62)

The structure 2,5-dihydroxy-1H-naphtho[2,1,8-*mna*]xanthen-1-one proposed for the natural pigment lachnanthofluorone (53) does not seem to be consistent with the published nmr data, and a synthetic sample of

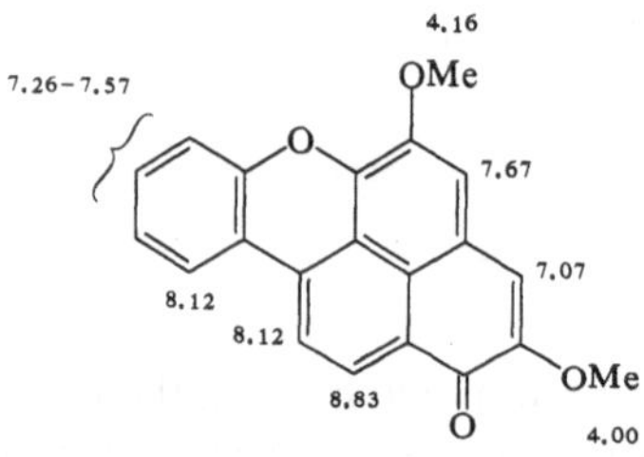

(59)

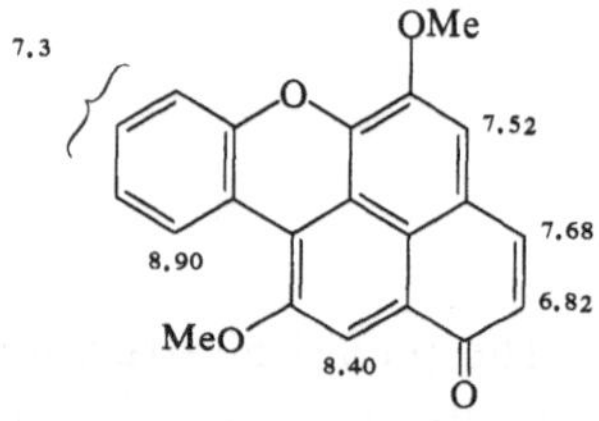

(60)

2,5-dimethoxy-1H-naphtho[2,1,8-*mna*]xanthen1-one (**59**) gave different results (*62*). The alternative structure (**60**) for lachnanthofluorone dimethyl ether was therefore considered to be more in accord with the observed nmr spectrum. Current work on total synthesis of these pigments should resolve the uncertainties.

X-ray crystallography has been used to establish the structure of lachnanthospirone (**63**), a novel dimeric phenalenone pigment. The single crystal analysis was made with the derivative (**64**).

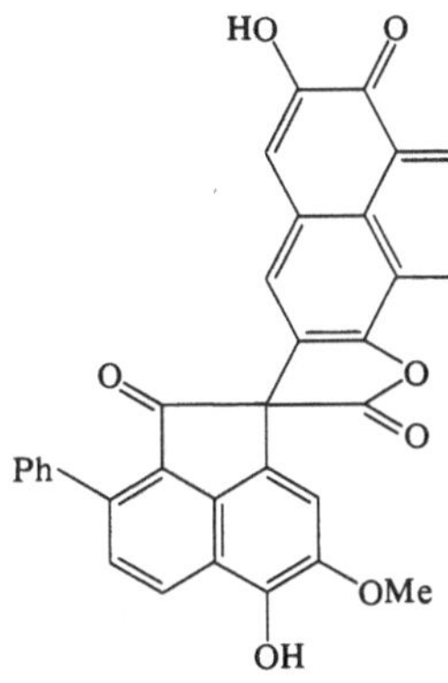

(**63**) 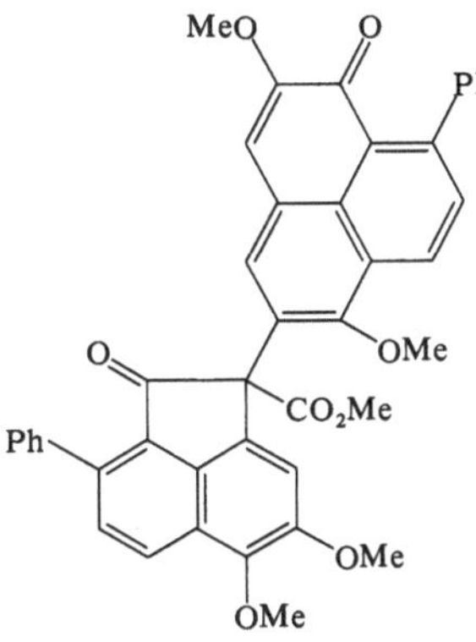   (**64**)

D. Synthesis

The first published methods for synthesis of natural arylphenalenones involved laborious multistep construction of a phenylnaphthalene intermediate from benzene derivatives and thence to the phenalenone (*50*) or, alternatively, the preparation of a dihydrophenalenone and then reaction with the arylmagnesium halide, dehydration and dehydrogenation (*58*). General approaches to the synthesis of simple phenalenones have been reviewed previously (*72*), and the most generally useful method for the synthesis of the natural arylphenalenones has been the 1,4-addition of arylmagnesium halides to simple phenalenones, followed by dehydrogenation of the intermediate to give, usually, only the 9-arylphenalenone (*45, 46, 59*). The introduction of a 2-hydroxy group can be accomplished when required by epoxidation at the 2,3-bond of a phenalenone with alkaline hydrogen peroxide solution (*45, 51*) or better with *tert*-butyl hydroperoxide and a quaternary ammonium hydroxide (*46, 59, 62*). The epoxide is isomerized easily by acid to give the 2-hydroxyphenalenone.

A third generally useful reaction is the selective acid hydrolysis of a 6-methoxy group, which may be regarded as a vinylogous ester group.

The application of the above reactions is illustrated in Scheme VIII by the synthesis of 2,6-dihydroxy-5-methoxy-9-phenyl-1H-phenalen-1-one (haemocorin aglycone) (59).

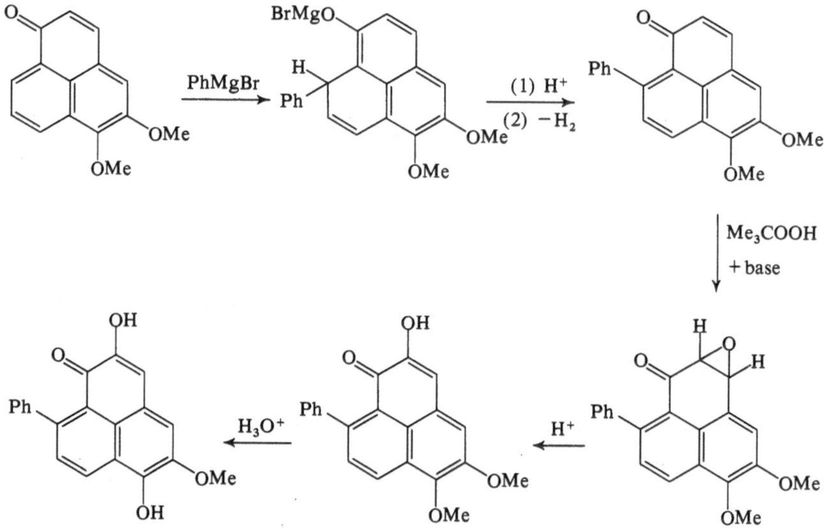

Scheme VIII. Synthesis of haemocorin aglycone

6-Hydroxy-9-arylphenalenones are also conveniently prepared by condensation of 2,7-dihydroxynaphthalene and its derivatives with 2,4-dioxo-4-arylbutanoic acids as shown in Scheme IX (71).

An unusual synthesis (52) of 2,6-dihydroxy-9-phenyl-1H-phenalen-1-one (lachnanthocarpone) was achieved by oxidation of 1-phenyl-7-(3,4-dihydroxyphenyl)-1,3-heptadien-5-one (65). It was assumed that the 1,2-

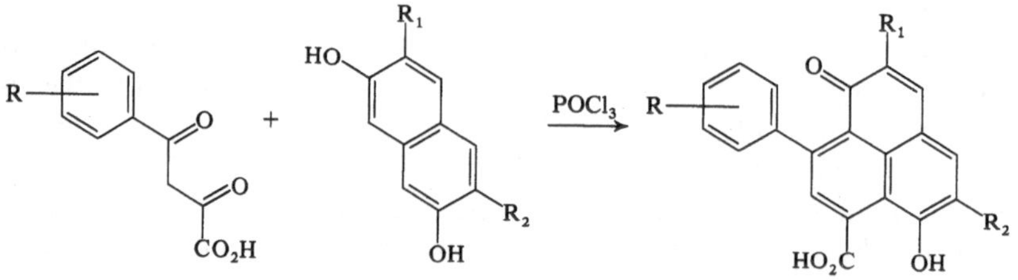

Scheme IX

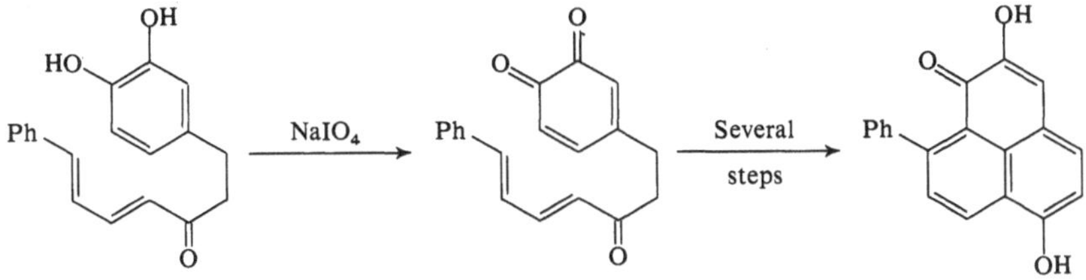

Scheme X. Synthesis of lachnanthocarpone

benzoquinone **(66)** so formed undergoes intramolecular Diels-Alder addition and then further oxidation steps produce the arylphenalenone (Scheme X).

Another approach which is useful for specific direct synthesis of 7-aryl-6-methoxyphenalenones is the treatment of a 1-aryl-3-naphthyl-propan-1-one **(67)** with polyphosphoric acid. Apparently cyclode-hydration is followed by selective demethylation, controlled by steric factors, and a dehydrogenation step *(70)*. This method is shown below.

The only approach to the synthesis of the natural arylnaphthalides has been the preparation of two fully methylated derivatives shown in Scheme XI *(65)*. A 2,7-diarylpent-4-enenitrile **(68)** was cyclized to the phenyltetralin **(69)**. Bromination and aromatization with N-bromosuccin-imide then gave the phenylnaphthalene derivative **(70)**. Finally basic hydrolysis produced the phenylnaphthalide **(71)**.

1,8-Naphthalic anhydrides are most conveniently obtained by oxi-dation of phenalenones or acenaphthenes *(45)*.

It has been demonstrated that photolysis of 6-hydroxy-9-phenyl-phenalenones is an efficient method for synthesis of 1H-naphtho-[2,1,8-*mna*]xanthen-1-one and its derivatives *(62, 69)*. They have also been prepared by acid catalyzed cyclization of 6-hydroxy-9-(2-methoxy-phenyl) phenalenones *(70, 71)*. For example cyclization and de-carboxylation convert 7-carboxy-2,5-dimethoxy-6-hydroxy-9-(2-methoxy-

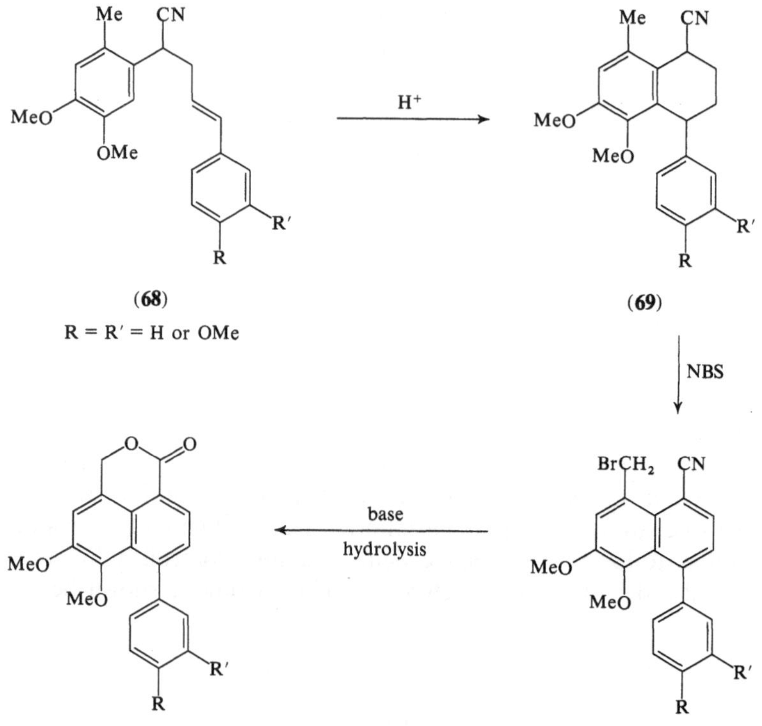

(68)

R = R' = H or OMe

(69)

(71)

(70)

Scheme XI. Synthesis of arylnaphthalides

phenyl)-1H-phenalen-1-one (72) into 2,5-dimethoxy-1H-naphtho[2,1,8-mna]xanthen-1-one. The product was identical with the material produced by the photochemical method (62) but its properties did not agree with those reported for the dimethyl ether of lachnanthofluorone (53).

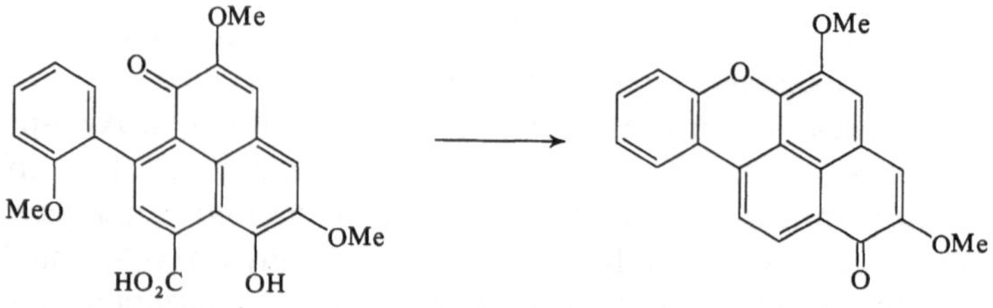

(72)

E. Biological Activity

The earliest reference to biological activity of a plant in the family Haemodoraceae seems to have been one by DARWIN (73) to the toxic action of *Lachnanthes* on white pigs and the immunity of the black ones. This has frequently been assumed by later writers to indicate a photodynamic action. Some experimental support for this belief has been obtained by the observation that extracts of *Lachnanthes tinctoria* induce phototoxicity in microorganisms (74). The Australian plant *Haemodorum corymbosum* has also been regarded as toxic to animals (75). Haemocorin, the phenalenone pigment from this plant, was shown to have antitumor activity (76) and a wide range of antibacterial activity (77).

F. Biosynthesis

The possible biosynthesis of the carbon skeleton of the 9-phenyl-phenalenones presents a complex problem; it has been studied experimentally, but the interpretation of the finding is by no means straightforward. The pattern of oxygenation observed in the haemodoraceous pigments does not suggest a polyketide origin; on the other hand there is no obvious way in which the C_{19} skeleton can be assembled from shikimate-derived precursors. THOMAS (78, 79) has proposed a possible biosynthesis involving the condensation of one molecule each of phenylalanine and tyrosine (or their metabolic equivalents) with one molecule of acetic acid, with loss of one of the three carboxyl groups, to yield a diarylheptanoid (73). This intermediate could then cyclize to the ring system of the 9-phenylphenalenones. Diarylheptanoids are fairly widely distributed in the plant kingdom (Zingiberaceae, Betulaceae, Leguminosae, Myricaceae, etc.); as yet, however, no compound of this type has been identified in a haemodoraceous plant.

The general features of the biosynthetic scheme, i. e. the incorporation of acetate, phenylalanine, and tyrosine into the aromatic system, have been substantiated both in *Lachnanthes* (80) and *Haemodorum* (81). Thus, both phenylalanine and tyrosine were incorporated (0.3 and 0.5%) into haemocorin by *Haemodorum corymbosum*, while 1 and 2-^{14}C acetate were incorporated to a lesser extent (0.1%). By means of a degradation of the aglycone (46) of haemocorin, based on the original chemical studies of the pigment by COOKE (vida supra), the specific incorporation of radioactivity from 2-^{14}C-tyrosine into C-5 of the pigment was also demonstrated (81). Very similar incorporations of the primary precursors into lachnanthoside aglycone (74) were observed in studies with *Lach-*

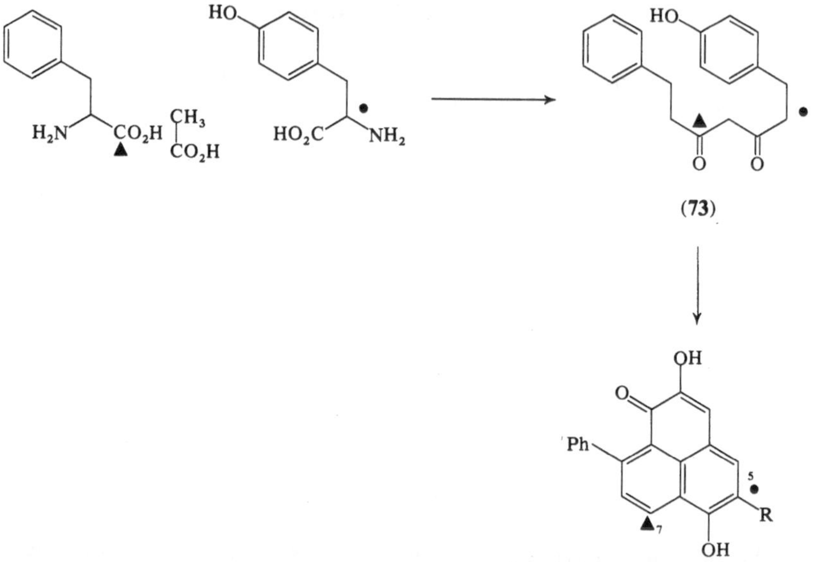

(73)

R = OMe Haemocorin aglycone (46)
R = OH   Lachnanthoside aglycone (74)

Scheme XII. Postulated biosynthesis of 9-phenylphenalenones (78, 79)

nanthes (80); in this work a remarkably high incorporation (3.5%) of phenylalanine was achieved, and the observation that in the case of both phenylalanine and tyrosine the 1- and 3-labelled amino acids were incorporated to the same extent, militates against the loss of the carboxyl group from these precursors in the biosynthesis. A ^{13}C-study (68) (one of the few to be achieved in a higher plant) has shown the specific incorporation of label from 1-^{13}C-phenylalanine into C-7 of (74). None of these findings has directly implicated a diarylheptanoid in the biosynthetic scheme, and the role of acetate is unclear since no specific incorporation has been demonstrated. Furthermore, Roughley and Whiting (82), after an investigation of the biosynthesis of curcumin (75), a well known constituent of the rhizomes of Curcuma longa, have

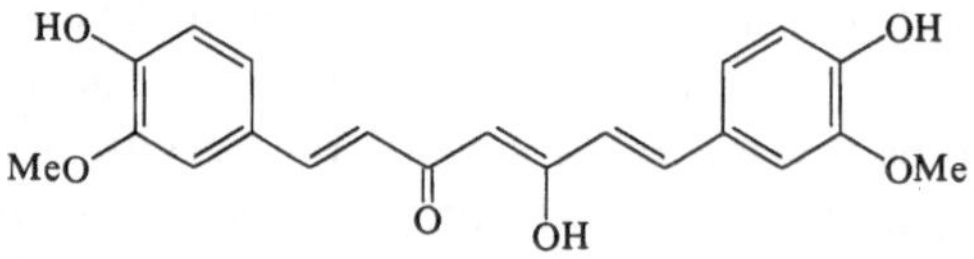

Curcumin (75)

concluded that this molecule is probably formed from one molecule of phenylalanine or tyrosine, and five acetate units. It is of interest that an investigation of the biosynthesis of 6-gingerol (**76**) by *Zingiber officinale* (*83, 84*) has indicated that its biosynthesis probably involves the condensation of one molecule each of dihydroferulic acid, acetic acid, and hexanoic acid, with loss of CO_2, in a manner exactly analogous to that proposed by THOMAS (*78*) for the formation of the phenalenone pigments.

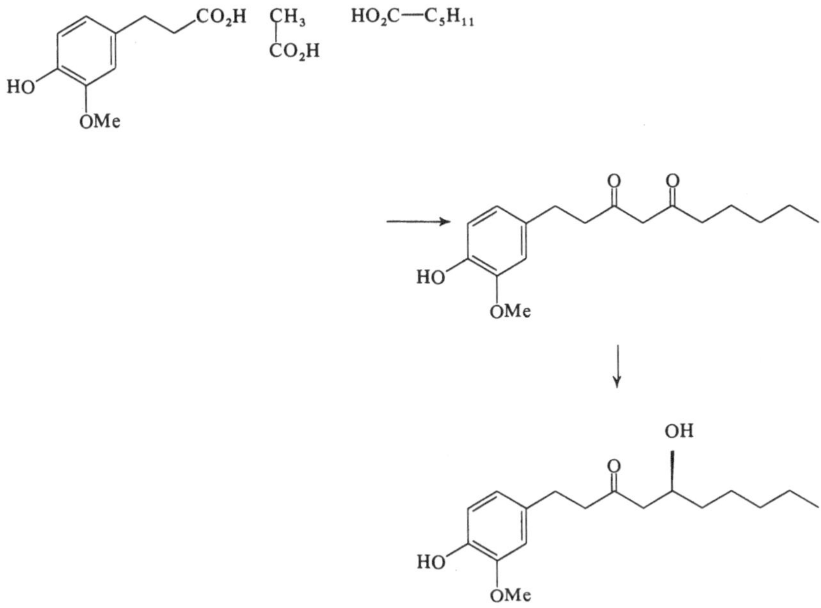

Scheme XIII. Biosynthesis of [6]-gingerol (*84*)

If the actual biosynthesis of the 9-phenylphenalenones is related to the scheme proposed by THOMAS, then the cyclization of the diaryl-heptanoid intermediate [e.g. (**73**)] could proceed through the biosynthetic equivalent of a Diels-Alder reaction. As was mentioned on page 181, the diarylheptadienone (**65**) has been synthesized and converted into lachnanthocarpone through the diarylheptanoid orthoquinone (**66**) *in vitro* (*52*). To the extent that this result shows that a 9-phenylphenalenone can be constructed from a suitable diarylheptanoid, it can be considered as adding support to THOMAS' hypothesis.

The nature of the compounds intermediate between the primary building blocks and the plant phenalenones is unknown; the formulation of diarylheptanoids in the biosynthetic scheme remains hypothetical, and suggestions as to the sequence of gain or loss of oxygen functions are mere speculation. Similarly there is as yet no evidence concerning the metabolism of the phenylphenalenones *in vivo*. The isolation of the 6-oxa and 6-aza compounds from *Lachnanthes*, and the presence of naphthalic anhydrides, naphthalides, and naphthoxanthenone pigments in the same plant, suggests a sequential oxidative metabolism (*53*) in this species.

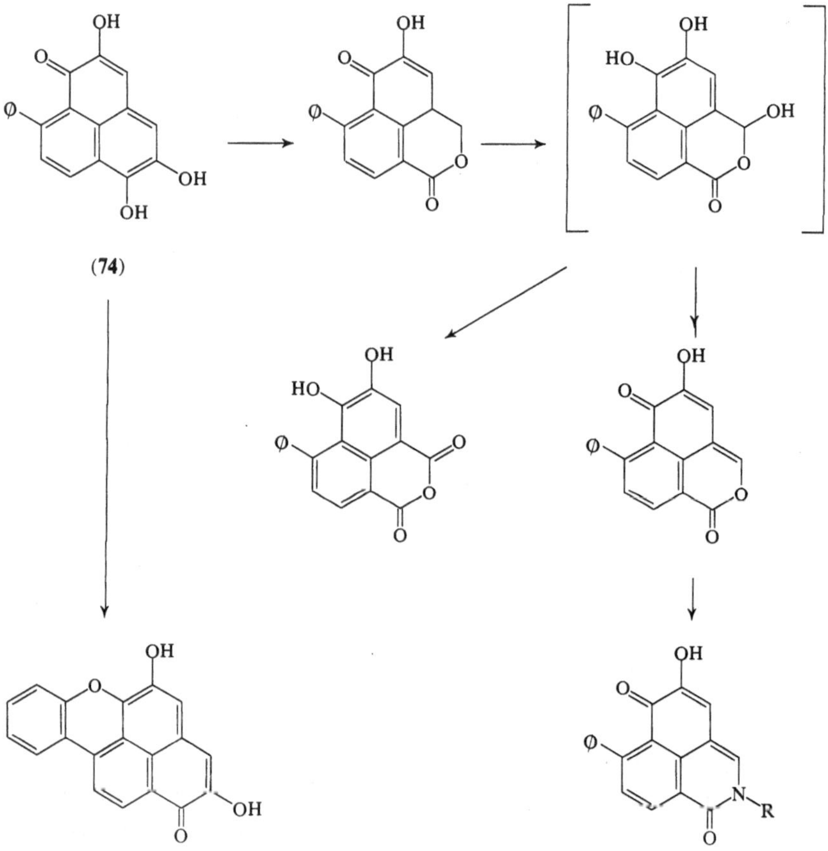

Scheme XIV. Possible biosynthetic relationship between *Lachnanthes* pigments

Acknowledgement. We would like to thank Professor G. A. MORRISON of the University of Leeds, and Professor R. THOMAS of the University of Surrey for reading parts of this review in manuscript.

The extensive work on *Lachnanthes* carried out over the past ten years was inspired by the enthusiasm of Dr. U. WEISS, NIH, whose collaboration is gratefully acknowledged.

References

1. STODOLA, F. H., K. B. RAPER, and D. I. FENNELL: Pigments of *Penicillium herquei*. Nature **167**, 773 (1951).
2. GALARRAGA, J. A., K. G. NEILL, and H. RAISTRICK: The Colouring Matters of *Penicillium herquei* Bainier and Sartory. Biochem. J. **61**, 456 (1955).
3. HARMAN, R. E., J. CASON, F. H. STODOLA, and A. L. ADKINS: Structural Features of Herqueinone, a Red Pigment from *Penicillium herquei*. J. Organ. Chem. (USA) **20**, 1260 (1955).
4. BARTON, D. H. R., P. DE MAYO, G. A. MORRISON, and H. RAISTRICK: The Constitutions of Atrovenetin and Some Related Herqueinone Derivatives. Tetrahedron **6**, 48 (1959).
5. NEILL, K. G., and H. RAISTRICK: Metabolites of *Penicillium atrovenetum* G. Smith, I. Atrovenetin, A New Crystalline Colouring Matter. Biochem. J. **65**, 166 (1957).
6. NARASIMHACHARI, N., and L. C. VINING: Studies on the Pigments of *Penicillium herquei*. Canad. J. Chem. **41**, 641 (1963).
7. KRIEGLER, A. P., and R. THOMAS: 5th International Symposium on the Chemistry of Natural Products, London. Abstract C 67, p. 172 (1968).
8. BROOKS, J. S., and G. A. MORRISON: Naturally Occurring Compounds Related to Phenalenone. The Structure of Herqueinone and Norherqueinone and their Relationships with Isoherqueinone and Isonorherqueinone. J. Chem. Soc. (London) Perkin 1 421 (1972).
9. NARASIMHACHARI, N., and L. C. VINING: Herqueichrysin, a New Phenalenone Antibiotic from *Penicillium herquei*. J. Antibiotics **25**, 155 (1972).
10. FROST, D. A., D. D. HALTON, and G. A. MORRISON: Naturally Occurring Compounds Related to Phenalenone. Structure and Synthesis of Demethylherqueichrysin. J. Chem. Soc. (London) Perkin 1 2443 (1977).
11. ROSSI, C., and R. UBALDI: Characterization of a Pigment Produced by *Fusicoccum putrefaciens*. Ann. Ist. Super. Sanita **9**, 320 (1973).
12. VAN EIJK, G. W.: A Naphtho[1,2-b]furan Derivative from the Fungus *Roesleria pallida*. Phytochem. **10**, 3263 (1971).
13. SHIBATA, S., Y. OGIHARA, N. TOKUTAKE, and O. TANAKA: Duclauxin, A Metabolite of *Penicillium duclauxi*. Tetrahedron Letters 1287 (1965); Y. OGIHARA, O. TANAKA, and S. SHIBATA. Tetrahedron Letters 2867 (1966).
14. KUHR, I., J. FUSKA, P. SEDMERA, M. PODOJIL, J. VOKOUN, and Z. VANĚK: An Antitumor Antibiotic Produced by *Penicillium stipitatum;* its Identity with Duclauxin. J. Antibiotics **26**, 535 (1973).
15. MC CORKINDALE, N. J., A. MCRITCHIE, and S. A. HUTCHINSON: Lamellicolic Anhydride — a Heptaketide Naphthalic Anhydride from *Verticillium lamellicola*. Chem. Commun. 108 (1973).
16. CHEXAL, K. K., C. TAMM, K. HIROTSU, and J. CLARDY: Gilmaniellin and Dechlorogilmaniellin, Two Novel Dimeric Oxaphenalenones. Helv. Chim. Acta **62**, 1785 (1979).
17. PAUL, I. C., and G. A. SIM: Fungal Metabolites. III. The Structure of Atrovenetin: X-ray Analysis of Atroventin Orange Trimethyl Ether Ferrichloride. J. Chem. Soc. (London) 1097 (1965).
18. NARASIMHACHARI, N., and B. S. RAMASWAMI: Pigments from *Penicillium herquei*. Current Sci. (India) 66 (1966).
19. NARASIMHACHARI, N., V. B. JOSHI, and S. KRISHNAN: Photolytic Decomposition of Perinaphthenone Derivatives. Experientia **24**, 538 (1968).
20. BARTON, D. H. R., P. DE MAYO, G. A. MORRISON, W. H. SCHAEPPI, and H. RAISTRICK: Some Observations on the Constitutions of Herqueinone and Related Compounds. Chem. and Ind. 552 (1956).

21. TURNER, A. B.: Quinone Methides in Nature. Fortschr. Chem. Organ. Naturstoffe **24** 288 (1966).
22. BROOKS, J. S., and G. A. MORRISON: Absolute Configuration of Atrovenetin and Related Compounds. Chem. Commun. 1359 (1971).
23. CASON, J., J. S. CORREIA, R. B. HUTCHINSON, and F. PORTER: The Structure of Trimethylherqueinone B. Tetrahedron **18**, 839 (1962).
24. CASON, J., C. W. KOCH, and J. S. CORREIA: The Structure of Herqueinone, Iso-herqueinone, and Norherqueinone. J. Organ. Chem. (USA) **35**, 179 (1970).
25. BROOKS, J. S., and G. A. MORRISON: The Constitution of Herqueinone and its Relationship to Isoherqueinone. Tetrahedron Letters 963 (1970).
26. HALTON, D. D., and G. A. MORRISON: The Structure and Synthesis of Desmethyl-herqueichrysin. Tetrahedron Letters 1443 (1975).
27. SIMPSON, T. J.: Carbon-13 Nuclear Magnetic Resonance Structural and Biosynthetic Studies on Deoxyherqueinone and Herqueichrysin, Phenalenone Metabolites of *Penicillium herquei.* J. Chem. Soc. (London) Perkin 1 1233 (1979).
28. OGIHARA, Y., Y. IITAKA, and S. SHIBATA: X-Ray Study of Monobromoduclauxin. Tetrahedron Letters 1289 (1965).
29. — — — Crystal and Molecular Structure of Monobromoduclauxin. Acta Crystallogr. **24B**, 1037 (1968).
30. SHIBATA, S.: Chemistry and Biosynthesis of some Fungal Metabolites. Chem. in Britain 110 (1967).
31. BYCROFT, B. W., and A. J. EGLINGTON: Synthetic Approaches to Some Naturally Occurring Phenalenones and Related Compounds. Chem. Commun. 72 (1968).
32. FROST, D. A., and G. A. MORRISON: Naturally Occurring Compounds Related to Phenalenone. Part VI. Synthesis of Atrovenetin and Related Compounds. J. Chem. Soc. (London) Perkin 1 2388 (1973).
33. THOMAS, R.: Studies in the Biosynthesis of Fungal Metabolites. Biochem. J. **78** 807 (1961).
34. — Biosynthesis of Phenalenones. Pure and Applied Chem. **34**, 515 (1973).
35. KRIEGLER, A. B., and R. THOMAS: Biosynthetic Interrelationships of Fungal Phenal-enones. Chem. Commun. 738 (1971).
36. GANDHI, R. N.: Biosynthesis of the Methyl Carbonate Unit in 4-O-Carbomethoxyl-amellicolic Anhydride. Indian J. Chem. **15B**, 482 (1977).
37. SANKAWA, U., H. TAGUCHI, Y. OGIHARA, and S. SHIBATA: Biosynthesis of Duclauxin. Tetrahedron Letters 2883 (1966).
38. NARASIMACHARI, N., K. S. GOPALKRISHNAN, R. H. HASKINS, and L. C. VINING: Pro-duction of the Antibiotic Atrovenetin by a Strain of *Penicillium herquei* Bainier and Sartory. Canad. J. Microbiol. **9**, 134 (1963).
39. NARASIMACHARI, N., B. N. VASAVADA, and S. VISWANATHAN: Antibiotic Activity of Deoxyherqueinone. Experientia **21**, 376 (1965).
40. FUSKOVÁ, A., B. PROKSA, and J. FUSKA: *In vitro* Effect of Duclauxin and Derivatives of Coumarin on Nucleic Acid and Protein Synthesis in Ehrlich's Ascites Carcinoma (EAC). Pharmazie. **32**, 291 (1977).
41. THOMAS, R.: Personal communication.
42. HUTCHINSON, J.: The Families of Flowering Plants, 3rd ed. Oxford: Clarendon Press. 1973.
43. EDWARDS, J. M., J. A. CHURCHILL, and U. WEISS: A Chemical Contribution to the Taxonomic Status of *Lophiola americana.* Phytochem. **9**, 1563 (1970).
44. HEGNAUER, R.: Chemotaxonomie der Pflanzen, Vol. 2, p. 228. Basel und Stuttgart: Birkhäuser Verlag. 1963.
45. COOKE, R. G., and R. L. THOMAS: Colouring Matters of Australian Plants. XVIII. Constituents of *Anigozanthos rufus.* Austral. J. Chem. **28**, 1053 (1975).

46. COOKE, R. G., and I. J. DAGLEY: Colouring Matters of Australian Plants. XX. Synthesis of Hydroxyanigorufone and Related Phenalenones. Austral. J. Chem. 31, 193 (1978).
47. COOKE, R. G. et al.: Unpublished data.
48. EDWARDS, J. M., and U. WEISS: Perinaphthenone Pigment from Fruit Capsules of Lachnanthes tinctoria. Phytochem. 9, 1653 (1970).
49. EDWARDS, J. M.: Phenylphenalenones from Wachendorfia Species. Phytochem. 13, 290 (1974).
50. LAUNDON, B., G. A. MORRISON, and J. S. BROOKS: Naturally Occurring Compounds Related to Phenalenone I. The Synthesis of Lachnanthocarpone. J. Chem. Soc. (London) C. 36 (1971).
51. FORTE, G. J., J. A. ZITO, and J. M. EDWARDS: The Synthesis of Lachnanthocarpone. Lloydia 39, 192 (1976).
52. BAZAN, A. C., J. M. EDWARDS, and U. WEISS: Synthesis of Lachnanthocarpone [9-Phenyl-2,6-dihydroxyphenalen-1-(6)-one] by Intramolecular Diels-Alder Cyclization of a 1,7-Diarylheptanoid Orthoquinone; Possible Biosynthetic Significance of Diels-Alder Reactions. Tetrahedron 34, 3005 (1978).
53. EDWARDS, J. M., and U. WEISS: Phenalenone Pigments of the Root System of Lachnanthes tinctoria. Phytochem. 13, 1597 (1974).
54. COOKE, R. G., and W. SEGAL: Colouring Matters of Australian Plants IV. Haemocorin: A Unique Glycoside from Haemodorum corymbosum. Vahl. Austral. J. Chem. 8, 107 (1955).
55. — — Colouring Matters of Australian Plants V. Haemocorin: The Chemistry of the Aglycone. Austral. J. Chem. 8, 413 (1955).
56. COOKE, R. G., B. L. JOHNSON, and W. SEGAL: Colouring Matters of Australian Plants VI. Haemocorin: The Structure of the Aglycone. Austral. J. Chem. 11, 930 (1958).
57. EDWARDS, J. M., M. MANGION, J. B. ANDERSON, M. RAPPOSCH, and G. HITE: Lachnanthospirone, A Dimeric 9-Phenylphenalenone from the Seeds of Lachnanthes tinctoria Ell. Tetrahedron Letters 4453 (1979).
58. LAUNDON, B., and G. A. MORRISON: Naturally Occurring Compounds Related to Phenalenone II. The Synthesis of Haemocorin Aglycone. J. Chem. Soc. (London) C. 1694 (1971).
59. COOKE, R. G., and I. J. RAINBOW: Colouring Matters of Australian Plants XIX. Haemocorin: Unequivocal Synthesis of the Aglycone and Some Derivatives. Austral. J. Chem. 30, 2241 (1977).
60. BICK, I. R. C., and A. J. BLACKMAN: Haemodorin — A Phenalenone Pigment from Haemodorum distichophyllum Hook. Austral. J. Chem. 26, 1377 (1973).
61. CREMONA, T. L., and J. M. EDWARDS: Xiphidone, the Major Phenalenone Pigment of Xiphidium caeruleum. Lloydia 37, 112 (1974).
62. COOKE, R. G., and I. J. DAGLEY: Colouring Matters of Australian Plants XXI. Naphthoxanthenones in the Haemodoraceae. Austral. J. Chem. 32, 1841 (1979).
63. BAZAN, A. C., and J. M. EDWARDS: Phenalenone Pigments of the Flowers of Lachnanthes tinctoria. Phytochem. 15, 1413 (1976).
64. COOKE, R. G.: Phenylnaphthalene Pigments of Lachnanthes tinctoria. Phytochem. 9, 1103 (1970).
65. COOKE, R. G., and R. A. H. FLETCHER: Phenylnaphthalene Pigments of Lachnanthes tinctoria II. Synthesis of Two Phenylnaphthalides. Austral. J. Chem. 24, 873 (1971).
66. EDWARDS, J. M., and U. WEISS: Quinone Methides Derived from 5-Oxa- and 5-Aza-9-Phenyl-1-Phenalenone in the Flowers of Lachnanthes tinctoria. Tetrahedron Letters 1631 (1972).
67. HIGHET, R. J., and J. M. EDWARDS: Analysis of the Carbon-13 NMR Spectrum of Phenalenones. J. Mag. Res. 17, 336 (1975).

68. Harmon, A. D., J. M. Edwards, and R. J. Highet: The Biosynthesis of 2,5,6-Trihydroxy-9-Phenylphenalenone by *Lachnanthes tinctoria*. Incorporation of 1-^{13}C-Phenylalanine. Tetrahedron Letters 4471 (1977).

69. Weiss, U., and J. M. Edwards: Pigments of *Lachnanthes tinctoria* Ell. (Haemodoraceae) I. Isolation and Photolysis of some 9-Phenylperinaphthenones. Tetrahedron Letters 4325 (1969).

70. Cooke, R. G., and I. J. Dagley: Synthesis and Spectroscopic Properties of 1H-Naphtho[2,1,8-*mna*]xanthen-1-one and its 8-Methoxy Derivative. Tetrahedron Letters 637 (1978).

71. Cooke, R. G., B. K. Merrett, G. J. O'Loughlin, and G. A. Pietersz: Colouring Matters of Australian Plants XXIII. A New Synthesis of Arylphenalenones and Naphthoxanthenones. Austral. J. Chem. **33**, 2317 (1980).

72. Reid, D. H.: The Chemistry of the Phenalenes Quart. Rev. (Chem. London) **19**, 274 (1965).

73. Darwin, C.: The Origin of Species, 13, 14. New York: Appleton and Company. 1895.

74. Kornfeld, J. M., and J. M. Edwards: An Investigation of the Photodynamic Pigments in Extracts of *Lachnanthes tinctoria*. Biochim. Biophys. Acta **286**, 88 (1972).

75. Webb, L. J.: Guide to the Medicinal and Poisonous Plants of Queensland, Bulletin No. 232, Council Sci. Indust. Res., Melbourne (1948).

76. Schwenk, E.: Tumor Action of Some Quinonoid Compounds in the Cheekpouch Test. Arzneim. Forsch. **12**, 1143 (1962).

77. Narasimhachari, N., V. B. Joshi, S. Krishnan, M. V. Panse, and M. N. Wamburkar: Antibacterial Properties of Perinaphthenone Derivatives. Current Sci. (India) **37**, 288 (1968).

78. Thomas, R.: Studies in the Biosynthesis of Fungal Metabolites. Biochem. J. **78**, 807 (1961).

79. Thomas, R.: The Biosynthesis of Phenalenones. Pure and Applied Chem. **34**, 515 (1973).

80. Edwards, J. M., R. C. Schmitt, and U. Weiss: Biosynthesis of a 9-Phenylperinaphthenone by *Lachnanthes tinctoria*. Phytochem. **11**, 1717 (1972).

81. Thomas, R.: The Biosynthesis of the Plant Phenalenone Haemocorin. Chem. Commun. 739 (1971).

82. Roughley, P. J., and D. A. Whiting: Experiments in the Biosynthesis of Curcumin. J. Chem. Soc. (London) Perkin 1 2379 (1973).

83. Denniff, P., and D. A. Whiting: Biosynthesis of [6]-Gingerol, Pungent Principle of *Zingiber officinale*. Chem. Commun. 711 (1976).

84. Macleod, I., and D. A. Whiting: Stages in the Biosynthesis of [6]-Gingerol in *Zingiber officinale*. Chem. Commun. 1152 (1979).

(Received June 2, 1980)

Molecular Mechanisms of Enzyme-Catalyzed Dioxygenation
(An Interdisciplinary Review)

By C. W. JEFFORD and P. A. CADBY, Department of
Organic Chemistry, University of Geneva, Switzerland

With 3 Figures

Contents

I. Introduction

Enzymes which catalyze the introduction of one or both atoms of a molecule of oxygen into an organic substrate are termed oxygenases. Further classification into mono- and dioxygenases in based on whether the enzyme causes transfer of just one atom to the substrate (S) (eq. 1) or whether the whole molecule is incorporated into a single substrate (eq. 2a) or is shared by a pair of substrates (S, S') (eq. 2b) (1, 2)[1].

$$S + O_2 \xrightarrow{\quad 2\,H^{\cdot}\quad} SO + H_2O \tag{1}$$

$$S + O_2 \longrightarrow SO_2 \tag{2a}$$

$$S + S' + O_2 \longrightarrow SO + S'O \tag{2b}$$

From a thermodynamic point of view, it might seem strange that there should even be a need for enzymes to catalyze these reactions. Molecular oxygen can be regarded as a high-energy molecule in the biochemical sense and its exothermicity of reaction with organic molecules is roughly equivalent to that of molecular chlorine. The fact that most organic molecules are stable in the presence of oxygen over long periods of time clearly points to a fundamental difference from chlorine. This difference lies in the energy of activation. Oxygen, but not chlorine, has a high barrier to activation. In other words, most organic molecules are kinetically stable to uncatalyzed oxygenation. The origin of this high energy of activation has stimulated much discussion. The general view is that the chief reason lies in the spin restriction arising from the reaction of ground-state organic molecules, which are nearly all singlet states, with ground-state molecular oxygen, which exists as a triplet (3—5). As most dioxygenases contain iron or copper co-factors or at least require iron or copper ions for activity, it has been argued that coordination of oxygen to these metals may circumvent either

---

[1] Strictly speaking, both the organic molecule that is dioxygenated and molecular oxygen itself are substrates of the dioxygenases. For the sake of simplicity, however, we shall only refer to the former as the substrate.

partially or completely this spin restriction, thereby resulting in the enzymatic activation of molecular oxygen (*3—7*). The notion of "oxygen activation" has largely been carried over from studies with the mono-oxygenases, in particular the cytochrome P-450 hemoproteins, where it is now established that oxygen is in fact activated by co-ordination to metal (*8*). In the present review this hypothesis will be examined in the light of the other ways which dioxygenases might adopt for catalyzing dioxygenation.

The biochemical properties of many of the dioxygenases have now been elucidated (*1, 2*). Moreover, much data has been obtained from the study of model systems in which the chemistry of oxygenated metal complexes (*9, 10*) and the reactions of the various forms of molecular oxygen with the substrates of the dioxygenases (*11*) have been studied. We shall draw on the evidence from these three areas in order to present a coherent mechanistic picture of enzyme-catalyzed dioxygenation.

II. Some Basic Chemistry of Molecular Oxygen

Oxygen possesses two characteristic properties, its paramagnetic nature and its electrophilic character (*3*). As a paramagnetic diradical, it tends to react readily with other paramagnetic species. It combines with organic compounds in their triplet state, with certain transition metal complexes, and even with itself (*12*). It also reacts with strained compounds which might have free radical character (*13, 14*) and readily with free radicals in the familiar process of chain autoxidation (*15, 16*).

On the other hand, as it is an electrophile, it can form charge-transfer complexes and can also undergo reduction by accepting one, two, three or four electrons which are conveniently accommodated in its half-filled π^* and empty σ^* molecular orbitals (*3*).

As we have already mentioned, the inertness of organic compounds toward oxygen has been attributed to the difference in the spin multiplicity between the reactants. While this is qualitatively true, it is not sufficient to account for the magnitude of the effect. Even singlet oxygen ($^1\Delta_g$) is not overly reactive and only molecules like olefins substituted with electron-donating substituents exhibit rapid oxygenation (*17, 18*). Consequently, the mere removal of the spin barrier by the heavy-atom effect in metal-oxygen complexes would in no way explain the remarkable degree of catalysis afforded by the dioxygenases.

It has been pointed out that the spin-forbidden nature of singlet-triplet reactions can be viewed as a problem of time-scale (*5*). The time needed for spin inversion is usually longer than the life-time of a collision complex. When reactions actually take place between organic molecules and molecular oxygen, they may do so just because the life-time of the collision complex is prolonged.

Recent studies (19) have shown that reaction of triplet molecular oxygen with carbanions in the gas-phase proceeds only if the electron affinity of the corresponding organic radical is less than 20 kcal. In such cases, spin inversion apparently occurs by the device of an electron jumping from the carbanion to the oxygen molecule to generate the radical pair which subsequently collapses to the singlet peroxy anion (eq. 3).

$$R^- + O_2 \longrightarrow R^{\boldsymbol{\cdot}} + O_2^{\boldsymbol{\overline{\cdot}}} \longrightarrow R-O-O^- \tag{3}$$

On the basis of this idea, we see that there are two distinct ways in which an enzyme may catalyze dioxygenation of an organic substrate. Initially, the enzyme converts the substrate to a carbanion and if needs be makes it "soft", if its highest occupied molecular orbital (HOMO) is too stabilized or "hard" (20). As a result, the frontier molecular orbitals of the organic carbanion and triplet oxygen will be close enough in energy, thereby permitting one of two processes to take place. The first will be the aforementioned electron jump from the anion to the oxygen molecule to create the radical pair which on combination furnishes substrate peroxide. The second is the formation of a charge-transfer complex between the organic anion and triplet oxygen which is sufficiently

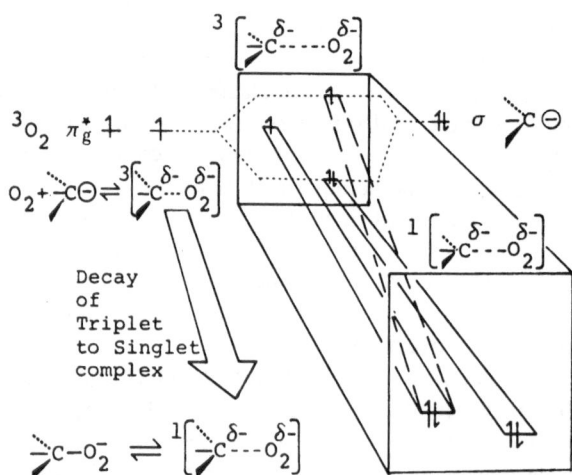

Fig. 1. Interaction of the HOMO of a "soft" or destabilized, tetrahedral carbanion with the half-filled π^* orbitals of triplet oxygen to give initially a triplet charge-transfer (CT) complex which decays to a singlet CT complex and finally to organic peroxide

long-lived to relax to the corresponding singlet charge-transfer complex. The latter then collapses to the same substrate peroxide (Fig. 1).

In principle, it is difficult to differentiate between these alternatives, but the course of oxygenation of phenolate anions (Scheme 1) has permitted a distinction to be made (21). Oxygenation of the hindered phenolate ion (1) gives the p-quinone peroxide (2). Here it seems that a long-lived charge-transfer complex is formed because the corresponding phenoxy radical (3) is not only reduced to starting phenolate ion (1) by superoxide radical anion, but further adds molecular oxygen to give a dimeric peroxide (4) and other products. As a general rule, all but the most destabilized anions should prefer charge-transfer complexation with molecular oxygen to donating an electron, a process which would be energetically more demanding (vide infra).

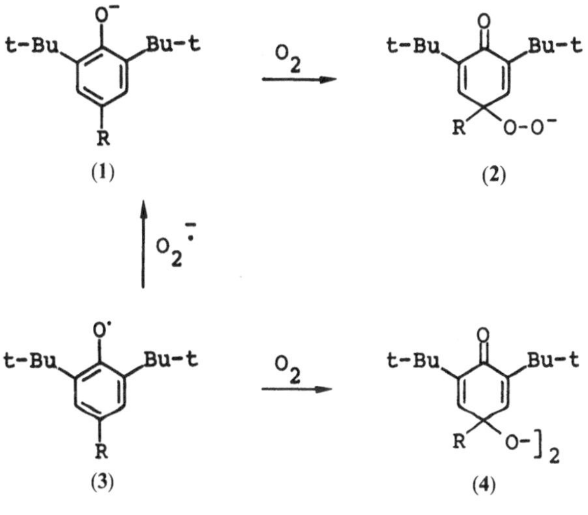

Scheme 1

Having now at this preliminary stage introduced the idea that "activation" of oxygen by the metal co-factor may be unimportant and that the substrate could be activated instead, we will first examine in some detail the precedents for "oxygen-activation" and later we will outline, where possible, from the evidence, mechanistic schemes for "substrate activation".

Before proceeding further, it is worth mentioning that the metal co-factor, if it does not "activate" oxygen, may well bind it, making it available when needed. Given the ionizing, hydrogen bonding, and hydrophobic capabilities of amino acid residues, binding of normal organic substrates is a simple matter. Binding molecular oxygen to a protein is a more difficult task. The solubility of oxygen is not greatly enhanced in hydrophobic organic solvents and the equilibrium constants for charge-transfer complex formation with oxy anions (22), amino functions (23) and for hydrogen bonded complex formation (24) are much too low to account for the measured oxygen binding constants (2). Therefore it is likely that metal co-factors, where necessary, present the only means of binding oxygen to the enzyme.

III. Precedents for Metal-Activation of Dioxygen

There is no doubt that transition metal complexes are able to catalyze the oxygenation and oxidation of organic substrates which otherwise would only react slowly or not at all with molecular oxygen. In most of these cases, however, the reaction is inhibited by free radical scavengers and shows an initial induction period, which is undoubtedly due to an initiating step typified by the Haber-Weiss process (eq. 4) (9, 15).

$$M^{n+} + R-O-OH \longrightarrow M^{(n+1)^+} + RO^{\cdot} + {}^{-}OH \qquad (4)$$

Although there may be one or two exceptions, there is no evidence, as revealed by kinetics, effects of anti-oxidants, electron paramagnetic resonance spectral data, etc. for the belief that dioxygenases exploit this form of catalysis. Unfortunately, precedents for bona fide non-radical chain catalyses are hard to find. Nevertheless, several examples are known of oxidation of ligands and substrates by dioxygen co-ordinated to metal.

1. How Would an Iron- or Copper-Protein Interact with Molecular Oxygen?

The thermodynamic stability of transition metal-oxygen complexes increases with the number of d-electrons in the metal. This means that Fe(II) and Cu(I) tend to form rather weak, reversible dioxygen complexes compared to rhodium and iridium for example (12). The electronic and geometrical properties of dioxygen complexes with Fe(II) are well documented (9, 10, 25—28). In general, the sigma-type superoxo complex (5) is formed, although when steric hindrance is absent, solutions of oxygenated ion salts eventually form binuclear μ-peroxo complexes (6) (29). It should be immediately stated that such binuclear peroxides are not involved in enzymic catalysis. With few exceptions, the non flavin-containing dioxygenases only possess one metal atom for one active site.

Another type of mononuclear dioxygen complex, the π-peroxo triangular complex (7) is known to exist, but is confined to metals having more d-electrons than iron or copper.

$$
\underset{(5)}{\overset{\overset{\text{II}}{\text{Fe}}}{\text{Fe}}\diagup\overset{O}{\diagdown}O}
\qquad
\underset{(6)}{\overset{\overset{\text{II}}{\text{Fe}}}{\text{Fe}}\diagup O\diagdown O\diagup\overset{\overset{\text{II}}{}}{\text{Fe}}}
\qquad
\underset{(7)}{O\diagdown_{M}\diagup O}
$$

Apart from the sheer bulk effect of the ligands attached to Fe(II), which will favor (5) over (6), the stability of the particular complex in the electronic state of the central metal atom will also depend on the nature of the ligands. As an illustration, the degree of interaction between Co(II) and oxygen, exemplified by the stability of the dioxygen complex, increases proportionally with the donating power of the other ligands on the metal (30). Proteins, on account of the variety of potential ligands they contain, are admirably suited for fashioning metal-dioxygen complexes of differing stabilities. An apt example is provided by the naturally occurring ferro-porphyrin proteins, hemoglobin and cytochrome P-450. The former is a reversibly oxygenating oxygen carrier, while the latter cleaves its dioxygen ligand into water and the highly reactive oxenoid function ($Fe^{3\pm}0$). The origin of this profound difference in chemical reactivity apparently resides in the different donating properties provided by the axially disposed histidine ligand in hemoglobin and the cysteine thiolate ligation of cytochrome P-450 (31).

It is to be expected then that metallo-proteins will form dioxygen complexes of great diversity. For example, those containing the same metal ion such as the ferrous-containing dioxygenases will not automatically share the same mechanisms of catalysis.

How transition metal-coordinated dioxygen might be more reactive than free molecular oxygen has been the subject of much speculation (9). The most commonly cited reason is that as co-ordinated dioxygen is often diamagnetic it should react with organic substrates with spin conservation. The suggestion that metal-bound dioxygen behaves like singlet molecular oxygen is without empirical basis. Moreover, it should not be forgotten that certain paramagnetic dioxygen complexes can also oxygenate organic molecules (9).

As the act of co-ordination brings about partial reduction of the bonded dioxygen molecule, the increase in electron density accruing on the oxygen atoms and the corresponding weakening of the oxygen-oxygen bond might, at first sight, lead to enhanced reactivity. However, such changes can scarely result in "activation" for purposes of enzymatic dioxygenation. In general, dioxygenase substrates are electron-rich and would prefer to react with dioxygen in which the electron density is diminished. Moreover, in the resulting products the oxygen-oxygen bond is retained.

In the light of this preamble, precedents will be examined for three different, possible modes of reaction of metal-activated oxygen.

2. Precedents for the Transfer of Dioxygen Within the Co-ordination Sphere

When both the substrate and oxygen are co-ordinated to the same metal ion there may be instances where their relative spatial relationship is propitious for reaction. This type of transfer of oxygen within the co-ordination sphere occurs when the ligands are readily oxidizable entities such as phosphines, sulfur dioxide, nitrous oxide, and isonitriles (6, 9, 32). A typical example is shown in Scheme 2 where triphenylphosphine is apparently oxidized in this way to its oxide.

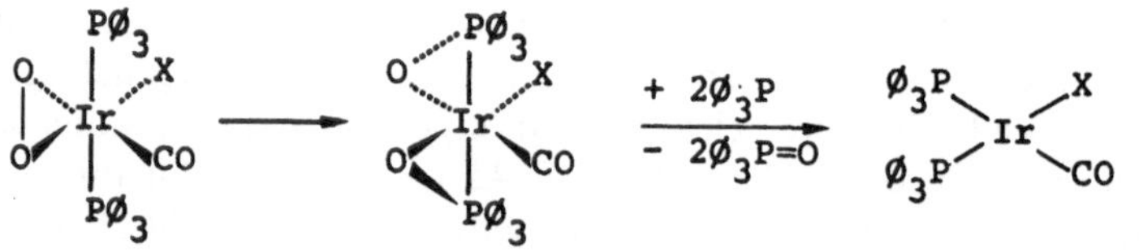

Scheme 2

Mechanisms have rarely been elucidated unambiguously, especially for easily oxidizable ligands, where alternative explanations are difficult to rule out. In fact, closer examination of some of these so-called transfer reactions has revealed that free hydrogen peroxide is the oxidant (33). Moreover, this type of reaction is restricted to the heavier transition metals which form π-type peroxo complexes (7) with dioxygen and are therefore scarcely models for the iron- and copper-containing dioxygenases.

Certain cobalt complexes which form binuclear dioxygen adducts (6) are also known to oxidize phosphine ligands (34, 35). However, ligated mono-oxy radicals appear to be implicated (Scheme 3) (*35*). For the oxidation of phosphines by Fe(III) and Fe(IV) complexes co-ordination-sphere dioxygen transfer has been ruled out (*36*).

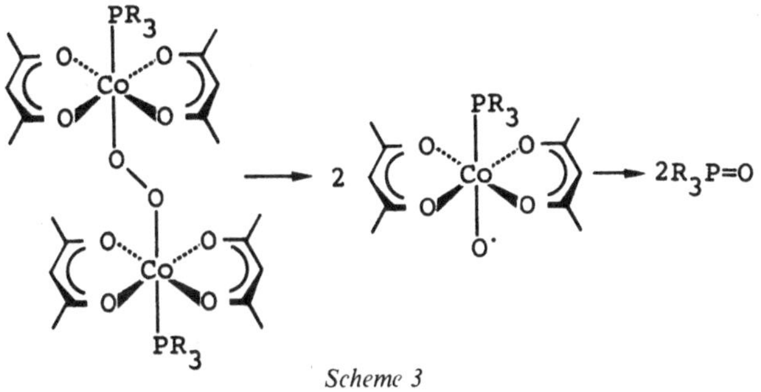

Scheme 3

There are genuine examples where co-ordination-sphere dioxygen transfer is likely, but here again these involve heavier metal complexes such as those of Rh(I). Three typical examples are shown in which different mechanisms may be operating. For cyclic olefins (Scheme 4), an oxygen atom inserts into the conveniently positioned allylic carbon-hydrogen bond (*37*). For acyclic olefins (*38, 39*) oxidation initially occurs *via* a sigma complex, followed by reduction with triphenyl-

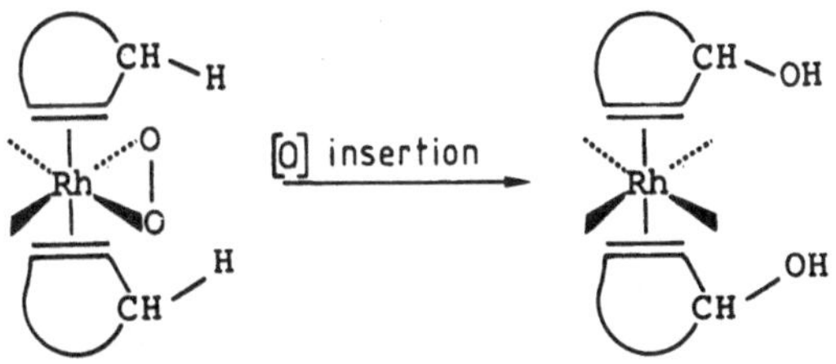

Scheme 4

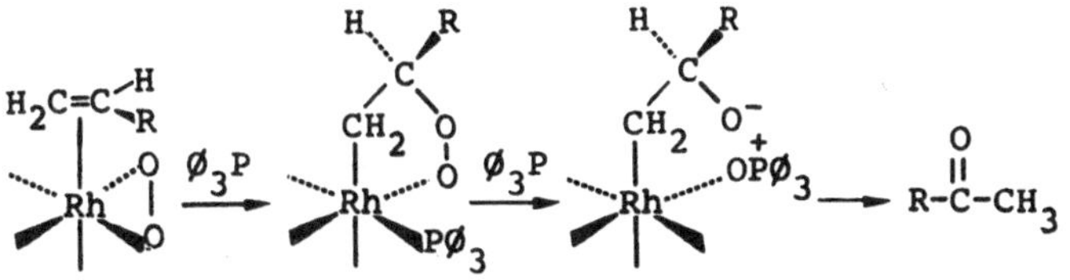

Scheme 5

phosphine to yield the corresponding ketone (Scheme 5). A third variant is encountered with styryl derivatives (40, 41) which undergo oxidative cleavage to benzaldehyde and a formyl derivative (Scheme 6).

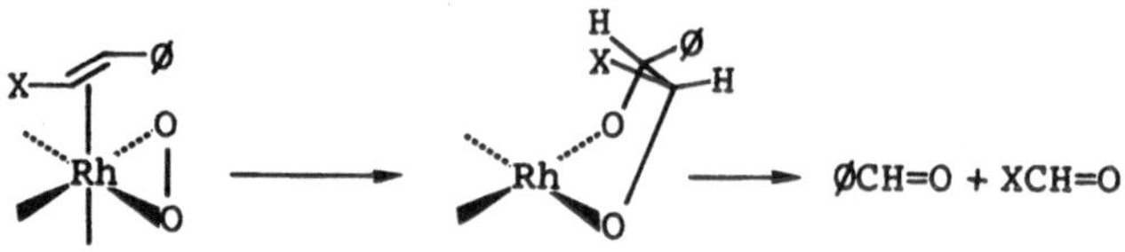

$$X = CH=O, \ OAc$$

Scheme 6

The cleavage of catechol by cuprous chloride-methanolic pyridine in the presence of oxygen was originally thought to involve electrophilic attack by co-ordinated oxygen (Scheme 7) (42). Re-investigation showed that a Cu(II) reagent was responsible with oxygen merely serving to re-oxidize the Cu(I) species so formed (43—45). A similar mechanism is probably operative for the "biomimetic" cleavage of indoles by the same reagent (46).

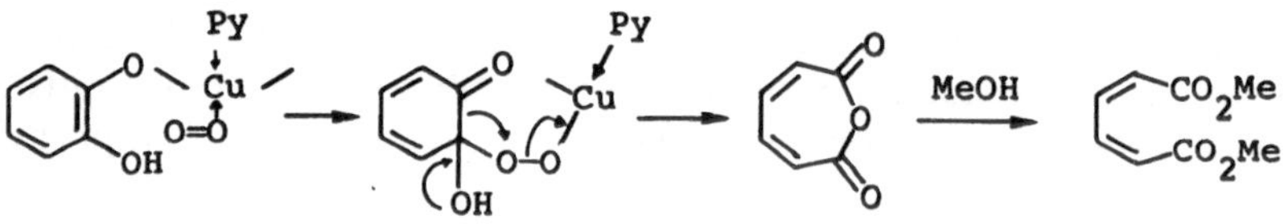

Scheme 7

A similar process mediated by a catecholato-bipyridyl (phenanthrolyl) Cu(II) complex (Scheme 8) was found to be oxygen-dependent in keeping with intra-complex dioxygen transfer (47). Once again, other mechanisms are possible. Oxygen could be reduced to hydroperoxide and react with the appropriate *o*-quinone outside the co-ordination sphere (48). The binuclear μ-peroxo complex (8) could also be implicated and, if so, it would behave similarly to the catalytically active species encountered in other oxygen-dependent cuprous chloride/pyridine oxidations (49).

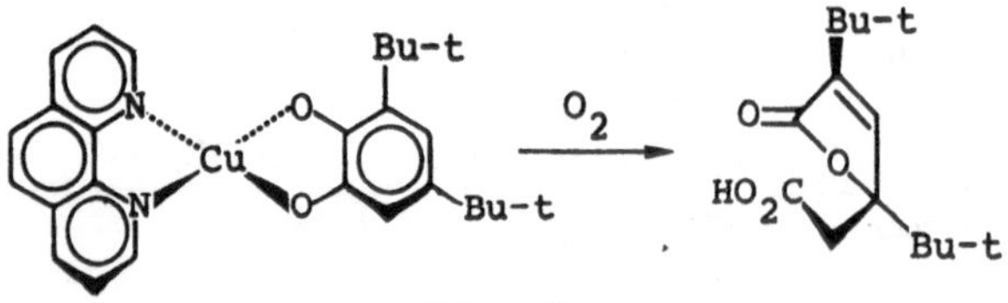

Scheme 8

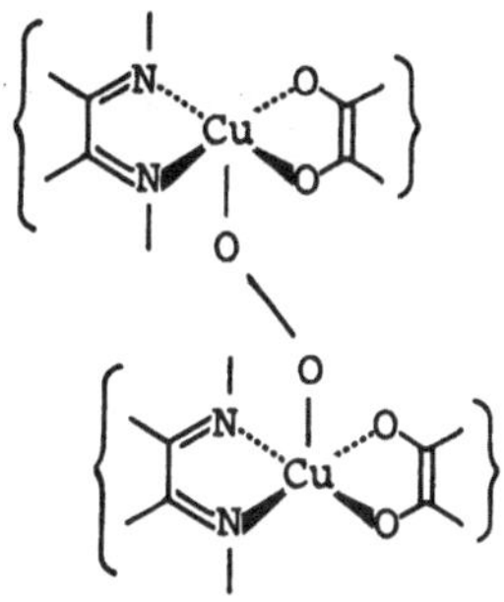

(8)

Another type of reaction occurring within the co-ordination sphere is the conduction of electrons by the metal atom to oxygen from a readily oxidizable ligand such as an ene-diol (Scheme 9) (7). These reactions are essentially oxidase-mimetic. It was originally thought that molecular oxygen only participated in a subsequent redox cycle in which the active, oxidized form of the metal (M_{ox}) was regenerated (arrow a) (48). Alternatively, both processes, substrate oxidation and oxygen reduction might occur simultaneously (arrow b) by means of the displacement of an electron pair or by two separate one-electron transfers within the co-ordination sphere (50, 51).

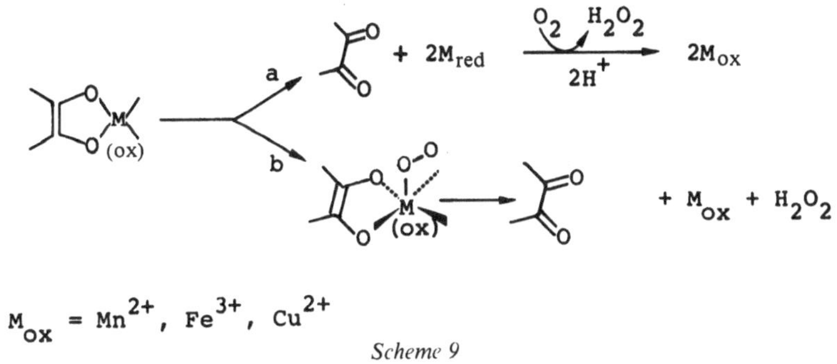

$$M_{ox} = Mn^{2+}, Fe^{3+}, Cu^{2+}$$

Scheme 9

In the case of the analogous dehydrogenation of copper-complexed ene-diamines, hydrogen abstraction probably occurs stepwise (52). Under these conditions, α-diketo and α-diimino intermediates arising from ene-diols and ene-diamines undergo subsequent biomimetic cleavage as the result of Baeyer-Villiger oxidation by free or ligated hydrogen peroxide (Scheme 10) (48).

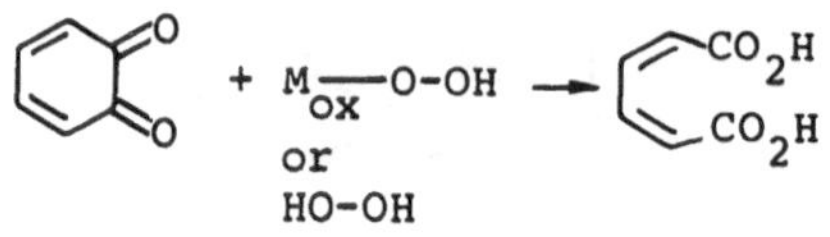

Scheme 10

3. Precedents for the Reaction of Co-ordinated Dioxygen with Free Substrates

Many metal-catalyzed oxygenations of readily oxidizable compounds such as phosphines, sulfur dioxide, nitrous oxide, etc., are believed to be due to the action of ligated dioxygen on the unco-ordinated substrate (6, 9, 32). They are not valid models for biological oxygenation as they only occur with metals which give stable π-peroxo complexes (7) and in any event their mechanisms are ambiguous owing to the ease of substrate oxidation.

There are several reports of superoxo metal complexes of type (5), in particular Co(II) oxygen complexes, which initiate autoxidation (6, 53), cause hydrogen abstraction (9, 54), catalyze the oxidation (54—57) or bring about the oxidative coupling (57, 58) of phenols and thiols (59). Clearly, free radical processes are involved, but it is difficult to attribute reactivity to the superoxo function because binuclear μ-peroxo complexes may well form and these are ready sources of free radicals (Scheme 3). Indeed, the Co(II)salen-catalyzed oxygenation of flavones

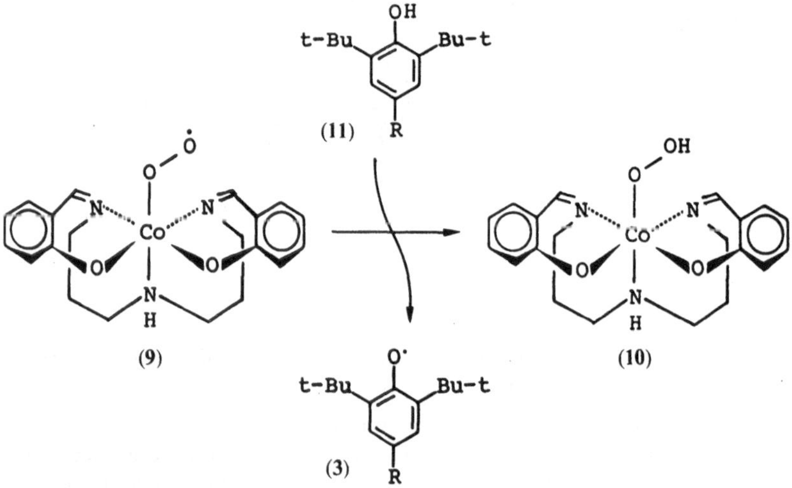

Scheme 11

only proceeds in solvents which favor the binuclear $Co-O_2-Co$ complex over the mononuclear superoxo form (*11, 56*).

The Co(II)salpr complex does not form a binuclear $Co-O_2-Co$ complex with molecular oxygen (*60*). Despite this, hydrogen abstraction from hindered phenols (**11→3**) still takes place (Scheme 11) (*61, 62*). The superoxo complex (**9**) must therefore have been responsible, but the analogy with dioxygenases goes no further as there was no ensuing reaction between (**10**) and (**3**) (Scheme 11) (*61*).

Perhaps the only reaction of ligated oxygen which mimics the dioxygenases inasmuch as both oxygen atoms are incorporated is its addition to double bonds to give the peroxy radical (Scheme 12) (*9*). To-date this addition has only been observed for cobaltous and manganous acetylacetonates and copper phthalocyanine with 2-phenylpropene. Unfortunately this result is not unambiguous as chain oxygenation also takes place (*63, 64*).

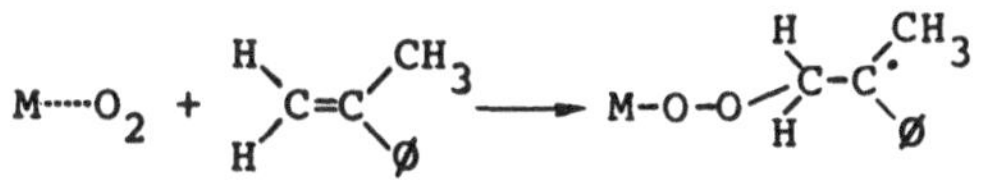

Scheme 12

4. Free Forms of Activated Dioxygen Generated by Metals

Oddly, this third possibility for activating oxygen has received more support than the preceding two, despite the rarity of enzymatic processes involving unbound substrates. In principle, a metallo-enzyme could generate free active dioxygen by reducing, oxidizing or exciting it electronically. The first and last forms, namely superoxide ion and singlet oxygen, have often been invoked by default for those cases where enzyme activity has been inhibited by singlet oxygen quenchers and scavengers and by superoxide dismutating enzymes.

From a biochemical point of view, the concept of free, activated species is unattractive, as there would be nothing to restrain their escape from the active site or to prevent them from reacting adventitiously with other substrates. Moreover, the leakage of superoxide ion or singlet oxygen would constitute a considerable energy loss to the enzyme and probably damage its structural units by non-selective reaction. Free reactive dioxygen species can also be discounted for good chemical reasons. Many of the dioxygenase-catalyzed reactions occur with high regio- and stereoselectivity, often in defiance of the availability of more reactive sites and more energetically favorable stereoelectronic modes of reaction.

The oxidation of oxygen by iron or copper enzymes, for example, to generate O_2^+ is highly unlikely since oxygen is a poor electron donor as its high ionization potential ($+12.1$ e.v.) indicates (65). In fact, only highly electronegative metal complexes such as $[PtF_6]^-$ can accomplish oxidation (66).

Reduction of oxygen by ferro- and cupro-enzymes on the other hand is energetically feasible, its electron affinity being -0.9 e.v. (67), and probably occurs with many dioxygenases as an undesired side reaction (68). This process would, of course, oxidize the metal co-factor thus necessitating its reduction before the commencement of a new catalytic cycle (eq. 5). It is important to know if the superoxide ion could actually carry out the type of reaction which the dioxygenases are supposed to catalyze. The superoxide ion is strongly basic (68) and nucleophilic (70), but is a poor acceptor of electrons and a weak abstractor of hydrogen atoms (68, 71). Since most dioxygenase substrates are electron-rich and easily oxidized, it is hard to see how superoxide ion could be much more reactive than triplet molecular oxygen. Consequently, caution should be exercised in interpreting inhibition of dioxygenases by superoxide-dismutating enzymes and copper chelates. In particular, reports of inhibition of intradiol and extradiol-cleaving dioxygenases (72), α-ketoglutarate-decarboxylating hydroxylases (73), ribulose bisphosphate oxygenase (74) and nitropropane dioxygenase (75) should not be taken as irrefutable evidence for the participation of superoxide ion. The case of galactose oxidase readily illustrates how superoxide dismutating agents will show a net inhibitory effect even when this species plays no role in the actual catalytic process (76).

$$M^{n+} + O_2 \longrightarrow M^{(n+1)+} + O_2^{\overline{\cdot}} \xrightarrow{\ S\ } SO_2^{\overline{\cdot}} \xrightarrow{\ M\,(n+1)+\ } SO_2 + M^{n+} \qquad (5)$$

Electronic excitation of oxygen by metallo-enzymes poses three special problems. The first concerns the provision of the 23 kcal/mol necessary for excitation, the second concerns the avoidance of undesired quenching of singlet oxygen before it encounters the substrate and the third concerns the recovery of excess energy from vibrationally excited products (77).

The catalytic, thermal generation of singlet oxygen by decomposition of the charge-transfer complex of strained acetylenes and ketenes with triplet oxygen has been proposed (2, 16, 78) and a similar mechanism has been suggested as a possibility for metallo-enzymes (11). Nevertheless, the central problem of generating and controlling a highly energetic, free species still remains and as yet, there are no unambiguous precedents for metal-catalyzed thermal conversion of triplet into singlet oxygen.

IV. Precedents for Metal Activation of Organic Substrates

If mechanisms implicating "oxygen-activation" are without precedent or otherwise unattractive, the only remaining alternative would be "substrate-activation". It turns out that dioxygenases act on three general classes of compounds, (i) electron-rich systems such as thiols, enols and enamines, (ii) homo-conjugated polyunsaturated hydrocarbons and (iii) certain α-ketocarboxylic acids. Apart from this last class (79, 80) there exists much information concerning their reactions with Fe(III) and Cu(II) ions. A key feature of these reactions is the metal-induced formation of radicals and radical cations (81, 82). In the absence of oxygen, typical free radical reactions occur such as dimerization, rearrangement and fragmentation, whereas under aerobic conditions the intermediate radicals combine with molecular oxygen and chain or non-chain autoxidation is observed (15, 16). In many cases, the dioxygenase substrates are pre-disposed to form resonance-stabilized radicals and the question arises: Will these radicals react sufficiently rapidly with molecular oxygen? We have already mentioned that phenoxy radicals may be formed in the base-catalyzed hydroperoxidation of phenols, but are not under any circumstances actually implicated in product formation (see Scheme 1) (21). It now appears that when the same type of reaction is mediated by a cobalt complex in the absence of base, the phenoxy radical still plays no active role. Interestingly, a quite different type of metal-substrate activation is exploited.

In the presence of molecular oxygen and Co(II) or Co(III)salpr complexes, hindered phenols (11) are converted into peroxy-o-quinolato (13) or peroxy-p-quinolato-Co(III) complexes (14), depending on the

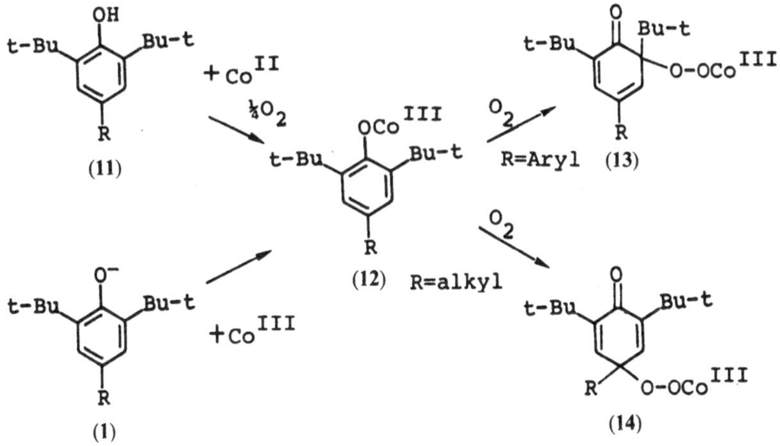

Scheme 13

nature of the *p*-substituent (Scheme 13) (*61, 62*). We described earlier how the oxygenated Co(II) salpr complex oxidizes the phenol (**1**) to its phenoxy radical (**3**) (Scheme 1), but it is important to realize that this particular reaction plays no direct part in peroxidation. The phenoxy radical (**3**) is rapidly reduced to its phenolate (**1**) by un-oxygenated Co(II)salpr and the overall effect is the generation of the Co(III) oxidation state with the consumption of one quarter equivalent of oxygen (*61, 62*). It does not matter if Co(II) or Co(III)salpr is used. In either case the reaction will pass through the same phenolato Co(III) complex (**12**) (*61, 62*).

As cobalt complex-mediated oxygenation of these hindered phenols (**11**) occurs 30 times faster than base-catalyzed oxygenation (*62*), the phenolato Co(III) complex (**12**) evidently possesses special "activation" for reaction with molecular oxygen. In keeping with the desirability of a localized, "soft" anionic center in an oxygenase substrate, we can only assume that the metal is able to localize the π-anionic charge into an orbital with more [sp^3] character.

An entirely similar obligatory intermediate appears to be required to account for the quercetinase-mimetic dioxygenation of 3-hydroxy-flavones (**15**) to the acid (**17**), which is induced by Co(II)salen (Scheme 14) (*11, 56*). Here again a preliminary redox cycle is necessary to produce the appropriately reactive Co(III) enolate (**16**). The intermediacy of the enoxy radical (**18**) is ruled out by the behavior of the latter when independently generated from (**15**) by manganese dioxide in the presence of oxygen. The result is dimerization to (**19**) rather than oxygenation (*11*).

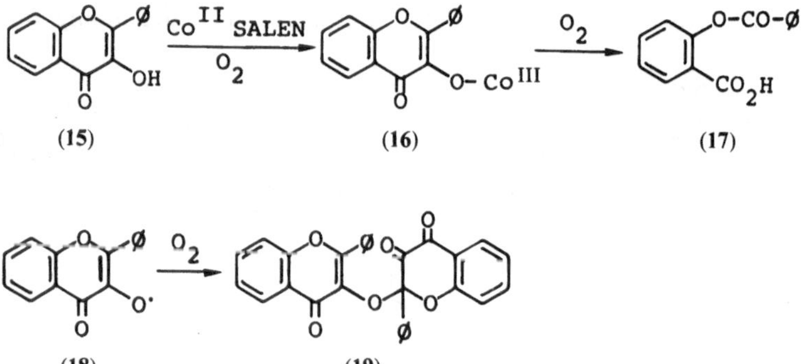

(15)   (16)   (17)

(18)   (19)

Scheme 14

The biomimetic oxygenation of indoles, *e. g.* (**20**), induced by Co(II)salen probably involves the initial formation of the Co(III) derivative (**21**) as an obligatory intermediate. Oxygenation at C-3 affords (**22**) which cleaves its C-2, C-3 bond to give (**23**) (Scheme 15) (*11*).

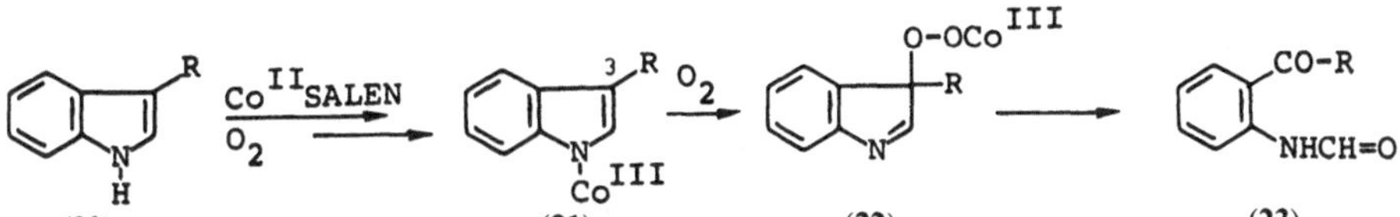

Scheme 15

The essential point which is demonstrated by these examples is that, first, the oxygen is not activated by the cobalt complex, but the substrate is, and second, the metal does something more than impart radical character. We maintain that "soft" anion character is engendered in these substrates. As we will see later, these examples constitute valid precedents for substrate "activation".

V. The Double Bond-Cleaving Dioxygenases

1. Ene-diol Cleaving Dioxygenases

With one exception, in all the substrates the ene-diol structure is part of an aromatic nucleus. Oxidative cleavage of the ene-diol yields a dicarboxylic acid which retains the geometry of the remaining double bonds. There are eight of these so-called intradiol dioxygenases (*1, 2, 83*).

(1) Pyrocatechase (E.C. 1.13.11.1 – Catechol:oxygen 1,2-oxidoreduct-ase (decyclizing)

$$(6)$$

(2) Protocatechuate intradiol dioxygenase (E.C. 1.13.11.3 – Protocate-chuate:oxygen 3,4-oxidoreductase (decyclizing)

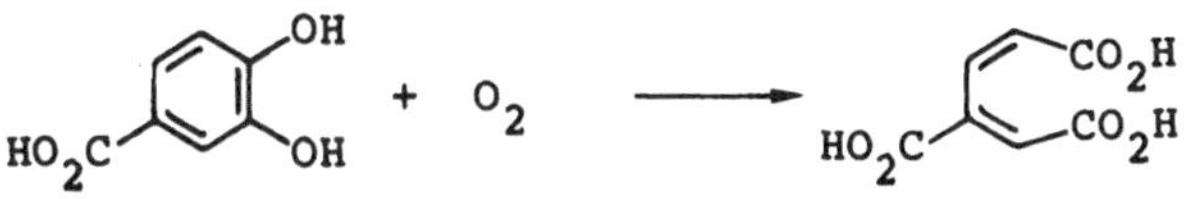

$$(7)$$

(3) Ascorbate dioxygenase (E.C. 1.13.11.13 – (L)-Ascorbate:oxygen 2,3-oxidoreductase (decyclizing)

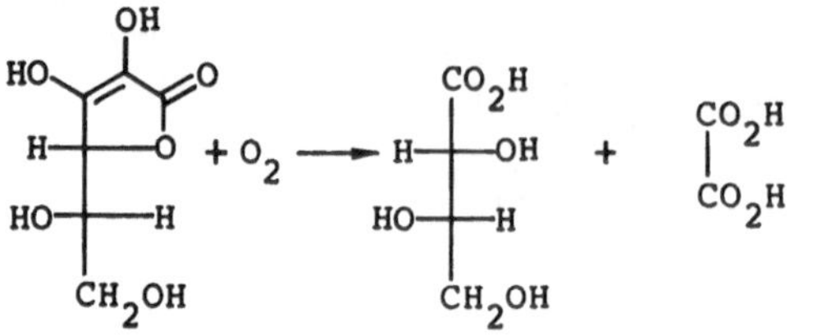

$$\text{(8)}$$

L-threonate   oxalate

(4) Caffeate dioxygenase (E.C. 1.13.11.22 – 3,4-Dihydroxy-*trans*-cinnamate: oxygen 3,4-oxidoreductase (decyclizing)

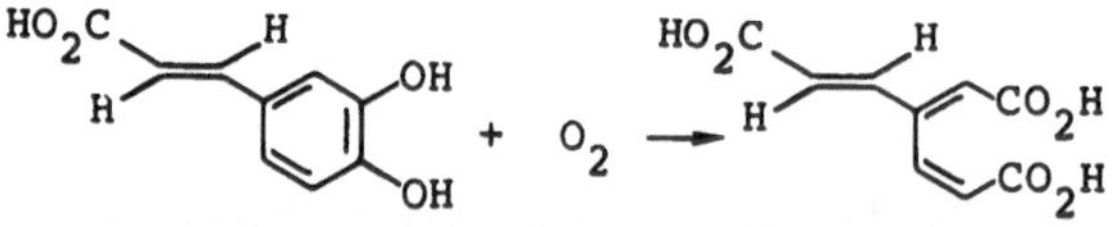

$$\text{(9)}$$

(5) Dihydroxyindole dioxygenase (E.C. 1.13.11.23 – 2,3-Dihydroxyindole: oxygen 2,3-oxidoreductase (decyclizing)

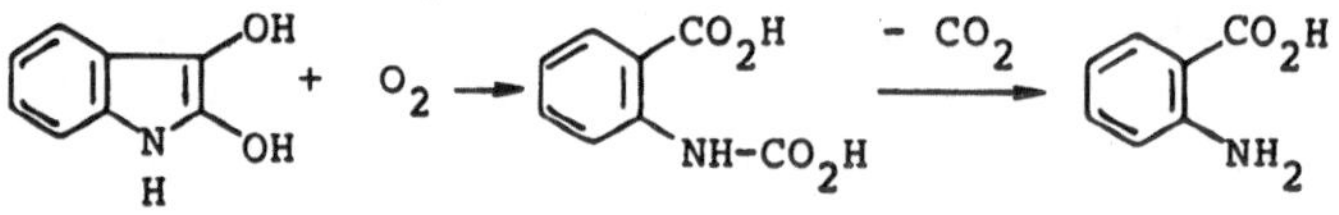

$$\text{(10)}$$

(6) Chloromethylcatechol dioxygenase (– 5-Chloro-3-methylcatechol: oxygen 1,2-oxidoreductase (decyclizing)

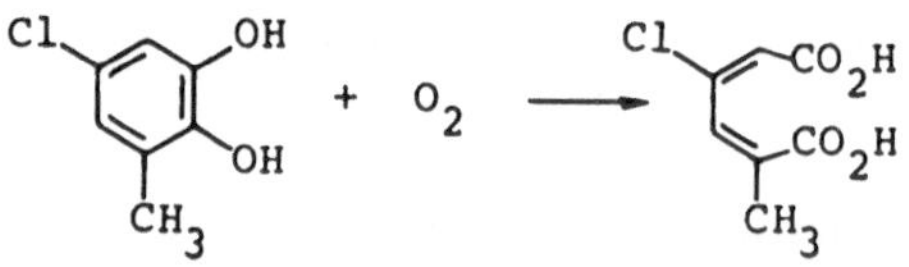

$$\text{(11)}$$

(7) Dihydroxybenzoate intradiol dioxygenase (E.C. 1.13.12.28 – 2,3-Dihydrobenzoate: oxygen 2,3-oxidoreductase (decyclizing)

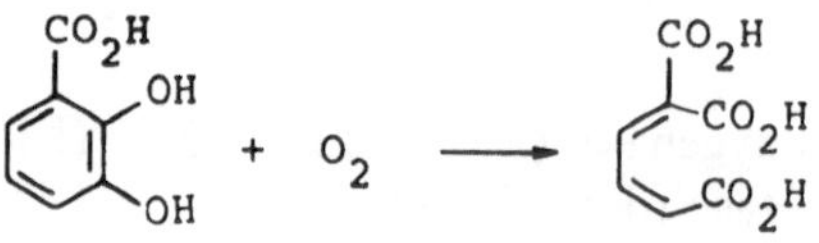

$$\text{(12)}$$

(8) Gallate dioxygenase (– Gallate: oxygen 3,4-oxidoreductase (de-cyclizing)

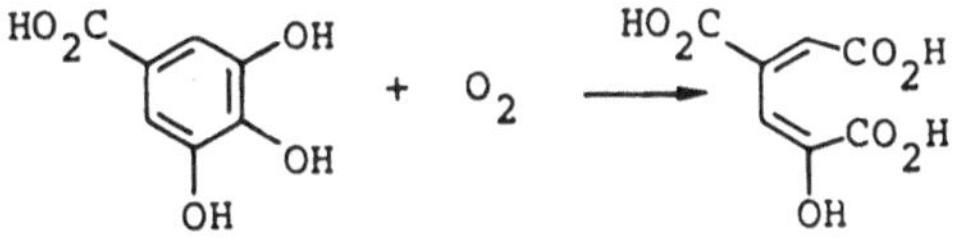

$$(13)$$

Only the first two enzymes have received close attention. Each possesses one atom of ferric iron at the active site (*2, 83*). The ascorbate (*84*) and chloromethylcatechol (*85*) dioxygenases also contain iron, but its oxidation state is unknown. Dihydroxyindole dioxygenase is probably a metallo-enzyme as it is inhibited by metal chelators (*86*). Dihydroxy-benzoate dioxygenase probably contains thiolate-bound metal as it is inhibited by sulfhydryl reagents (*87*). Analysis for co-factors has not yet been attempted for caffeate (*88*) and gallate (*89*) dioxygenases.

The dioxygenases which cleave catechol and protocatechuate are also the only ones for which the kinetics have been elucidated and the different stages examined spectroscopically (*2, 83, 90—91*). Both exhibit sequential kinetics with an obligatory order of substrate addition (eq. 14) (*2, 83, 91*).

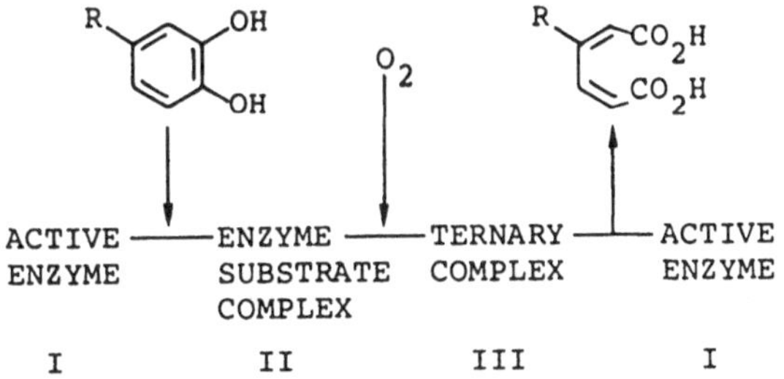

$$(14)$$

Initially, the active enzyme I binds the ene-diol substrate to give the E–S complex II which then binds oxygen to produce a ternary complex III. The ternary complex then regenerates the active enzyme and liberates the product. The active enzyme I and the two reaction complexes (II and III) and various enzyme-inhibitor complexes have been thoroughly examined using U. V., Raman, E. P. R. and Mössbauer spectroscopic techniques. A recent review (*92*) has summarized these observations. It is reported that the active enzyme I contains a high-spin ferric ion in a rhombic environment with double ligation by tyrosinate (*92—96*). Under aerobic conditions, but in the absence of substrate, there is no spectral change, thereby confirming the order of substrate addition shown in eq. 14. Under anaerobic conditions, but in the presence of substrate a new high-spin E.P.R. signal appears with minor changes in other spectral properties (*90, 92*). Under no circumstances does the enzyme substrate complex II resemble the ferro-enzymes generated in the presence of dithionite (*83, 90*). In other words, the substrate is not reducing the metal co-factor. Furthermore, the tyrosine co-ordination is not altered by substrate binding (*95*). Spectral comparison with synthetic catecholato complexes suggests that both phenolic oxygens are chelating with the metal in II in pyrocatechase (*97*) and the proto-catechuate dioxygenases (*95, 96, 98*). Studies of enzyme-inhibitor complexes, on the other hand, suggest that only one of the phenolic oxygens, namely the para-oxygen in protocatechuate dioxygenase co-ordinates with the metal co-factor (*92*).

The ternary complex III is also in the high-spin ferric state (*90, 92*) and still possesses a strong tyrosinate-iron charge transfer interaction (*92*). Moreover, the ternary complex resembles enzyme-bound product rather than enzyme-bound reactants (*96, 99*).

Hence, it appears that when the enzyme binds the substrate and oxygen the ligand field environment of the metal co-factor is changed, but not its oxidation state. Furthermore, oxygen adds after the substrate has interacted with the metal attaching itself either to the ferric iron or to a site on the substrate made reactive by interaction with the metal.

The resemblance of these observations to the simpler model systems of Scheme 13 is striking. It was previously seen that oxygen reacts with the Co(III)-phenolate complex (**12**) more rapidly than with any phenoxy radical or phenolate anion derived from it. As there are no vacant sites left over for co-ordination of oxygen, this means that the Co(III) ion induces special dioxygen binding at the ring positions ortho and para to the phenolic hydroxy group, thereby giving (**13**) and (**14**).

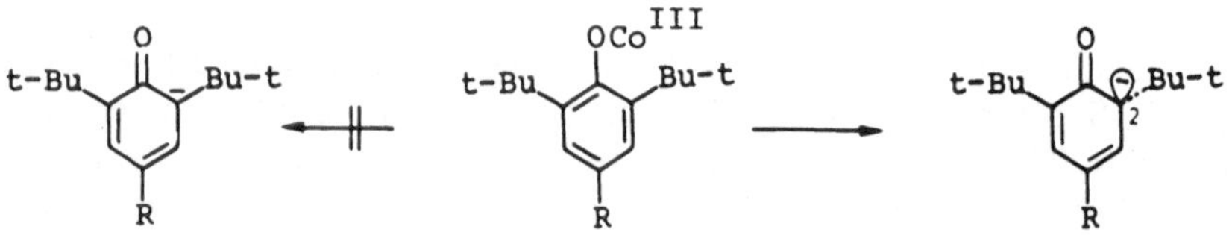

Scheme 16

As molecular oxygen can interact favorably with carbanionic centers, the logical conclusion is that complexation of a phenol with Fe(III) in the enzyme, for example, will develop a similar center. Of course, the hindered phenol (11), when complexed with Co(III) as in (12) constitutes a special case in that the bulky t-butyl groups prevent the phenolate or the enolate entity from achieving planarity. Consequently, the negative charge is no longer delocalized as in (24) of Scheme 16 but is rather localized at C-2 in an sp³ hybridized orbital (25). Nevertheless, molecules of the catechol type, as ene-diols, when bound to the enzyme,

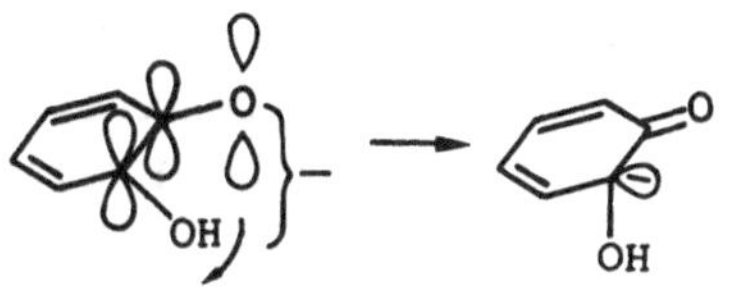

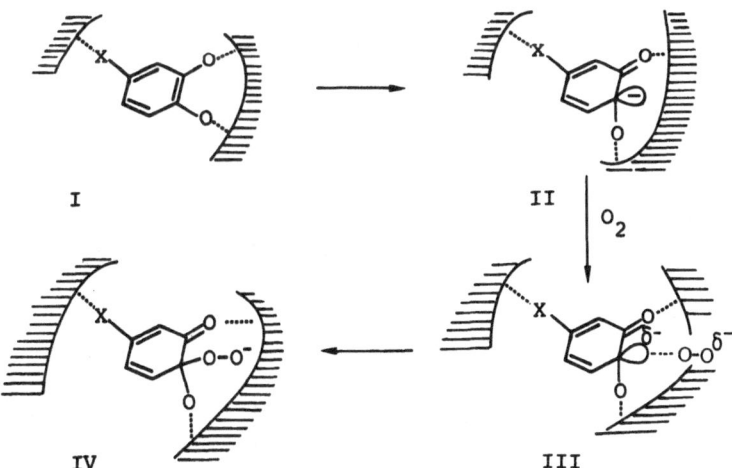

(26)   (27)

Fig. 2. Orbital tautomerism. Forcing the HO-substituent out of the plane of the ring destabilizes the phenolate ion 26 and gives 27 containing the tetrahedral carbanionic center

also present special cases. It does not really matter if both or only one of the catechol oxygen atoms are co-ordinated to metal. The essential point is that conformational constraints or the demands of metal co-ordination induce orbital tautomerism. One of the oxygen atoms is pushed out of the plane of the ring. Consequently, the intially planar

I   II

IV   III

Fig. 3. Enzyme converts planar catecholate anion (I) to tetrahedral form (II). Oxygen complexes (III) to finally give peroxide (IV)

π-delocalized phenolate anion (26) is forced to adopt a non-planar conformation, thereby placing the negative charge on a tetrahedral carbon
atom (27) (Fig. 2). The resulting electronically destabilized carbanion is
thus rendered "softer" than its phenolate precursor. Since its HOMO is
raised, it would be able to interact more efficiently with the π_g^* orbital
of molecular oxygen. The charge transfer complex so obtained will
be long-lived and permit the necessary spin inversion to take place
(Fig. 3).

The other possibility that the "soft" carbanion transfers an electron
completely to oxygen is ruled out by the cobalt precedent and also
for energetic reasons (see Scheme 13).

The second mechanistic problem concerns the cleavage of the α-
peroxy-carbonyl intermediate (28) (Scheme 17). Labelling studies have
shown that both oxygen atoms from the oxygen molecule are incorporated
into the dicarboxylic product (31) implying, without proof on account
of the low concentrations of ^{18}O used, that they are located one each
in both carboxyl groups (100). This assumption has led to the proposal
of two mechanims involving cyclization to a dioxetane (29) and ring
enlargement to an anhydride (30), both of which eventually afford the
diacid (31) (100).

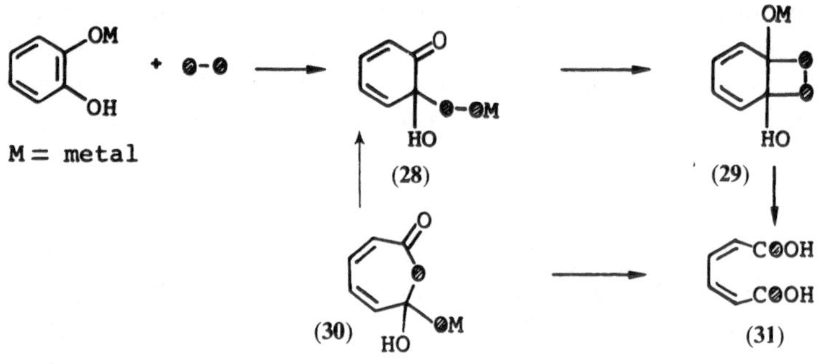

Scheme 17

The dioxetane pathway is mechanistically unsatisfactory. It has rightly
been pointed out that dioxetanes are thermodynamically improbable
under the circumstances (5). Ring enlargement is mechanistically more
plausible, α-hydroperoxycarbonyl compounds prefer to undergo Baeyer-
Villiger type rearrangements rather than give dioxetanes (101—107).
Migration of the acyl group (28→32) could occur as shown in Scheme 18
retaining the leaving group ($^-$OM) at the reaction site, so that labelled
oxygen is prevented from escaping under enzymic conditions. These
mechanisms are still at the speculative stage, as it is not known for certain
if both atoms are generally retained. To-date, labelling studies have
only been performed on pyrocatechase and caffeate dioxygenase (88).

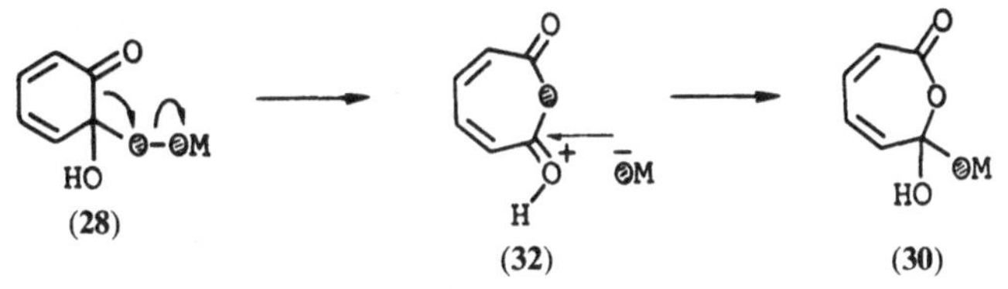

Scheme 18

A second class of ene-diol cleaving enzymes, the extradiol dioxygenases, cleaves the adjacent bond (Scheme 19). This change in cleavage

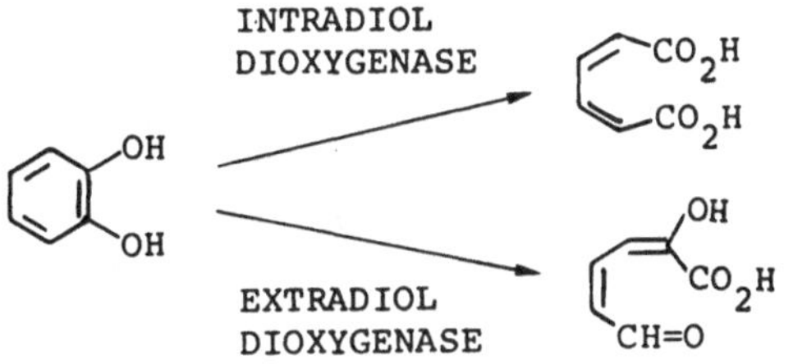

Scheme 19

mode could be realized in three different ways. Firstly, assuming the same initial intermediate is formed, the enzyme could catalyze rearrangement of the peroxyquinol (33) (Scheme 20) so as to favor vinyl over

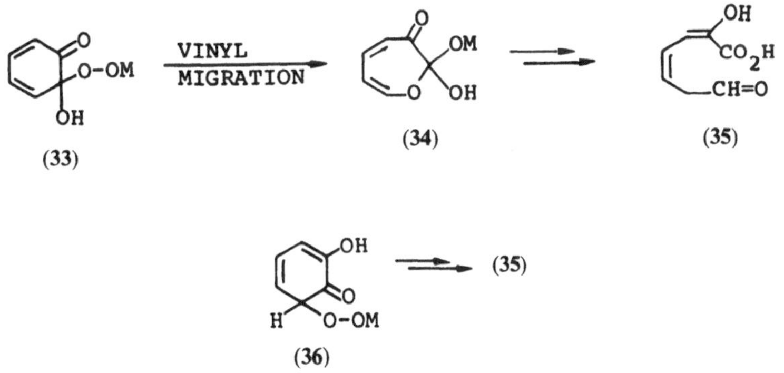

Scheme 20

acyl migration (33→34). Secondly, oxygenation could be induced to give the isomeric peroxyquinol (36). As the change in regioselectivity of base-catalyzed oxygenation of phenols is dependent on solvent and counter-

ions, enzymes might be expected to have the same capabilities (108, 109).
Thirdly, an entirely different mechanism of catalysis could be operating.
From a consideration of the chemistry of α-hydroperoxycarbonyl compounds, it is quickly appreciated that acid-catalyzed (101), base-
catalyzed (102), free-radical promoted (110) and normal Baeyer-Villiger
conditions (107), all exhibit a preference for acyl migration. Consequently,
the reason for the change in cleavage mode must be sought elsewhere.

A possibly significant finding is that pyrocatechase gave a mixture
of intradiol and extradiol cleavage products when 6-methyl- or 6-
methoxy-catechol was used as substrate (92, 111). It is unlikely that the
same enzyme would employ different mechanisms to oxygenate analogous
substrates (37) (Scheme 21). Consequently, in these cases both intradiol
and extradiol cleavage could occur by the same sequence of "anion-
softening", oxygenation and acyl migration (38→40 vs. 41→43), but
they will differ in the initial point of oxygen attachment (38 vs. 41). As
will be seen shortly, however, the uniquely extradiol-cleaving dioxygenases
differ from their intradiol-cleaving counterparts in certain fundamental
properties and the possibility that they employ different mechanisms for
oxygenating and cleaving their substrates cannot be ruled out.

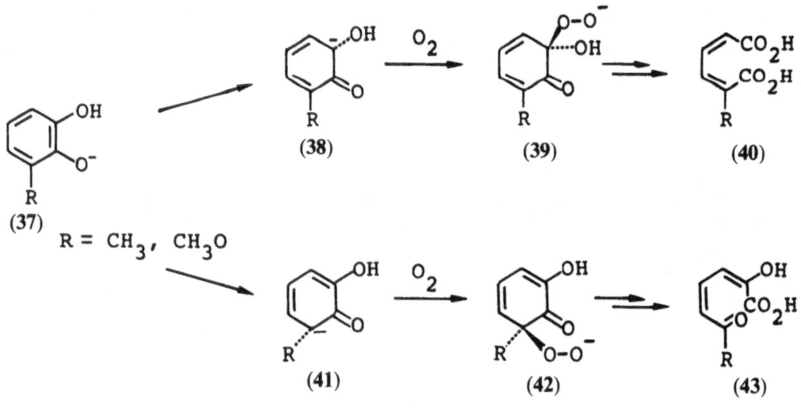

Scheme 21

2. Extradiol-Cleaving Dioxygenases

In addition to the classical extradiol cleaving dioxygenases, the
dioxygenases which cleave o-amino phenols (44) and p-diphenols (45)
(Scheme 22) also appear to share the same catalytic properties.

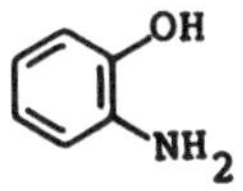

(44)

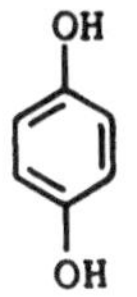

(45)

Scheme 22

(1) Metapyrocatechase (E.C. 1.13.11.2 – Catechol: oxygen 2,3-oxido-reductase (decyclizing)

$$+ O_2 \longrightarrow \qquad\qquad (15)$$

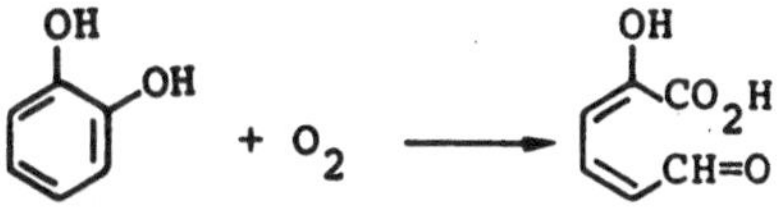

(2) Protocatechuate extradiol dioxygenase (E.C. 1.13.11.8 – Protocate-chuate: oxygen 4,5-oxidoreductase (decyclizing)

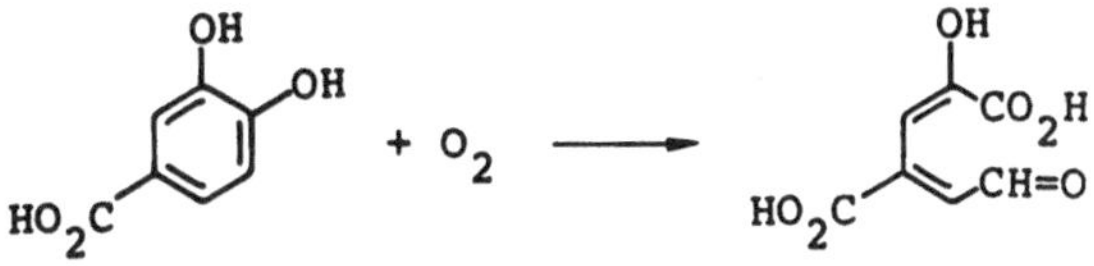

$$+ O_2 \longrightarrow \qquad\qquad (16)$$

(3) Dihydroxykynurenate dioxygenase (E.C. 1.13.11.10 – 7,8-Dihydro-xykynurenate: oxygen 8,8a-oxidoreductase (decyclizing)

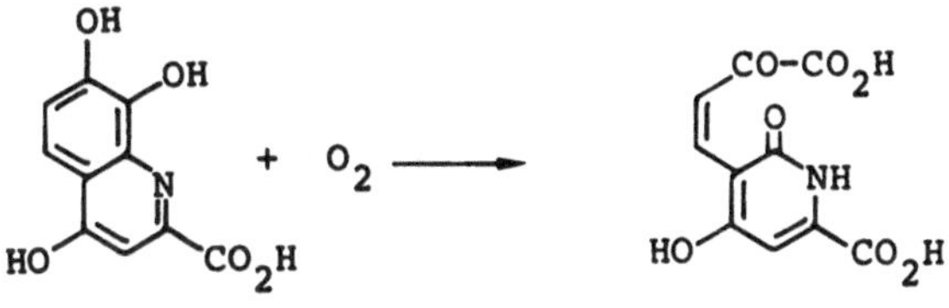

$$+ O_2 \longrightarrow \qquad\qquad (17)$$

(4) Dihydroxybenzoate extradiol dioxygenase (E.C. 1.13.11.14 – 2,3-Dihydroxybenzoate: oxygen 3,4-oxidoreductase (decyclizing, de-carboxylating) (*112*)

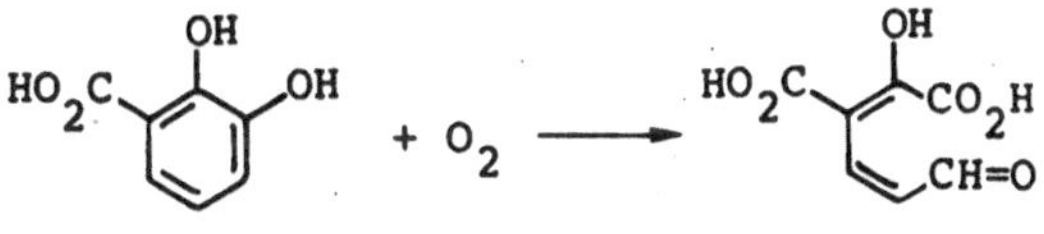

$$+ O_2 \longrightarrow \qquad\qquad (18)$$

(5) Homoprotocatechuate dioxygenase (E.C. 1.13.11.15 – 3,4-Dihydroxyphenylacetate: oxygen, 2,3-oxidoreductase (decyclizing)

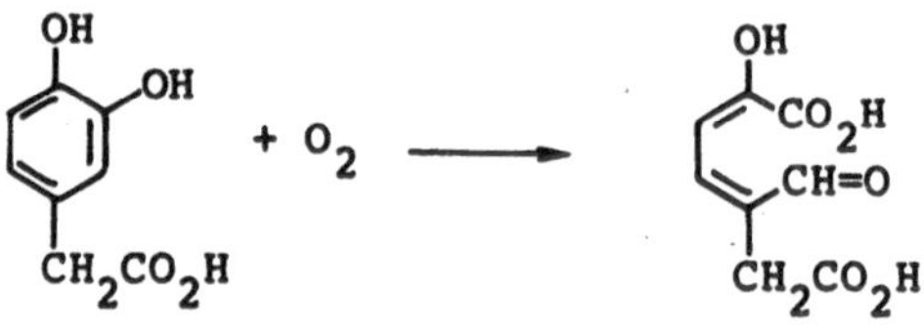

$$+ O_2 \longrightarrow \qquad\qquad (19)$$

(6) Carboxyethylcatechol dioxygenase (E.C. 1.13.11.16 – β-(2,3-Dihydroxyphenyl)propionate: oxygen 1,2-oxidoreductase (decyclizing)

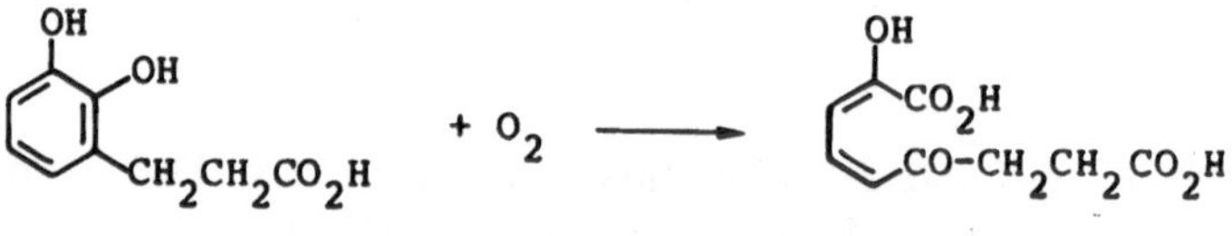

$$+ O_2 \longrightarrow \qquad\qquad (20)$$

(7) Steroid dioxygenase (E.C. 1.13.11.25 – 3,4-Dihydroxy-9,10-seco-androsta-1,3,5(10)-triene-9,17-dione: oxygen 4,5-oxidoreductase (decyclizing)

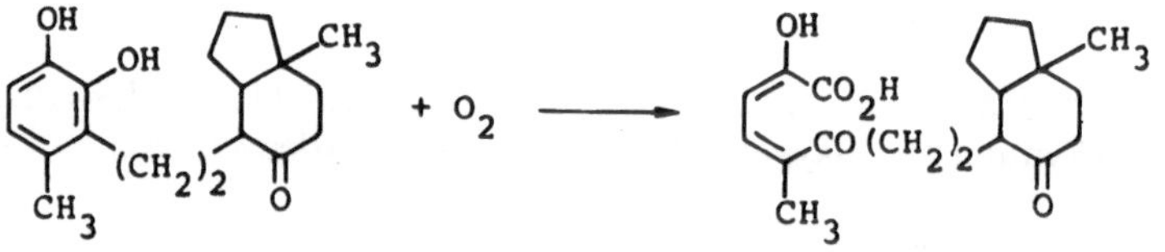

$$+ O_2 \longrightarrow \qquad\qquad (21)$$

(8) Gentisate dioxygenase (E.C. 1.13.11.4 – Gentisate: oxygen 1,2-oxidoreductase (decyclizing)

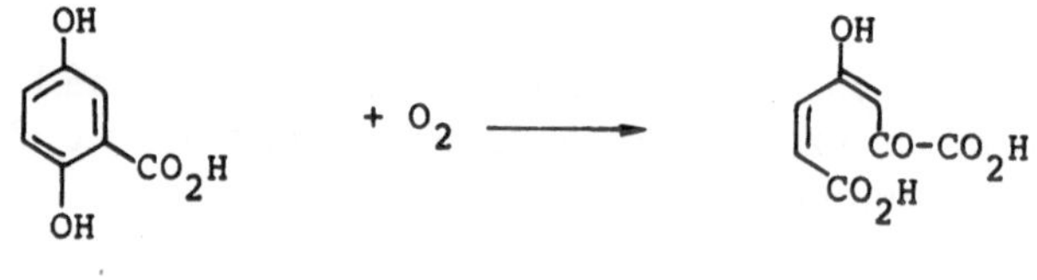

$$+ O_2 \longrightarrow \qquad\qquad (22)$$

(9) Homogentisate dioxygenase (E.C. 1.13.11.5 – Homogentisate: oxygen 1,2-oxidoreductase (decyclizing)

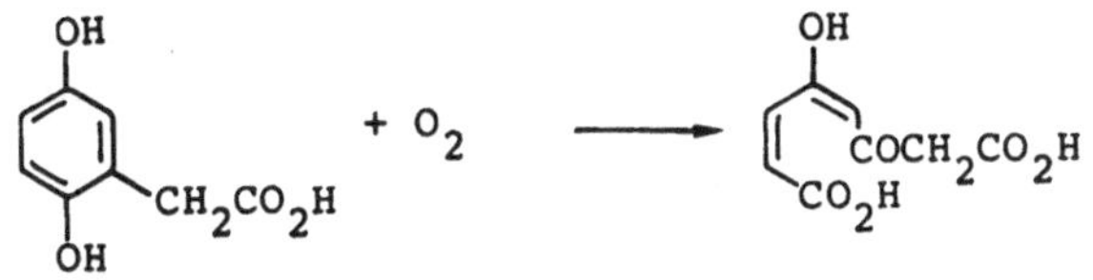

$$\text{(structure)} + O_2 \longrightarrow \text{(structure)} \qquad (23)$$

(10) Hydroxyanthranilate dioxygenase (E.C. 1.13.11.6 – 3-Hydroxy-anthranilate: oxygen 3,4-oxidoreductase (decyclizing)

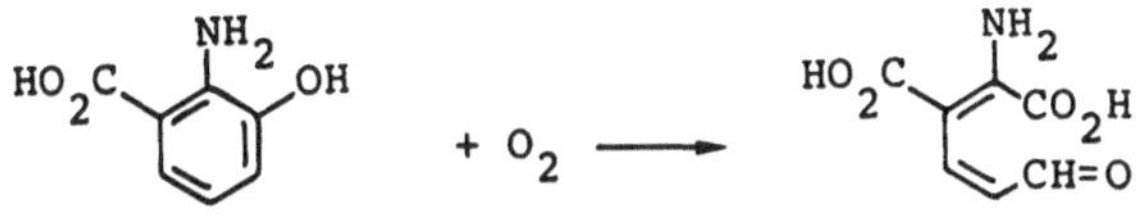

$$\text{(structure)} + O_2 \longrightarrow \text{(structure)} \qquad (24)$$

(11) Dihydroxypyridine dioxygenase (E.C. 1.13.11.9 – 2,5-Dihydroxy-pyridine: oxygen 5,6-oxidoreductase (decyclizing, amide hydrolyzing) (113)

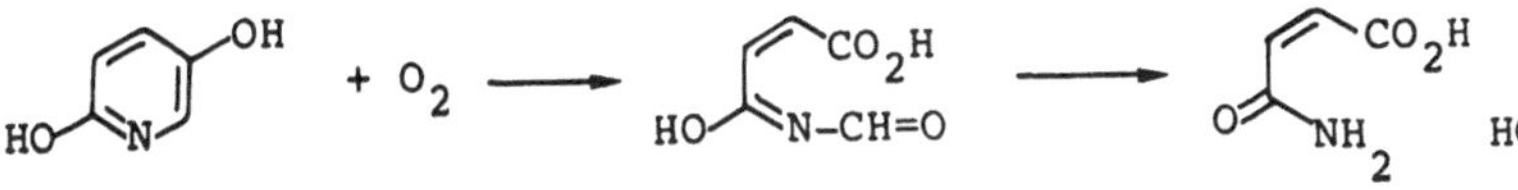

$$\text{(structure)} + O_2 \longrightarrow \text{(structure)} \longrightarrow \text{(structure)} \quad HCO_2H \qquad (25)$$

All the foregoing enzymes are assumed to be dioxygenases, although labelling studies demonstrating incorporation of both atoms of oxygen have only been carried out on homogentisate (114), hydroxyanthranilate (115), and dihydroxypyridine (113) dioxygenases.

Co-factor studies have been performed for all the enzymes listed, except for dihydroxybenzoate extradiol dioxygenase. Tests for dependence on ferrous ions and reducing agents and of inhibition by ferrous chelators and oxidizing agents indicate the presence of iron in the ferrous state in the active forms of metapyrocatechase (116), protocatechuate (83, 117), dihydroxykynurenate (2), homoprotocatechuate (118), carboxy-

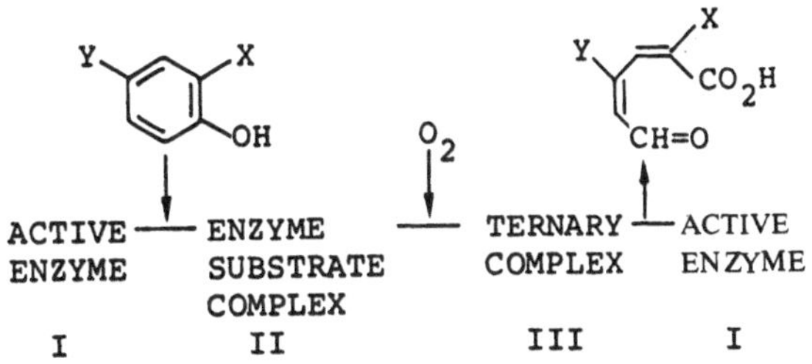

$$\text{(structure)} \qquad O_2 \qquad \text{(structure)} \qquad (26)$$

ACTIVE ENZYME	ENZYME SUBSTRATE COMPLEX	TERNARY COMPLEX	ACTIVE ENZYME
I	II	III	I

ethylcatechol (*119*), steroid (*120*), gentisate (*121*), homogentisate (*122, 123*), hydroxyanthranilate (*123, 124*) and dihydropyridine (*113, 125*) dioxygenases.

Notwithstanding this abundance of information, only five of these enzymes have been studied in detail (*2, 83*). They show a sequential mechanism with an obligatory order of substrate binding. It transpires that metapyrocatechase- and probably protocatechuate- and hydroxy-anthranilate-cleaving dioxygenases bind the substrate before oxygen (eq. 26), whereas steroid and probably homoprotocatechuate dioxygenases bind their substrates only after the enzyme has been oxygenated (eq. 27) (*2, 83*). E.P.R. spectroscopy of the active forms of metapyrocatechase and protocatechuate extradiol dioxygenase and homoprotocatechuate dioxygenase show no peaks around $g = 4$, consistent with the ferrous oxidation state of the metal atom. Mössbauer spectroscopy of the active form of the protocatechuate cleaving enzyme shows it to contain low-spin ferrous ion which changes to high-spin ferric ion in the ternary complex III (eq. 26) (*2, 83*). E.P.R. spectroscopy of the same ternary complex III and of the homoprotocatechuate dioxygenase complex III (eq. 27) shows a signal at $g = 4.3$ indicative of high spin ferric ion.

$$(27)$$

Although the spectral information is incomplete, it emerges that some of the extradiol dioxygenases are different from the intradiol dioxygenases in two respects; (i) they contain ferrous and not ferric iron in the active enzyme and (ii) the obligatory binding order is different. Naturally, speculation on mechanisms is without much empirical basis. However, in view of the consistent superior migratory aptitude of acyl over the vinyl group, a reasonable working hypothesis is that the enzyme causes an anionic center to form at the C-3 position (**46→47**) (Scheme 23). Subsequent addition of an oxygen molecule (**48**) followed by Baeyer-Villiger type rearrangement (**48→49**) ultimately leads to the aldehydic carboxylic acid (**50**).

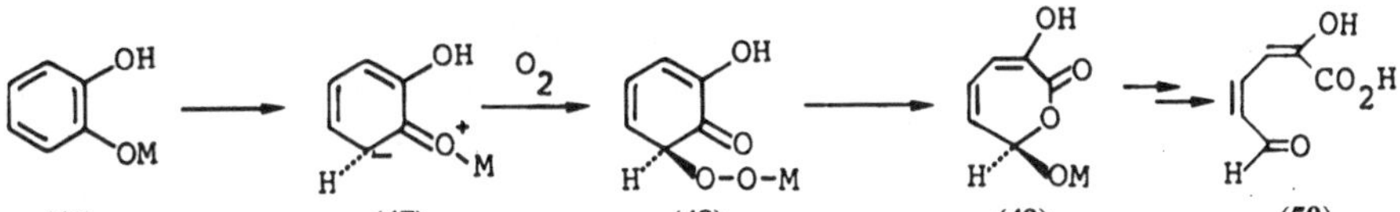

Scheme 23

3. Other Double Bond-Cleaving Dioxygenases

There is another class of enzymes which may resemble mechanistically the extradiol dioxygenases. None of their substrates however can form quinones. They also show diversity in their co-factors and include cupro- and ferro-proteins as well as two heme-containing enzymes.

(1) Peptidyl tryptophan dioxygenase (E.C. 1.13.11.26 – Peptidyl (L)-Tryptophan : oxygen 2,3-oxidoreductase (decyclizing) requires iron salts and a reducing agent (*i.e.* ferrous-requiring) (*126*).

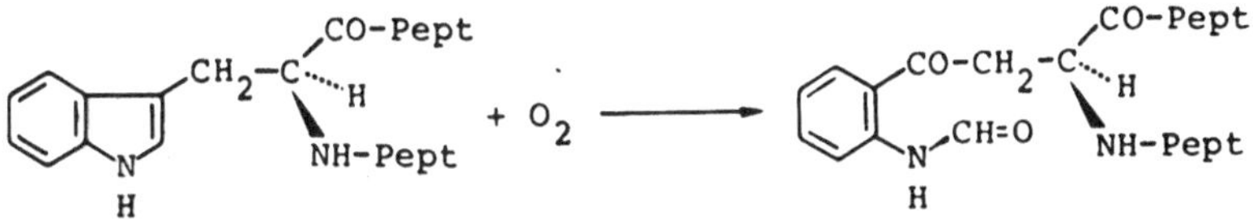

(28)

(2) Indole dioxygenase (E.C. 1.13.11.17 – Indole : oxygen 2,3-oxido-reductase (decyclizing). It appears to contain cuprous ions (*2*).

(29)

(3) Tryptophan dioxygenase (E.C. 1.13.11.11 – (L)-Tryptophan : oxygen 2,3-oxidoreductase (decyclizing) contains ferro-heme prosthetic groups as the only metallic co-factors of catalytic importance (*2*, *127*, *128*).

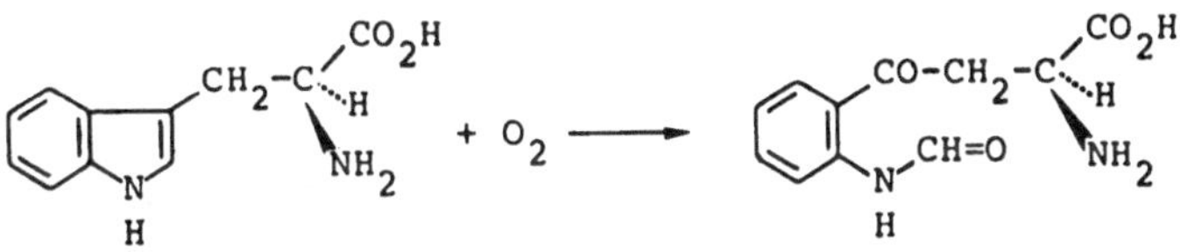

(30)

(4) Indolamine dioxygenase (– Tryptamine: oxygen 2,3-oxidoreductase (decyclizing). This enzyme which is either a ferroheme protein which uses oxygen as dioxygen substrate or a ferriheme protein using superoxide ion (129, 130) is non-specific in its choice of substrates (2, 127).

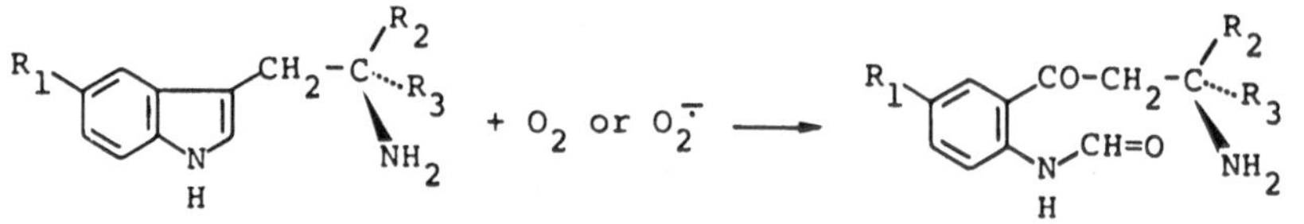

(31)

(5) Hydroxytryptophan dioxygenase (– (L)-5-Hydroxytryptophan: oxygen 2,3-oxidoreductase (decyclizing). The identity of this enzyme as being distinct from the preceding dioxygenase, is not certain. Its co-factor requirements are not yet known (131).

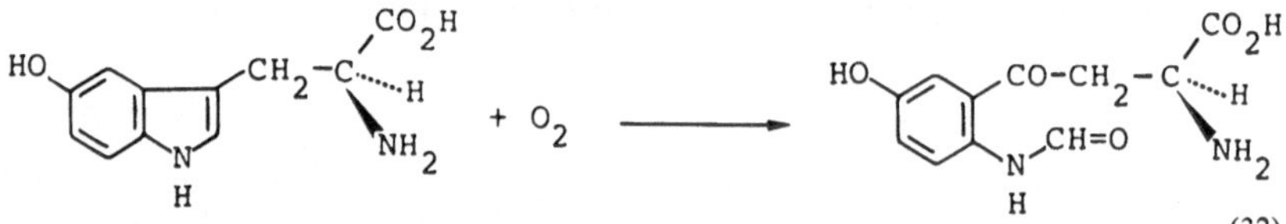

(32)

(6) Quercetinase (E.C. 1.13.11.24 – Quercetin : oxygen 2,4-oxidorectuct-ase (decyclizing, decarbonylating) which is copper-containing and incorporates two atoms from molecular oxygen into the 2- and 4-positions, suggesting the mechanism shown below (132).

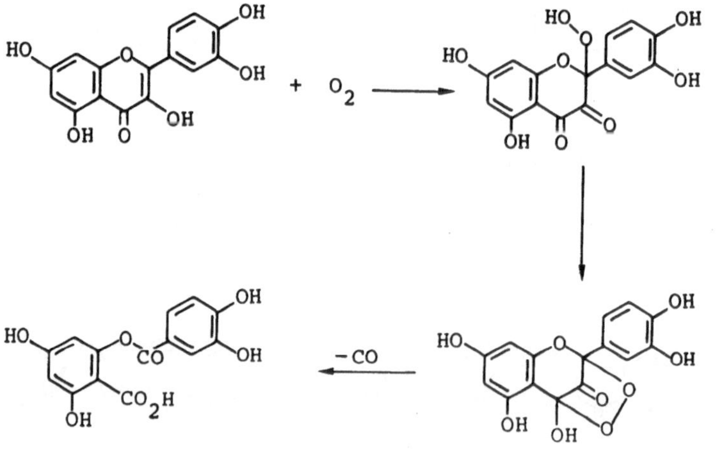

(33)

(7) Heme oxygenase (E.C. 1.14.99.3 – Heme, hydrogen donor: oxygen oxidoreductase (α-methene oxidizing, decyclizing). This enzyme-complex contains a dioxygenase and a mono-oxygenase which does not use cytochrome P-450 (*133*). Initially, oxyheme is formed (*134*). Subsequently, dioxygenase-catalyzed oxygenation and decarbonyl-ation occurs. Labelling studies reveal that only one of the amide carbonyl functions derives from dioxygen while the other comes from water (*133, 135, 136*). The same reaction can also be accomplished with mixtures of ferrous ion, ascorbate and oxygen to give the same labelling pattern (*134—138*).

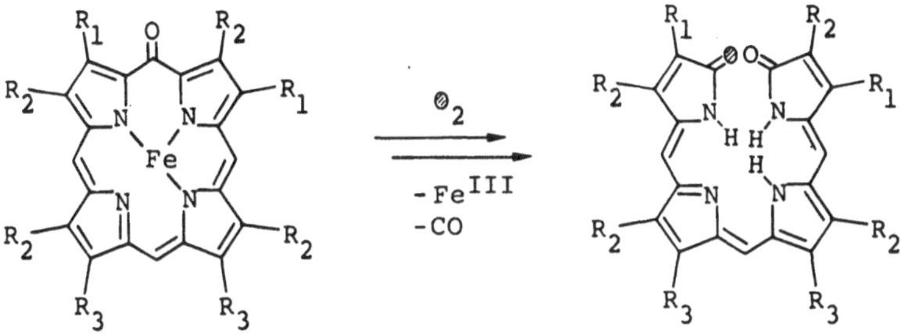

(34)

All the preceding substrates, even oxyheme, are expected to form carbanions readily. In fact, under solution conditions, in a mixture of *t*-butanol, dimethylformamide and base, the indole and flavonol derivat-ives undero oxygenation biomimetically to cleave the C-2, C-3 bond. Like indole itself, the C-3-substituted indoles, on enzymic catalysis, are structurally pre-disposed to yield tetrahedral carbanionic centers at C-3, suitably "soft" for easy and efficient addition to triplet dioxygen (*139*). The resulting peroxide (**53**) (Scheme 24) could possibly cleave its C-2, C-3 bond by the expedient of hydration (**54**) followed by cleavage to the keto-aldehyde (**55**).

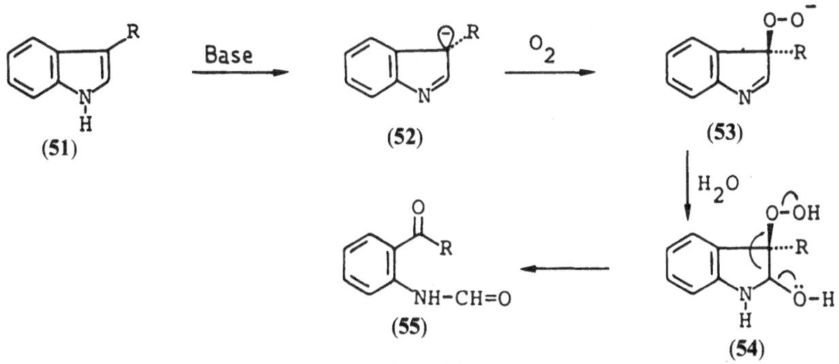

Scheme 24

Oxygenation of flavonol-type molecules (see eq. 33) proceeds similarly. However, cyclization to the five-membered ring lactone occurs which fragments liberating carbon monoxide. Oxygenation of oxyheme (eq. 34) clearly occurs at the carbon atom adjacent to the carbonyl group. The labelling pattern excludes the formation of a similar cyclic lactone and decarbonylation probably is caused by a linear fragmentation process (*105*).

VI. The Luciferases

There are three distinct classes of luciferases which bring about bioluminescence in fireflies, in certain aquatic species (both dioxygenases), and in bacteria (mono-oxygenase) (*140, 141*).

(1) Firefly luciferase (E.C. 1.13.12.7 – Photinus luciferin : oxygen 4-oxidoreductase (decarboxylating, ATP-hydrolyzing).

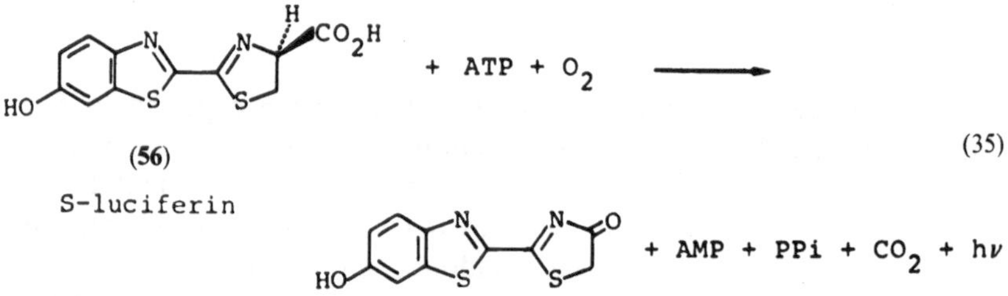

(56)

S-luciferin

(57)

(35)

This enzyme has been the subject of considerable study (*142, 143*). Its main catalytic function is probably to transfer the adenyl group from ATP to the carboxyl group of the luciferin (**56**) to give (**58**) (Scheme 25). Such functionalization is of prime importance as it creates a good leaving group which is necessary for the key formation of the light-producing element, the dioxetanone ring. Subsequently, oxygenation followed by decarboxylation produces inorganic pyrophosphate (PPi), oxyluciferin (**57**) and chemiluminescence.

The enzymes shows no requirement for metal ions, and there are indications that the active site is extremely hydrophobic (*142*). In fact, the luciferyl adenylate (**58**) undergoes rapid, spontaneous oxygenation in organic solvents to give a hydroperoxide (**60**) which eventually decomposes decarboxylatively to the oxyluciferin (**57**), accompanied by luminescence (*142*). There is no need for oxygen activation and the enzyme may catalyze oxygenation by simply creating a "soft" anionic center in the substrate $(58 \to 59)$. The resulting hydroperoxide (60) cyclizes to dioxetanone (61), thanks to the efficiency of the leaving group (144, 145). Scission then occurs spontaneously to carbon dioxide and oxyluciferin (57) with light emission (Scheme 25).

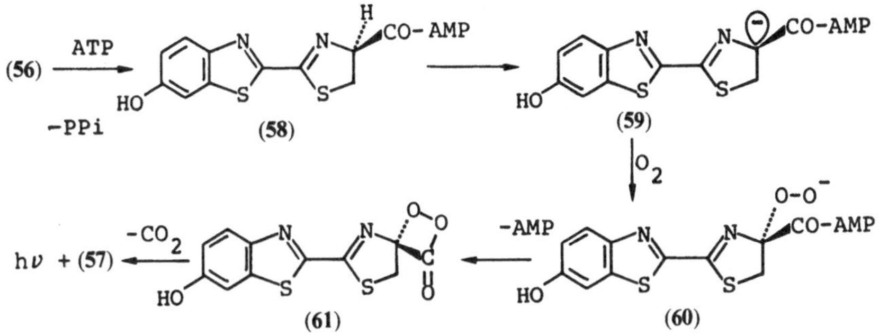

Scheme 25

A detailed mechanism has recently been advanced to explain how decarboxylation of (61) efficiently generates singlet-excited oxyluciferin (57) (146, 147).

(2) *Renilla* and *Cypridina* luciferases (E.C. 1.13.12.5 – *Renilla* luciferin: oxygen 2-oxidoreductase (decarboxylating) and E.C. 1.3.12.6 – *Cypridina* luciferin: oxygen 2-oxidoreductase (decarboxylating)).

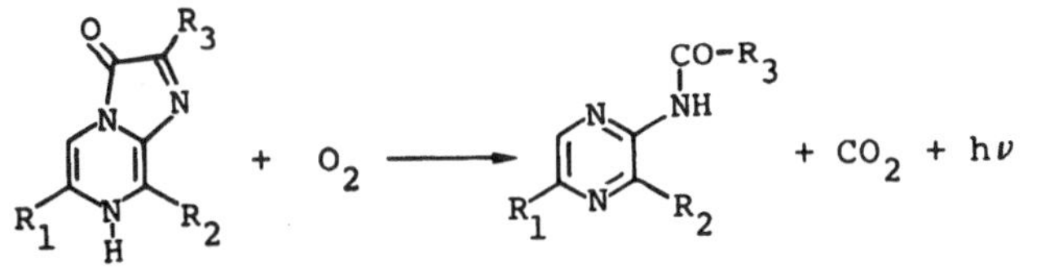

	$R_1 =$	$R_2 =$	$R_3 =$
(62) Renilla luciferin			
(63) Cypridina luciferin			

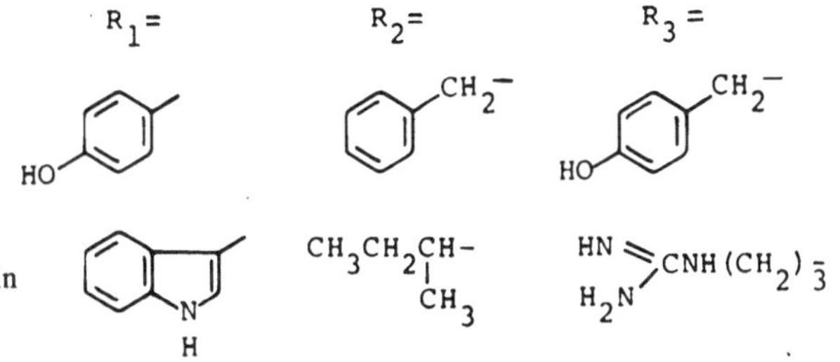

The anthozoan sea pansy, *Renilla reniformis* (*148*), the decapod shrimps *Oplophorus spinosus* (*149*), *Oplophorus gracilorostris* (*150*) and *Heterocarpus laevigatus* (*149*) as well as the myctophid fish *Neoscopelus* (*151*) all contain the same luciferin (**62**). The luciferin of the squid *Watasenia scintillans* also has the same structure, but possesses terminal sulfate ($-OSO_3H$) instead of hydroxyl groups (*152*). The luciferin of the crustacean *Cypridina* (**63**) contains a chiral center and different side-chains (*153*), but still possesses, as do all these luciferins, the same central, condensed 1,4-dihydropyrazo-imidazolone structure. Chemi-luminescence is observed as a consequence of the oxygenation and decarboxylation with rupture of the five-membered ring of the luciferins (eq. 36). Clearly, the same mechanism is followed by all of them.

Studies with *Cypridina* (*154*), *Oplophorus* (*150*) and *Renilla* (*155*) luciferases have uncovered the following steps (Scheme 26) (*156*). First, the appropriate luciferase deprotonates the luciferin (**62**) to its tetra-hedral carbanion (**64**). Reaction of the latter with oxygen yields the hydroperoxide (**65**). Cyclization affords the so-called high energy inter-mediate, the dioxetanone (**66**), which on cleavage loses carbon dioxide to give the amide (**67**) accompanied by chemiluminescence.

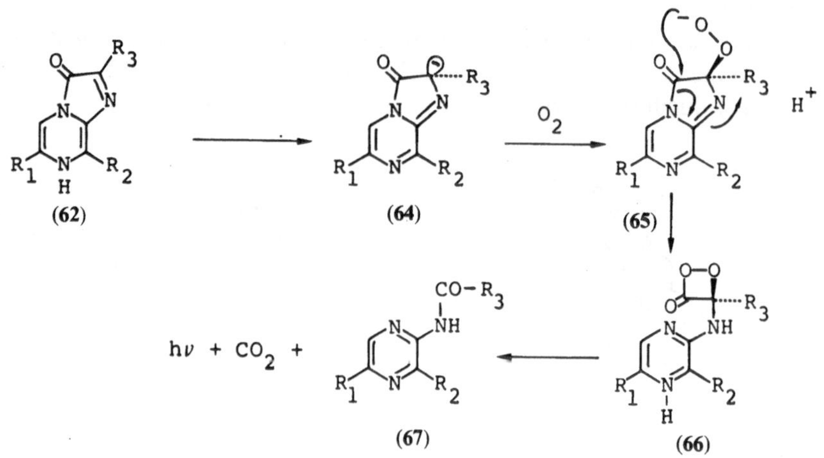

Scheme 26

Just as for the firefly luciferyl derivative (**58**), these luciferins (*e.g.* **62**) undergo rapid oxygenation even in non-aqueous solvents decomposing thereafter with chemiluminescent decarboxylation (*156, 157*). Again, there is no evidence for the involvement of metal ion co-factors. This is hardly surprising as there is no need since the 1,4-dihydropyrazine ring is predisposed to form a "soft" carbanion.

Similar behavior is observed with some of the flavoprotein mono-oxygenases, which also do not use a metal co-factor. The reduced flavin (68) has a structure which resembles that of luciferin (62) and reacts readily with molecular oxygen, through the intermediacy of its carbanion which forms charge-transfer intermediates leading to the hydroperoxide ion rather than to superoxide radical ion and pyrazyl radicals (158). Although the precise point of attachment of oxygen is controversial, the principle remains the same, namely the formation of a non-delocalized carbanion (69 or 70) (159—161).

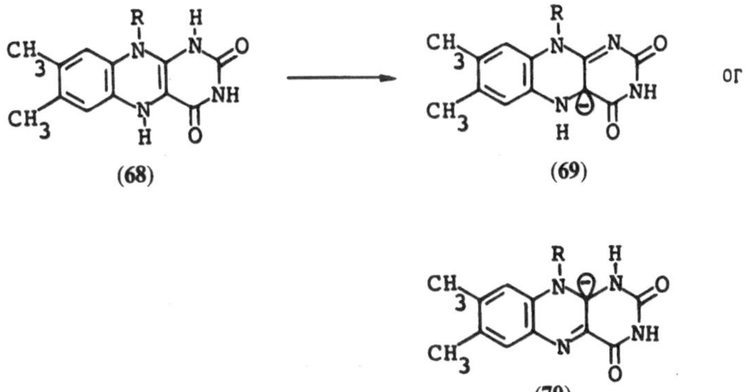

(68)   (69)   or

(70)

Scheme 27

VII. Peroxidizing Dioxygenases

Although there are a great many naturally occurring peroxides and hydroperoxides which presumably arise from dioxygenation of un-saturated precursors (162), only two of the enzymes responsible have been classified so far (163).

(1) Lipoxygenase (E.C. 1.13.11.12 – Linoleate: oxygen oxidoreductase).

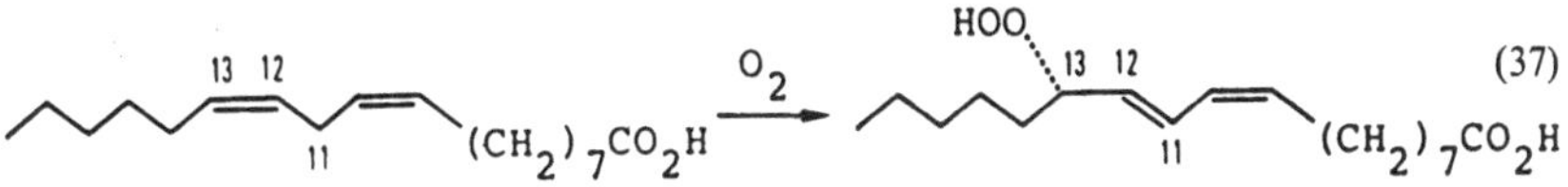

$$(37)$$

This enzyme is found in a wide variety of plants and exists in various iso-enzymic forms having different regioselectivities (163).

Two of the four iso-enzymes of soybean, lipoxygenase-1 (eq. 37) and -2 give the 13-hydroperoxy- and a mixture of the 9- and 13-hydro-peroxydienoic acids, respectively (163). Another iso-enzyme gives the same two products, but in a ratio which is dependent on pH (*164*). It appears that these enzymes share a common catalytic mechanism, but give different products reflecting minor conformational differences.

Any valid proposal for the mechanism of catalysis by lipoxygenase-1 must take into account the following observations (*163, 165, 166*).

(i) The native enzyme is inactive and contains a non-heme, low-spin ferrous ion.

(ii) The hydroperoxide product converts the native enzyme into its active, high-spin ferric form.

(iii) This activation is reversed by the binding of the substrate, linoleate.

(iv) The hydroperoxide product undergoes catalyzed decomposition, especially in the absence of oxygen. Under anaerobic conditions, mixtures of linoleic acid and its hydroperoxide give products strongly suggestive of free radical intermediates (*167, 168*).

(v) Spin-trapping studies (*166, 168, 169*), experiments with radical scavengers (*169*) and autoxidizable compounds (*166*) indicate that linoleic acid free radicals are involved and escape from time to time from the active site. Under anaerobic conditions, the linoleic acid radicals convert the inactive ferro-enzyme to the ferri-enzyme substrate complex (*170*).

(vi) There is a large primary kinetic isotope effect for removal of the C-11 hydrogen atom.

(vii) Only long-chain fatty acids possessing a *cis,cis*-1,4-pentadiene unit separated from the terminal methyl group by a chain of four saturated carbon atoms are peroxidized. This suggests specific hydrophobic binding.

Another important feature of the reaction is the stereochemistry of hydroperoxidation shown in eq. 37. Oxygen addition to the sp^2 hybridized C-13 atom is enantioface-differentiating (*171*), giving only the (*S*)-hydroperoxide. In addition, labelling studies have shown that C-11 hydrogen atom removal is enantiotopos-differentiating as only the pro-*S* hydrogen (*163, 165*) is removed. To attain this degree of stereocontrol, the substrate must obviously remain in a fixed conformation during hydrogen abstraction and oxygen addition. As the migrating double bond becomes *trans*, it follows that the pro-*S* substituent must be removed from the

$$(38)$$

opposite face of the substrate to which oxygen is added. Even if these two events are not synchronous, the net result is nonetheless antarafacial (eq. 38). This finding invalidates earlier proposals of singlet oxygen participation. Both oxygen addition and hydrogen abstraction occur on the same face of the olefin in the "ene" reaction of singlet oxygen (17).

There are also other mechanistic implications. As each lipoxygenase contains only one, tightly bound iron atom, the stereochemistry of lipoxygenation must preclude any mechanism which requires direct metal participation in both acts of hydrogen abstraction and oxygen addition. Moreover, it is generally agreed that the reaction is free radical in nature (165, 172, 173) and that the enzyme is activated to its ferric oxidation state by the hydroperoxide product in a Haber-Weiss reaction (eq. 39) (174).

$$E-Fe^{II} + R-O-OH \longrightarrow Fe^{III} + RO^{\cdot} + {}^-OH \longrightarrow \text{other products} \qquad (39)$$

GIBIAN and GALAWAY (165) believe that the enzyme is unlikely to have at its disposal a base of sufficient strength to remove the pro-S hydrogen substituent as a proton, nor would the ferric ion have sufficient oxidizing power to abstract it as a hydrogen radical. Instead, they propose the initial oxidation of the C-12; C-13 double bond by an electron transfer to the ferric ion (eq. 40). The resulting radical cation (Stages I→II of Scheme 28) would have a lower pK_a for its C-11-

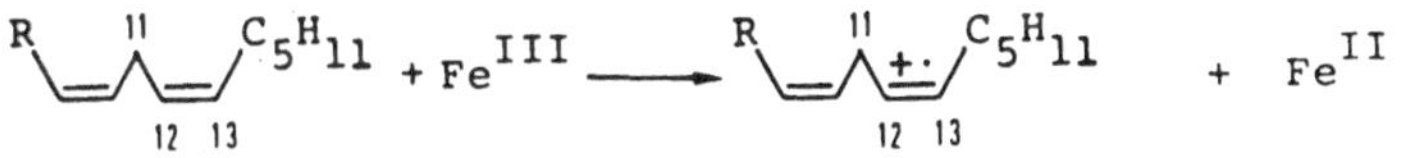

$$\underset{12\ \ 13}{R \diagdown\!\!\diagup\!\!\diagdown\!\!\diagup^{C_5H_{11}}_{11}} + Fe^{III} \longrightarrow \underset{12\ \ 13}{R\diagdown\!\!\diagup\!\!\diagdown\!\!\diagup^{C_5H_{11}}_{11 \quad \pm\cdot}} + Fe^{II} \qquad (40)$$

hydrogen substituent, thereby permitting its removal as a proton by a nitrogenous base in the rate-determining step (Stages II→III). Meanwhile, the ion co-factor, having been reduced to its ferrous state, is now ready

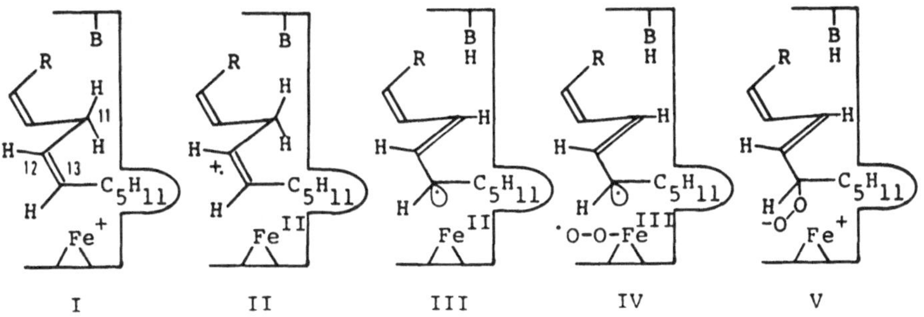

Scheme 28

to co-ordinate with an incoming dioxygen molecule (Stage IV) and to place it on the nearby terminus of the substrate allylic radical (Stage V).

This scheme is consistent with the biochemical and stereochemical properties of lipoxygenation and also explains the specificity for the 13-hydroperoxy product in the Haber-Weiss reactivation of the ferro-enzyme (eq. 39). This hydroperoxide will adopt almost the same spatial relationship with Fe(II) of the inactive enzyme as does the peroxide product with Fe(III) in Stage V. Once again, it is to be noted that this mechanism represents catalysis by substrate rather than oxygen activation. The role of ferrous ion in Stage IV is not to activate oxygen, but to simply hold it ready for reaction with the allylic radical. A minor points concerns the nature of the substrate in Stage III. Here it is shown as an allylic radical, but it could be regarded as a tetrahedral carbanion attached to ferric ion.

(2) Prostaglandin cyclooxygenase (E.C. 1.14.99.1 – 8,11,14-Eicosa-
   trienoate: oxygen oxidoreductase).

$$ (41) $$

This enzyme has recently been isolated from the prostaglandin synthetase complex (175) and has three C-20 fatty acids as substrates similar to the C-18 substrates of lipoxygenases, but containing one, two or three extra non-conjugated cis double bonds (163). The overall trans-formation (eq. 41) resembles lipoxygenation, except that double dioxy-genation occurs, giving an endo-peroxy hydroperoxide (163, 176).

Although lipoxygenases are found only in plants, and prostaglandins are limited to the animal kingdom, there are several striking similarities between the two enzyme systems.

(i) Both contain a non-heme iron atom (177). Both require hydro-peroxides for activation. Co-oxidation and autoxidizable compounds occur (178) and they share common inhibitors (175, 179).

(ii) Prostaglandin cyclooxygenase has trienoic acids as natural sub-strates, but also converts dienoic acids to mono-hydroperoxides in the same way as lipoxygenase (180). Furthermore, it does so with the same regio and enantiotopic selectivity (181). Lipoxygenase-1 gives a double hydroperoxide of a tetraenoic acid with the same regioselectivity as prostaglandin cyclooxygenase (182, 183) and with methyl linoleate gives cyclic peroxides (*184*). Lipoxygenase-2, on the other hand, converts a tetraenoic acid substrate of prostaglandin cyclooxygenase into an *endo*-peroxyhydroperoxide with the same regioselectivity, if not the same stereoselectivity as the latter enzyme (*185*).

(iii) The stereochemistry of cyclo-oxygenation and lipoxygenation are clearly similar. In cyclo-oxygenation the pro-*S* hydrogen at C-13 is specifically removed (*163, 176*).

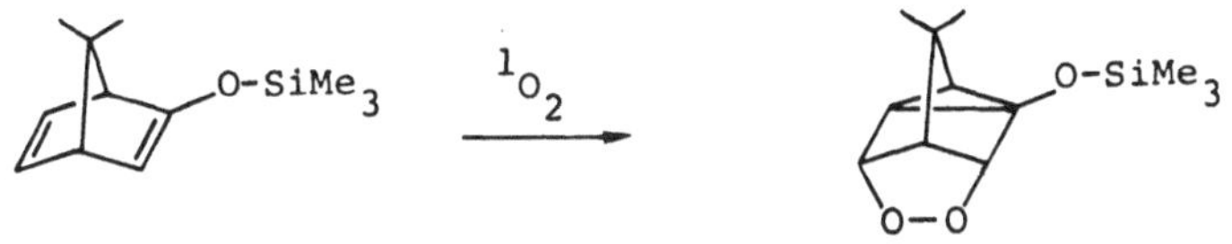

Scheme 29

The mechanism of endoperoxide formation is clearly puzzling and has led many authors to propose the participation of singlet oxygen (*176*). To the best of our knowledge, however, there is only one example of a homo-Diels-Alder addition of singlet oxygen. This is shown in Scheme 29 (*186*) and is the consequence of a polar substituent and steric hindrance which is absent in the substrates of cyclo-oxygenase. Precedents are better for a free radical cyclization (*11, 187*). We believe that a sequence similar to that proposed for lipoxygenation (*188*) would be more consistent with the biochemical facts. The main difference from lipoxygenation is that two molecules of oxygen add to different parts of the substrate. As in lipoxygenation, ferrous ion binds the molecule at C-11. The second site of oxygen attachment (C-15) is probably too

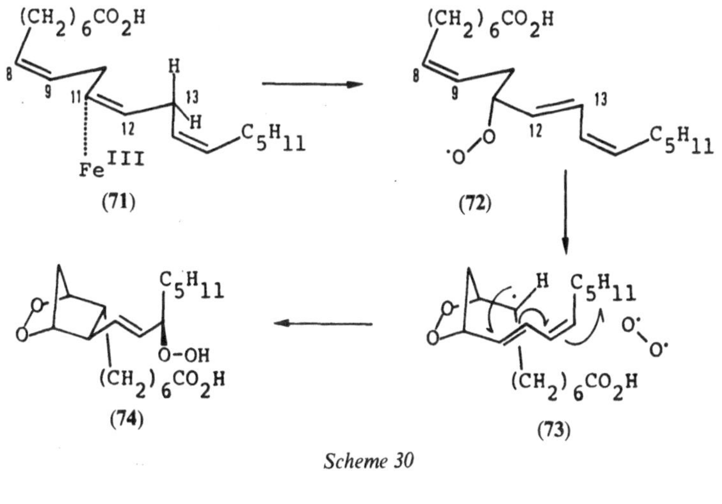

Scheme 30

far removed from this metal ion for the purposes of binding. However, a second type of metal co-factors exists in prostaglandin cyclo-oxygenase in the form of heme-bound iron (*177*), which may function as a binding site for the second oxygen molecule. This would explain the lack of regio-selectivity in polyperoxidation by lipoxygenases-1 and -2 which contain only one non-heme iron atom per active site.

Thus oxygenation occurs at C-11 of the activated substrate (**71**) (Scheme 30), giving the peroxy radical (**72**) which promptly cyclizes to the carbon radical (**73**). This last radical cyclizes in turn by conjugate addition to the diene moiety capturing a molecule of oxygen to generate the *trans* substituted prostaglandin (**74**).

VIII. Miscellaneous Dioxygenases

There are seventeen assorted dioxygenases which have not been studied sufficiently. Consequently mechanistic proposals are tentative.

1. External Flavoprotein Dioxygenases (*189*)

These are pyridine nucleotide-linked flavoprotein complexes that require ferrous ions for activity and which catalyze the double hydroxy-lation of aromatic nuclei in a highly stereoselective manner (*190*). Often the resulting diol undergoes dehydrogenation to the corresponding *o*-diphenol by the oxidized pyridine nucleotide (eq. 42) (*2*). Occasionally, the diol decomposes due to the inherent instability created by substi-tuents.

$$R \quad + NAD(P)H + O_2 \quad \xrightarrow{H^+} \quad \overset{R\ (R)}{\underset{H\ (S)}{}} \quad + NAD(P)^+ \quad \longrightarrow \quad R\text{-}OH \quad + NAD(P)H \qquad (42)$$

The most studied members of this class are listed below.

(1) Benzene dioxygenase (E.C. 1.14.12.3 – Benzene, NADH : oxygen 1,2-oxidoreductase. This enzyme catalyzes double hydroxylation of benzene and some alkyl benzenes. It requires ferrous ions and NADH (*191, 192*) and generally stops at the *cis*-1,2-diol stage (*193*).

$$\bigcirc \quad + NADH + O_2 \quad \xrightarrow{H^+} \quad \overset{H}{\underset{H}{\bigcirc}}\text{-}OH \quad + NAD^+ \qquad (43)$$

(2) Naphthalene dioxygenase (E.C. 1.14.12. – – Naphthalene, NAD(P)H: oxygen 1,2-oxidoreductase. Labelling studies have shown that both hydroxyl oxygens are derived from molecular oxygen (*194*). A tightly bound ferrous ion is also present (*195*).

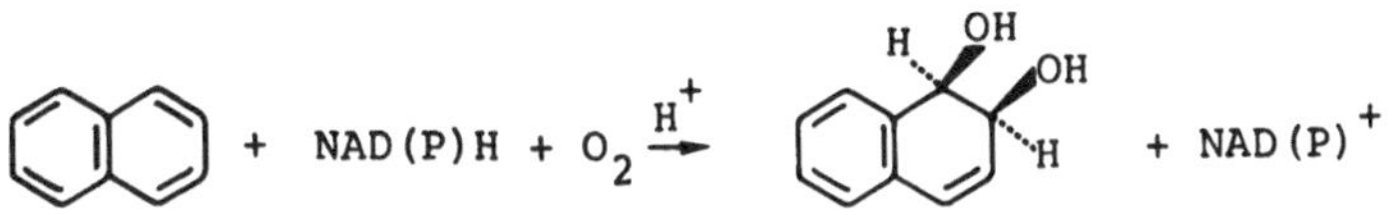

$$+ \ NAD(P)H \ + \ O_2 \ \xrightarrow{H^+} \ \ \ \ \ \ \ \ + \ NAD(P)^+ \quad (44)$$

(3) Toluene dioxygenase (E.C. 1.14.12. – – Toluene, NADH: oxygen 2,3-oxidoreductase (*196, 197*).

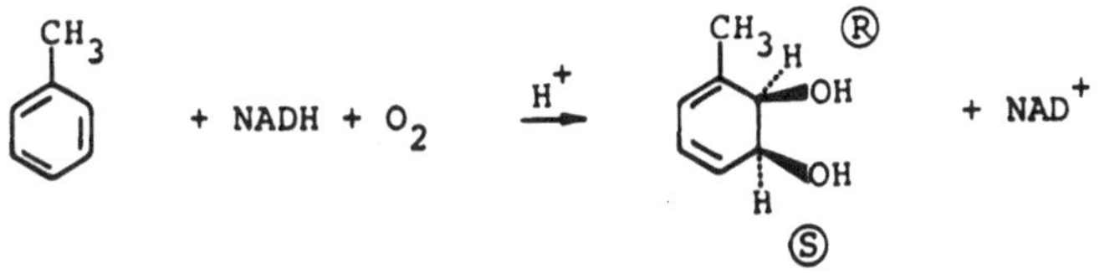

$$+ \ NADH \ + \ O_2 \ \xrightarrow{H^+} \ \ \ \ \ \ \ \ + \ NAD^+ \quad (45)$$

(4) Pyrazon dioxygenase (E.C. 1.14.12. – – 5-Amino-4-chloro-2-phenyl-2H-pyridazin-3-one, NADH: oxygen 2′,3′-oxidoreductase) (*198*).

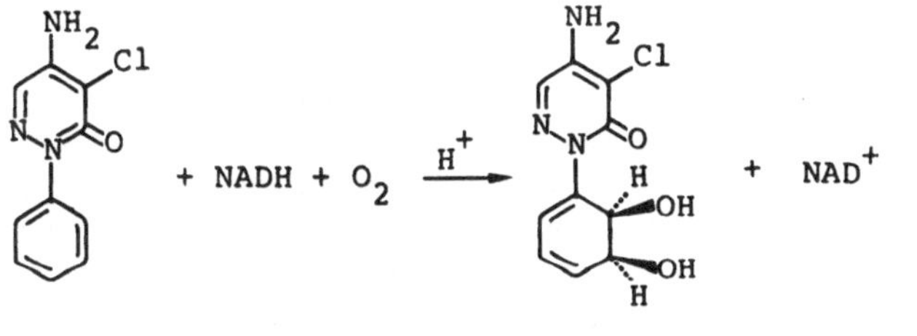

$$+ \ NADH \ + \ O_2 \ \xrightarrow{H^+} \ \ \ \ \ \ \ \ + \ NAD^+ \quad (46)$$

(5) Kynurenate 7,8-hydroxylase (E.C. 1.14.99.2 – Kynurenate, NADH: oxygen 7,8-oxidoreductase) (*199*).

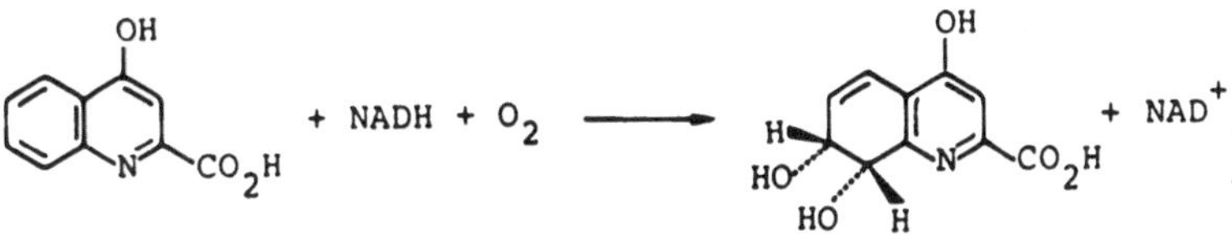

$$+ \ NADH \ + \ O_2 \ \longrightarrow \ \ \ \ \ \ \ \ + \ NAD^+ \quad (47)$$

(6) Benzoate 1,2-hydroxylase (E.C. 1.13.99.2 – Benzoate, NADH: oxygen, 1,2-oxidoreductase (decarboxylating). The intermediate diol has been isolated and labelling studies have verified that molecular oxygen is the source of the two hydroxyl groups (*200*).

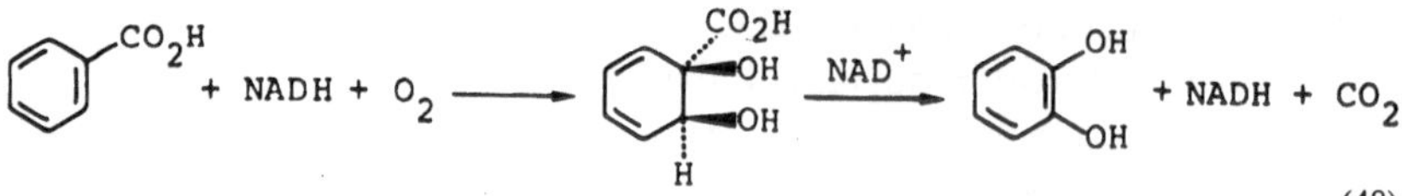

$$\text{(benzoic acid)} + NADH + O_2 \longrightarrow \text{(cyclohexadiene diol)} \xrightarrow{NAD^+} \text{(catechol)} + NADH + CO_2$$

(48)

(7) Anthranilate 1,2-hydroxylase (E.C. 1.14.12.1 – Anthranilate, NAD(P)H:oxygen 1,2-oxidoreductase (deaminating, decarboxylating) (201—203).

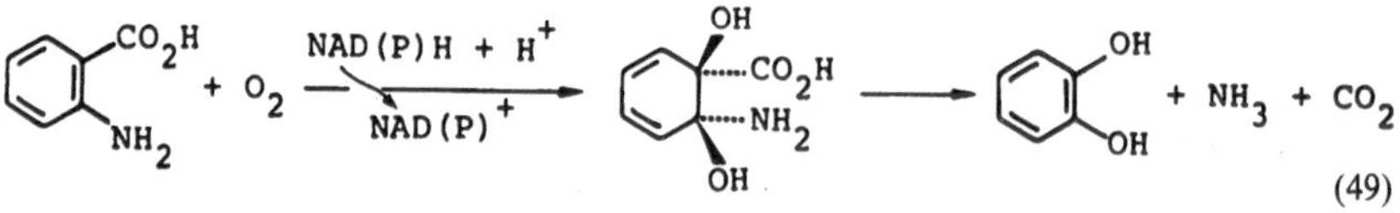

$$\text{(anthranilate)} + O_2 \xrightarrow[\text{NAD(P)}^+]{NAD(P)H + H^+} \text{(intermediate)} \longrightarrow \text{(catechol)} + NH_3 + CO_2$$

(49)

(8) Anthranilate 2,3-hydroxylase (E.C. 1.14.12.2 – Anthranilate, NADPH:oxygen 2,3-oxidoreductase (deaminating) (204, 205).

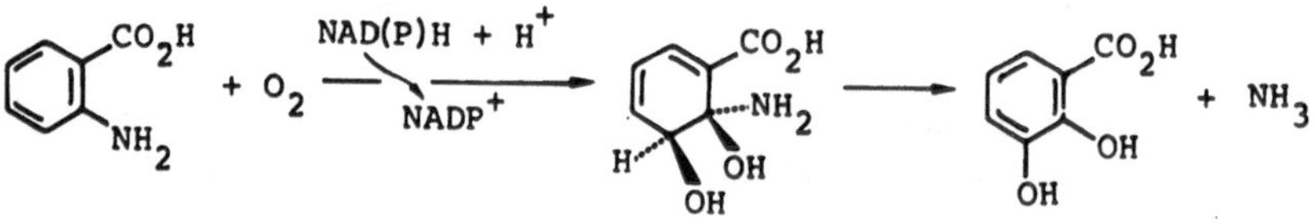

$$\text{(anthranilate)} + O_2 \xrightarrow[\text{NADP}^+]{NAD(P)H + H^+} \text{(intermediate)} \longrightarrow \text{(2,3-dihydroxybenzoate)} + NH_3$$

(50)

Recent studies with toluene dioxygenase (196, 206), benzene dioxygenase (207, 208), benzoate hydroxylase (209) and pyrazon and naphthalene dioxygenases (198), indicate the presence of three protein components in the enzyme. Together they make up an electron-transfer (Scheme 31) chain which is reminiscent of that found in certain mono-oxygenase systems (8, 189, 210). In benzene dioxygenase, the intermediate electron-carrying protein and the terminal dioxygenase contain one and two iron-sulfur clusters respectively (208). It is not known whether this last

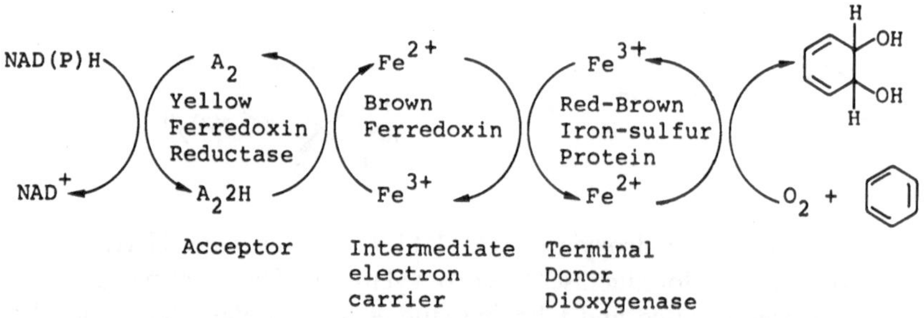

Acceptor	Intermediate electron carrier	Terminal Donor Dioxygenase

Scheme 31

pair of clusters occupies one active site, thereby permitting transfer of an electron pair or two active sites necessitating two sequential one-electron transfers (208). The aromatic nucleus can be conceived as being initially reduced to a radical anion, which then undergoes consecutive addition of molecular oxygen, a proton, an electron and finally a proton after an undefined formal attachment of the hydroxy cation to the new carbanionic center (Scheme 32). The observation of a small negative Hammett reaction constant for the dioxygenation of substituted benzoates is in agreement with this mechanism (211).

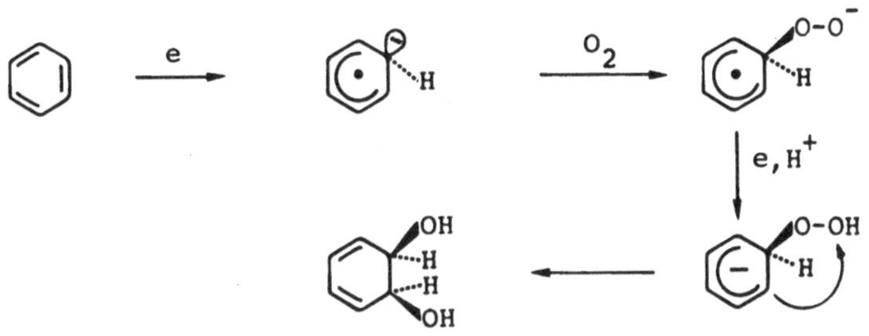

Scheme 32

Although labelling studies have shown that both hydroxyl groups are derived from molecular oxygen, a mechanism of double mono-oxygenation cannot be ruled out until the stoichiometry is accurately determined. The existence of such a possibility is revealed by benzoate dioxygenase which gives a mono-oxygenated product (76) when o-toluic acid (75) is used as substrate (Scheme 33) (200).

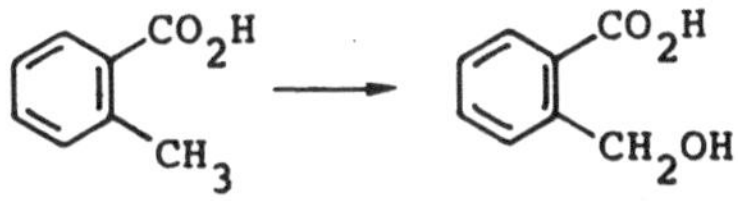

(75)   (76)

Scheme 33

There are two additional pyridine nucleotide-requiring flavo-enzymes for which the metal ion requirements are unknown (212) (see 9 and 10 below). Both cleave the pyridine nucleus at the bond adjacent to the hydroxyl substituent. However, the similarity with phenol-cleaving dioxy-genases is fortuitous. The reaction course derives from the inherent instability of the intermediate cis-diols rather than from a similar catalytic mechanism. Two different mechanisms have been advanced. The first, which is based on kinetic studies (213, 214), consists of reduction of the substrate to its anion followed by its oxygenation (eq. 51). The

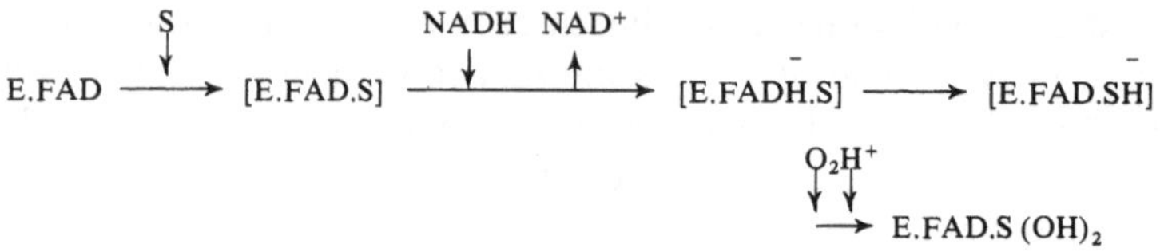

$$\text{(51)}$$

second involves the flavin (Fl)-mediated reduction of molecular oxygen

$$Fl^- + O_2 \rightleftharpoons Fl-O-O^- \rightleftharpoons Fl\cdot + O_2^{\overline{\cdot}} \qquad (52)$$

$$Fl\cdot + S^- \rightleftharpoons Fl^- + S\cdot \xrightarrow{\;O_2^{\overline{\cdot}}\;} S(O_2)^- \qquad (53)$$

as well as the flavin-mediated oxidation of the substrate. Combination of the two resulting radicals affords the hydroperoxide (eqs. 52, 53) (215). In both mechanisms, the outcome is the formation of a diol.

(9) Methylhydroxypyridinecarboxylate dioxygenase (E.C. 1.14.12.4 – 2-Methyl-3-hydroxypyridine-5-carboxylate, NADH: oxygen 2,3-oxidoreductase (decyclizing).

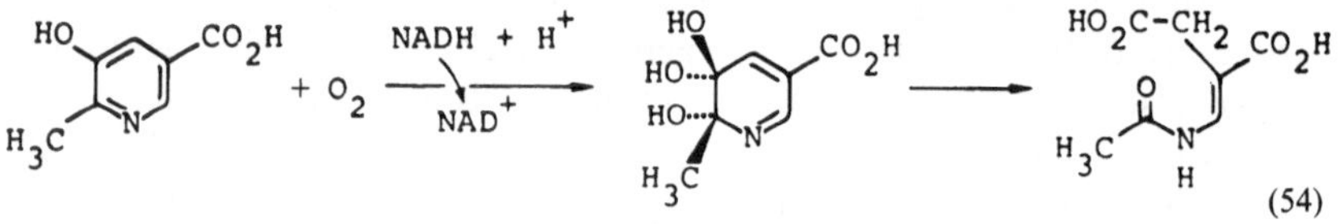

$$\text{(54)}$$

(10) Pyridoxate dioxygenase (E.C. 1.14.12.5 – 5-Pyridoxate, NADPH: oxygen 2,3-oxidoreductase (decyclizing).

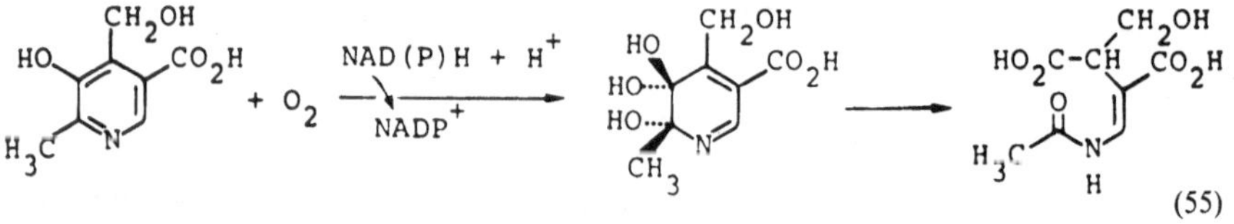

$$\text{(55)}$$

2. Sulfur Oxidizing Dioxygenases

There are three dioxygenases that oxidize sulfur.

(1) Sulfur dioxygenase (E.C. 1.13.11.18 – Sulfur: oxygen oxidoreductase) (216, 217).

$$2\,S + O_2 \longrightarrow H_2S_2O_3 \qquad\qquad (56)$$

Thiosulfate

(2) Cysteamine dioxygenase (E.C. 1.13.11.19 – Cysteamine : oxygen oxi-doreductase) (*218, 219*).

$$\begin{matrix} CH_2SH \\ | \\ CH_2NH_2 \end{matrix} + O_2 \longrightarrow \begin{matrix} CH_2SO_2H \\ | \\ CH_2NH_2 \end{matrix} \qquad\qquad (57)$$

Hypotaurine

(3) Cysteine dioxygenase (E.C. 1.13.11.20 – (L)-Cysteine : oxygen oxido-reductase (*220—222*).

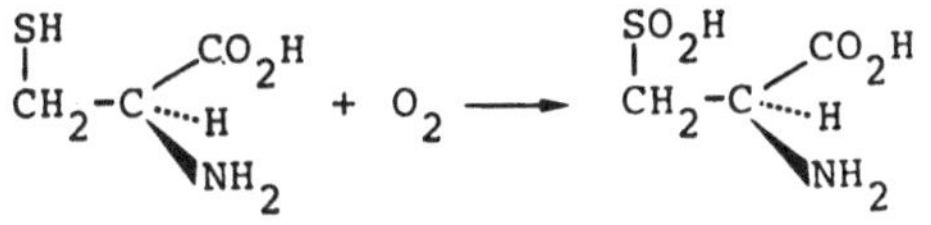

$$\begin{matrix} SH \\ | \\ CH_2-C \end{matrix}\!\!\begin{matrix} CO_2H \\ \diagdown \!\! H \\ NH_2 \end{matrix} + O_2 \longrightarrow \begin{matrix} SO_2H \\ | \\ CH_2-C \end{matrix}\!\!\begin{matrix} CO_2H \\ \diagdown \!\! H \\ NH_2 \end{matrix} \qquad (58)$$

Despite the similar functions of the preceding enzymes, they may well make use of different mechanisms (*1, 2*). Although all require iron, cysteine dioxygenase requires the ferrous state and NADPH (*220*), whereas cysteamine dioxygenase uses high-spin ferric ion in the actual catalysis (*219*). The fact that the E.P.R. spectrum of the last enzyme is modified by addition of the substrate, namely cysteamine, under anaerobic conditions (*219*) demonstrates that it is the substrate which interacts with the metal co-factor.

3. Inositol Dioxygenase
(E.C. 1.13.99.1 – myo-Inositol: oxygen oxidoreductase)

This enzyme contains one gram atom of ferrous ion per mole of enzyme and has a specific requirement for cysteine (*123, 223—225*).

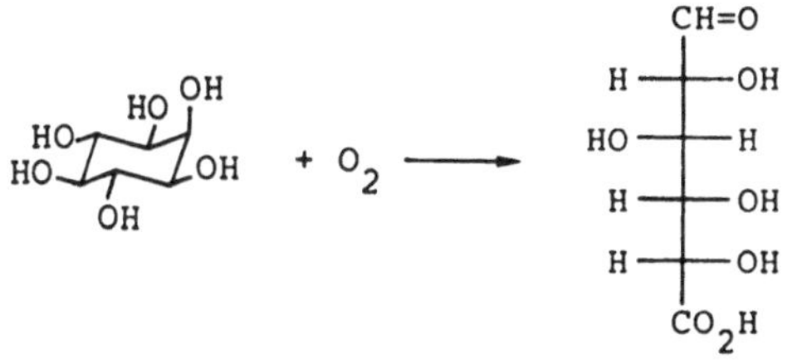

$$ (59) $$

D-Glucuronate

The carboxyl group of D-glucuronate only contains one oxygen atom derived from molecular oxygen (223). This does not invalidate dioxygenation since the aldehyde oxygen atoms undergo rapid exchange with those of water (123).

4. Nitropropane Dioxygenase
(E.C. 1.13.11. – – 2-Nitropropane: oxygen oxidoreductase)

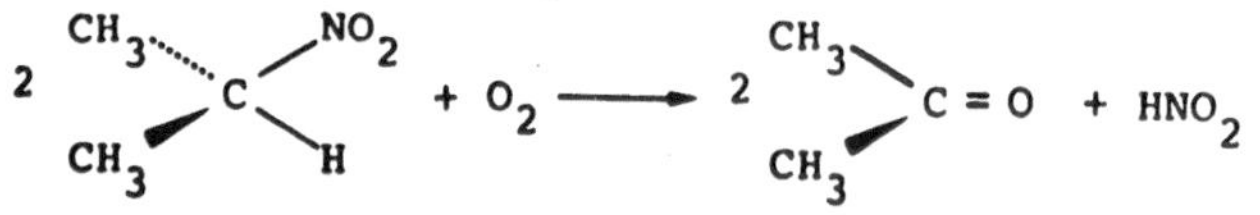

$$2 \quad \underset{CH_3}{\overset{CH_3}{>}}C\underset{H}{\overset{NO_2}{<}} + O_2 \longrightarrow 2 \quad \underset{CH_3}{\overset{CH_3}{>}}C=O + HNO_2 \qquad (60)$$

This flavo-enzyme contains one gram atom of non-heme ferric ion per mole of enzyme. Both atoms from molecular oxygen are incorporated into the acetone products (226). Arguments in favor of the intermediacy of superoxide ion have been suggested (75). Nevertheless, a carbanionic mechanism is equally feasible. The "softened" nitrocarbanion (77), on oxygenation, gives the hydroperoxide ion (78) which reacts with the parent molecule (79) to give the peroxide (80). Elimination of a molecule of nitrous acid yields the products (81) (Scheme 34).

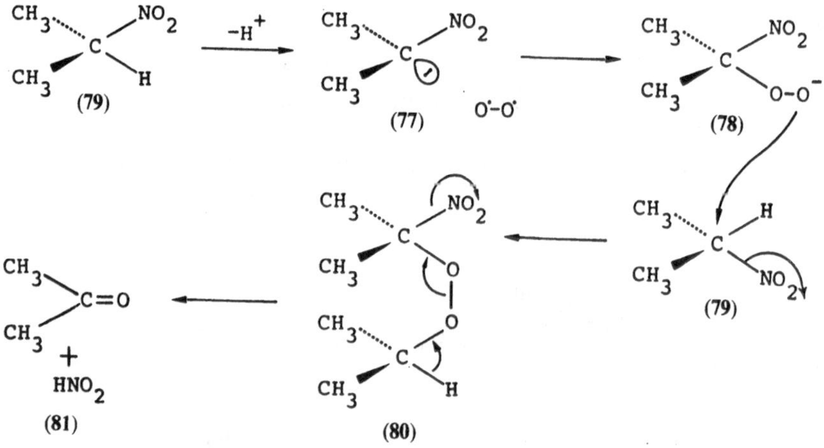

Scheme 34

5. Carotene Dioxygenase

(E.C. 1.13.11.21 – β-Carotene-oxygen 15,15′-oxidoreductase (bond cleaving). A ferrous requiring enzyme (227)

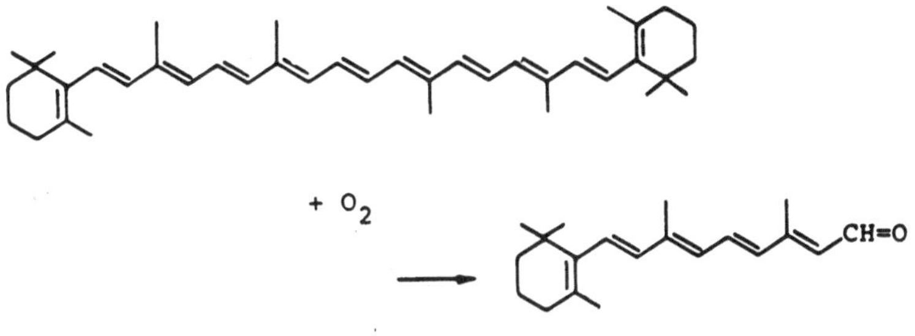

$$+ O_2 \qquad\longrightarrow\qquad \text{...CH=O} \qquad (61)$$

Retinal (vitamin A)

6. Ribulose Biphosphate Carboxylate Oxygenase

(E.C. 4.1.1.39 – 3-Phospho-D-glyceratecarboxylase (dimerizing)

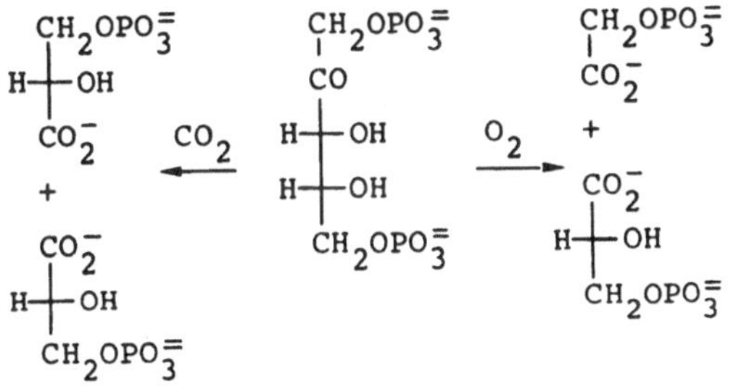

$$(62)$$

This enzyme constitutes an important case for our argument as it catalyzes oxygenation as well as carboxylation (228). Indeed, labelling studies have revealed that the former process is the major pathway for glycolate synthesis by plants (229). Although the native enzyme requires Mg^{2+} for activity (230), the oxygenase function becomes more competitive in the presence of Mn^{2+} (231—233) and Co^{3+} ions (232) which probably occupy allosteric sites on the protein. The detection of Cu^{2+} ion as a co-factor for oxygenation (234) may not be general for enzymes from all species of plants (235). Moreover, the inhibition of oxygenase, but not carboxylase activity, by hydroxylamine (236) does not necessarily indicate that there are separate active sites for each function. In any event, both functions are inhibited by cyanide ion (237).

As this enzyme would have certainly evolved millions of years before the advent of aerobiosis, it can be assumed that it possesses only one active site which generates the carbanionic center in the substrate, originally intended for attack on carbon dioxide, but which attacks molecular oxygen in its new function (238).

In view of the earlier discussion on orbital tautomerism, it is tempting to speculate that conformational differences in the enzyme would dictate the relative reactivities of the anionic center toward carbon dioxide and oxygen. The planar enolate ion, which is "hard", would react preferentially with the "hard" electrophile, carbon dioxide, whereas the tetrahedral carbanion would prefer attack on oxygen, as both participants are "soft" (Scheme 35).

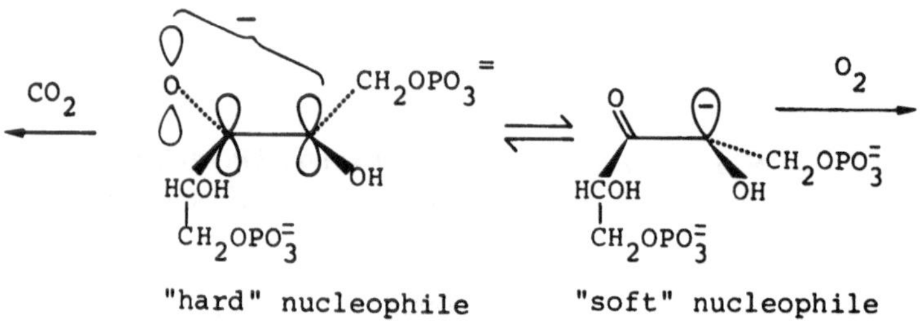

Scheme 35

The mechanistic details of both reaction courses have been charted by studies with ^{18}O-labelled ribulose biphosphate (239, 240) and with ^{18}O-labelled molecular oxygen and water (237) (Scheme 36).

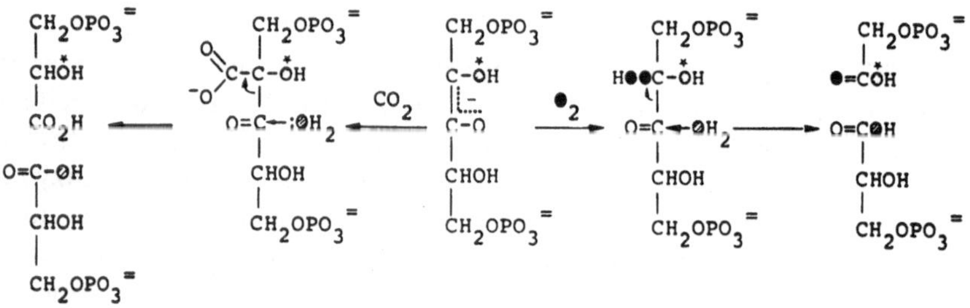

Scheme 36

IX. α-Keto Carboxylic Acid Decarboxylating Dioxygenases

Chemically speaking these dioxygenases have to carry out a difficult task as they catalyze the mono-hydroxylation of substrates that are often unreactive towards nucleophilic, electrophilic and free radical reagents. In this respect they resemble the mono-oxygenases (8) with the important difference that the other atom of oxygen is not reduced to water, but is incorporated into the new carboxyl group which arises from the α-keto acid on oxidative decarboxylation (eq. 63) (2).

$$S-H + O_2 + R-CO-CO_2H \longrightarrow SOH + RCO_2H + CO_2 \tag{63}$$

There are eight dioxygenases in this class, and a ninth which constitutes a special case.

(1) Butyrobetaine hydroxylase (E.C. 1.14.11.1 – 4-Trimethylamino-butyrate, 2-oxoglutarate: oxygen oxidoreductase (3-(R)-hydroxylating) (2).

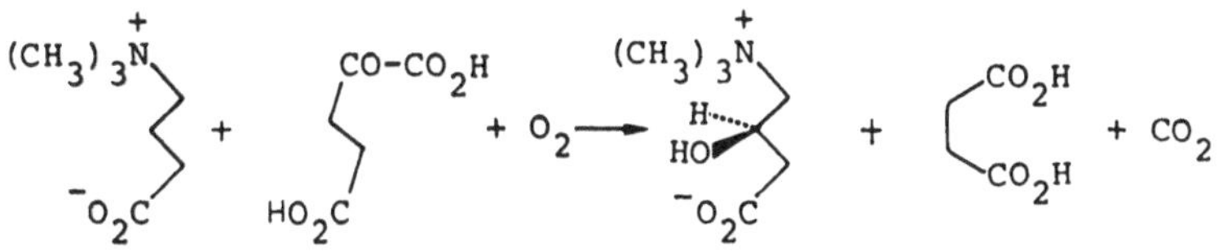

$$\tag{64}$$

(2) Peptidyl proline-4-hydroxylase (E.C. 1.14.11.2 – (L)-Prolyl-glycyl peptide, 2-oxoglutarate: oxygen oxidoreductase (4-(R)-hydroxylating) (2, 241).

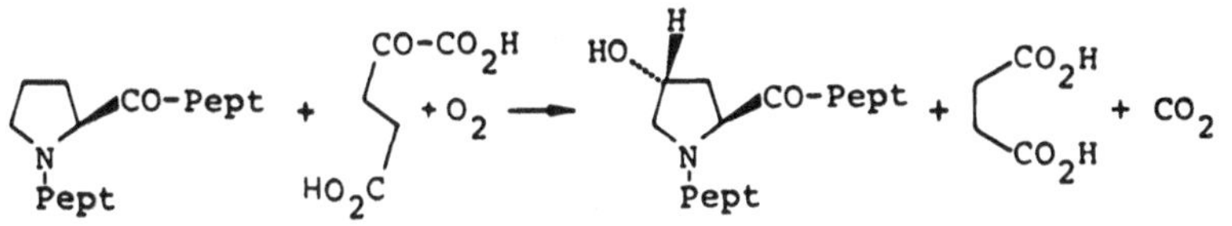

$$\tag{65}$$

(3) Peptidyl proline-3-hydroxylase (E.C. 1.14.11. – – (L)-Prolyl peptide, 2-oxoglutarate: oxygen oxidoreductase (3-(R)-hydroxylating) (242, 243).

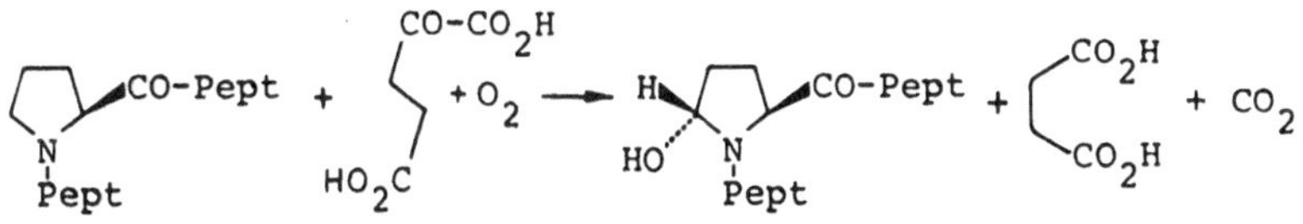

$$\tag{66}$$

(4) Thymidine-2′-hydroxylase (E.C. 1.14.11.3 – Thymidine, 2-oxoglutarate: oxygen oxidoreductase (2′-(S)-hydroxylating) (2).

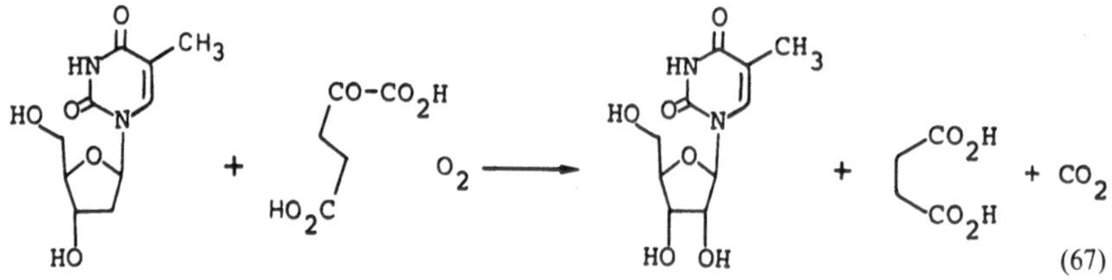

(67)

(5) Peptidyl lysine-5-hydroxylase (E.C. 1.14.11.4 – (L)-Lysyl peptide, 2-oxoglutarate: oxygen oxidoreductase (5-(R)-hydroxylating) (2, 224, 245).

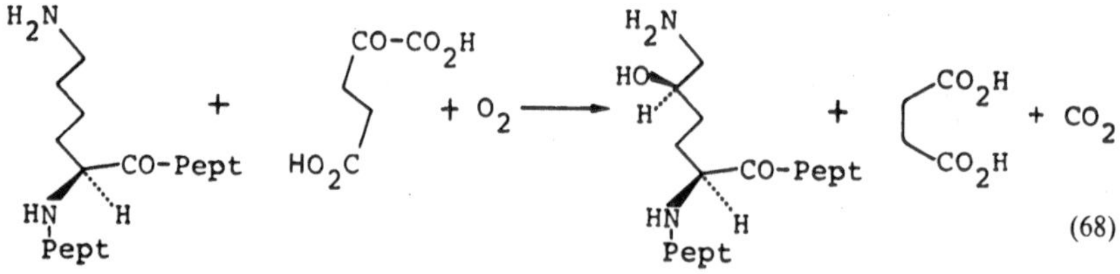

(68)

(6) Trimethyllysine 3-hydroxylase (E.C. 1.14.11 – ε-Trimethyl-(L)-lysine, 2-oxoglutarate: oxygen oxidoreductase (3-hydroxylating (246, 247).

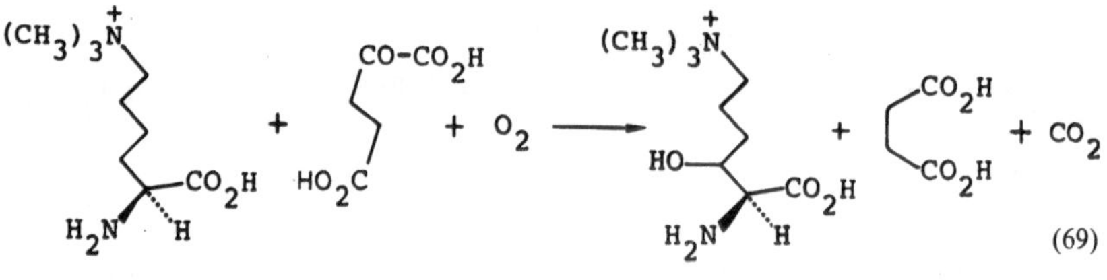

(69)

(7) Thymine-5-oxygenase (E.C. 1.14.11.6 – 1-(5)-alkyl (5-alkanol, 5-alkanal) uracil, 2-oxoglutarate : oxidoreductase (1-(7)-hydroxylating. This enzyme was originally classified as two separate enzymes (E.C. 1.14.11.5 – 5-Hydroxymethyluracil (5-oxygenating)) and (E.C. 1.14.11.6 – Thymine (7-hydroxylating)) (2). A recent study has shown however that both functions are performed by the same enzyme (248).

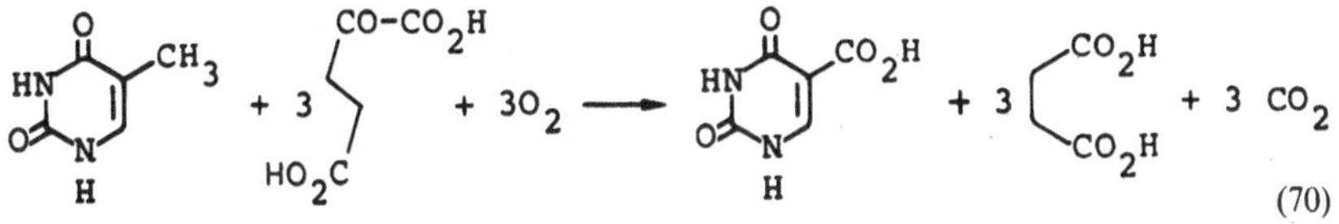

$$\text{(70)}$$

(8) Cephalosporin hydroxylase

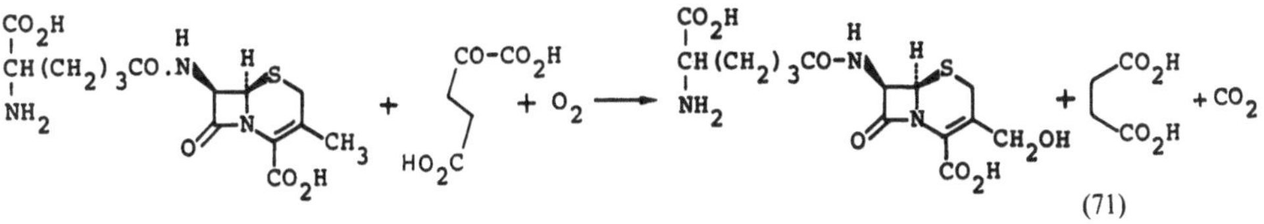

$$\text{(71)}$$

Enzymes from different living sources (249—251) show slight varia-
tions in activation properties and have yet to be studied in detail

(9) 4-Hydroxyphenylpyruvate dioxygenase (E.C. 1.13.11.27 – 4-Hydro-
xyphenylpyruvate : oxygen oxidoreductase (hydroxylating, decarboxylating) (2).

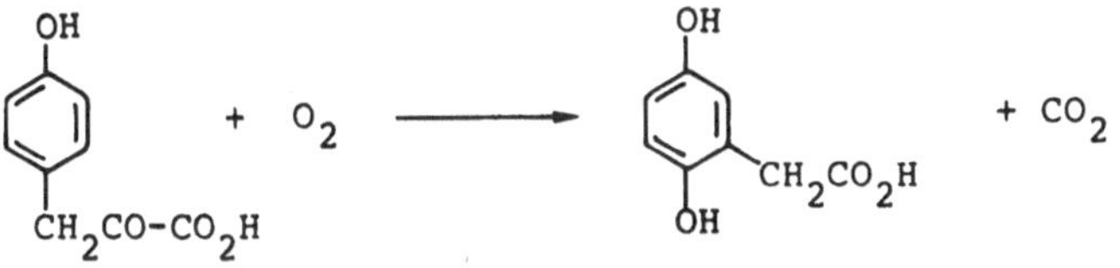

$$\text{(72)}$$

This last dioxygenase is clearly different from the others in that
the substrate and co-substrate form part of the same molecule. The
essential feature is that decarboxylation of the side chain is accompanied
by its migration to an adjacent ring position while the new hydroxyl
group occupies the vacated position. Furthermore, hydroxylation has
occurred at a center that is apparently activated by the p-hydroxyl
substituent. The electronic implication is negated by the observation
that 4-fluoro and unsubstituted phenylpyruvic acids are similarly hydroxylated by the same enzyme (252).

All the foregoing enzymes have been intensively studied and their
biochemical properties are well documented (2, 241, 253). Although minor
differences exist, they share the same important chemical properties indicating a common mechanism of catalysis. First, hydroxylation is coupled
to decarboxylation. A 1:1 stoichiometric relation exists between the two
processes, and if the oxoglutarate co-substrate is omitted or replaced
by a structurally similar analogue such as oxaloacetate, hydroxylation
is not observed (2, 253). On the other hand, the reverse is not true,

namely decarboxylation does not depend on hydroxylation. When substrate is absent or substrate inhibitors are added, decarboxylation of oxo-glutarate still occurs, albeit slowly, when catalyzed by proline-4-hydro-xylase (241, 254—256), lysyl hydroxylase (257) and thymine-7-hydro-xylase (258).

All the enzymes contain ferrous iron. Ferrous ions are essential for the catalytic activity of butyrobetaine hydroxylase (259), proline-4- (254) and 3- (242, 243), hydroxylases, thymidine hydroxylase (260), lysine hydroxylase (261), trimethyllysine hydroxylase (247), thymine oxygenase (262—264), cephalosporin hydroxylase (249), and hydroxy-phenylpyruvate (265—267). In addition, the enzymes show a non-stoichio-metric requirement for ascorbic acid which is supposed to maintain the ion co-factor in the ferrous oxidation state (2, 242, 243, 254). Labelling studies with butyrobetaine hydroxylase (268), proline-4-hydroxylase (269), thymine oxygenase (270), cephalosporin hydroxylase (251) and hydroxy-phenylpyruvate hydroxylase (271) all indicate a common pattern of oxygen incorporation. One atom of oxygen becomes the new hydroxyl group while the other is located in one of the carboxylic acid groups of succinic acid.

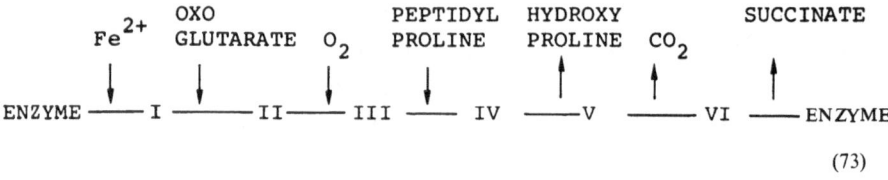

$$(73)$$

The kinetics of proline-4 hydroxylase (272), thymine oxygenase (273), lysyl hydroxylase (257, 274), and hydroxyphenylpyruvate dioxygenase (275) are all similar. There is an obligatory order of substrate binding and product release with the oxocarboxylate substrate being bound first. For purposes of discussion, the case of proline-4 hydroxylase will be taken, as it has been the most thoroughly studied until now (see eq. 73). There is direct evidence for the participation of Fe^{2+} in the catalysis of prolyl hydroxylation (254) and circumstantial evidence for its oxidation to Fe^{3+} in the intermediate stages (276, 277). The ferrous ion binds reversibly to the enzyme which is inhibited when Zn^{2+} or Pd^{2+} are substituted for Fe^{2+} (254). The ferrous ion probably binds to –SH groups at the active site on account of the inhibition of proline-4 hydroxylase activity observed by sulfhydryl reagents (278, 279). Oxalo-acetate und pyruvate inhibit the enzyme competitively with respect to oxoglutarate, but non-competitively with respect to Fe^{2+} (254). Oxalo-acetate and Zn^{2+} are also non-competitive inhibitors with respect to the peptidyl proline substrate (254). Consequently, we conclude that the sites of initial attachment of Fe^{2+}, oxoglutarate and peptidyl proline are distinct and spatially separate. The same general conclusions have been reached for the case of lysyl hydroxylase where oxygen and ascorbate are also found to have their own distinct binding sites and their own competitive inhibitors (257, 274).

There is a final characteristic of these enzymes that makes them unique among the dioxygenases. All the enzymes discussed so far catalyze the dioxygenation of substrates such as enols, phenols, enamines, etc., all of which react readily with molecular oxygen on free radical initiation, base or metal ion catalysis or electronic excitation (11). The substrates of the α-ketoacid decarboxylating hydroxylases, on the other hand, are inert towards the usual oxidants. Furthermore, the hydroxylations are highly stereoselective. This is nicely exemplified by the formation of single enantiomers, (e.g. (R)-carnitine) and diastereoisomers (e.g. 4-(R)-hydroxyprolines, thymine-2'-(S) nucleoside and 5-(R)-hydroxylysine). Moreover, hydroxylation differentiates between enantiotopic sites. Studies with 3-(S) and 3-(R) tritiated butyrobetaine show that the pro-R hydrogen atom is selectively removed to give the (R)-hydroxy product (280).

This stereoselective hydroxylation of an unactivated, saturated carbon atom is reminiscent of mono-oxygenase catalyzed hydroxylation where highly reactive oxygen-donating intermediates have been invoked (8). In order to retain configuration at the site of hydroxylation, it is likely that the oxidant inserts an atom of oxygen into the pro-R C–H bond rather like a carbene. For this reason, such highly reactive donors of ambiphilic oxygen atoms have been termed "oxenoid intermediates" (4, 5).

Apart from the high stereoselectivity, hydroxylation also occurs with high regioselectivity attacking unactivated methylene groups that are chemically indistinguishable from neighboring methylene groups as is the case in butyrobetaine, proline and lysine. It is well to note that hydroxylation of peptidyl lysine at C-3 and C-4 is due to the operation of two separate enzymes and is not due to inaccurate site-recognition by the oxidant (242, 281).

Therefore, the most important mechanistic questions concern (i) the nature of this highly reactive oxenoid intermediate, (ii) how the enzyme catalyzes its formation, and (iii) the role played by the α-keto acid.

$$(74)$$

Although many authors have maintained that substrate oxygenation precedes decarboxylation of the keto acid (eq. 74) (*2*), it is now recognized that the keto acid reacts with oxygen first to give a highly reactive oxenoid intermediate which subsequently brings about hydroxylation (eq. 75). This is substantiated by the following observations.

$$R\text{-}CO\text{-}CO_2H + O_2 \longrightarrow \begin{matrix} \text{OXENOID} \\ \text{REAGENT} \end{matrix} \qquad RCO_2H \qquad (75)$$

(i) Decarboxylation of the α-ketocarboxylate occurs in the absence of a hydroxylatable substrate. (ii) The enzymes show high specificity for oxoglutarate and are inhibited when close structural analogues or even homologues such as 2-oxoadipate and oxaloacetate are substituted. This is more consistent with the keto acid being converted into an oxenoid species which reacts with high stereo- and regioselectivity, than it being a non-selective reductant of substrate hydroperoxide.

In order to see how an α-keto carboxylic acid and molecular oxygen could combine to give a powerful hydroxylating reagent (oxenoid species), we must first look at the chemistry of α-keto carboxylic acids and in particular at their reactions with various dioxygen species.

There are several precedents for oxidative decarboxylation of α-keto carboxylic acids by hydrogen peroxide (*106*), sodium peroxide (*282*), potassium superoxide (*283*) and oxygenated iron phthalocyanine (*284*). In none of these reactions, however, were any products or intermediates detected which showed oxenoid character. Only the photo-oxygenation of α-keto carboxylic acids appeared to offer a clue. While not wishing to infer that the enzyme actually generates singlet molecular oxygen, we believe that the reaction of singlet oxygen with an α-keto carboxylic acid serves as a useful model in clarifying our understanding of the type of oxenoid species which might be expected in the enzyme-catalyzed reaction. Our own studies (*285, 287, 286*) have shown that oxidative decarboxylation gives initially the peroxy acid (**82**) (eq. 76). Normally, the peroxy acid is consumed by unreacted starting material. In an enzyme-controlled reaction however such destruction would not occur and peroxy-succinate (**83**) and *p*-hydroxyphenylperacetate (**84**) could act as discrete oxene donors (*4*).

$$R\text{-}CO\text{-}CO_2H \; + \; {}^1O_2 \longrightarrow R\overset{O}{\overset{\|}{C}}\text{-}O\text{-}OH \xrightarrow{\;RCOCO_2H\;} 2RCO_2H + CO_2 \qquad (76)$$

$$(\mathbf{82})$$

While peroxy acids clearly possess some oxenoid character, the suggestion that they participate in oxoglutarate and hydroxyphenylpyruvate-mediated dioxygenations presupposes that they can hydroxylate the substrate. In the former case, (83) should be capable of hydroxylating the saturated secondary centers of butyrobetaine, proline and lysine. In the latter case, (84) should undergo intramolecular oxene transfer to give an arene oxide (85) or zwitterion (86), both of which should rearrange to give the product of the enzyme-catalyzed reaction, namely homogentisate (87) (288).

Despite the attractiveness of this proposal, peroxy acids such as (83) and (84) are unlikely to function as the critical oxenoid intermediates in the enzymic process. Peroxysuccinic acid has been added to both the proline-4-hydroxylase and thymidine-2'-hydroxylase systems, but it does not replace the requirement for oxygen and oxo glutarate (253). Furthermore, it is unlikely that such a peroxy acid would have sufficient reactivity to carry out the specified hydroxylation. Moreover, we have prepared the peroxy acid (84) and have found that it too is incapable of performing the required intramolecular oxene transfer. No products like (85), (86) or (87) are formed (287).

The possibility that the iron co-factor in the enzyme interacts with and enhances the oxenoid character of the α-keto carboxylate-oxygen reagent cannot be ignored. HAMILTON (4) has already suggested that a ferrous peroxycarboxylate may be the oxenoid intermediate and that it is directly responsible for substrate mono-oxygenation. The prospects offered by this proposal are indicated by the report that ferrous ions will reduce peroxycarboxylates to give an Fe(IV) species (289), which when appropriately co-ordinated is capable of hydroxylating unactivated, saturated carbon atoms by a two-step free radical mechanism (290) (eq. 77). Such a two-electron reduction by iron would normally occur when the substituent hydrocarbon grouping is strongly electron-withdrawing. We have found, that the peroxy acid (84), in the presence of ferrous perchlorate prefers to undergo a one-electron reduction (287, 291) (eq. 78). It is unclear how an enzyme would be able to favor and bring about the two-electron reduction process.

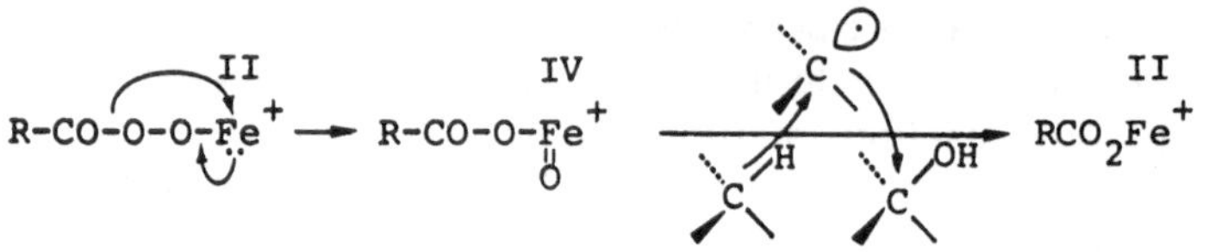

$$R\text{-}CO\text{-}O\text{-}O\text{-}Fe^+ \longrightarrow R\text{-}CO\text{-}O\text{-}Fe^+ \longrightarrow RCO_2Fe^+ \quad (77)$$

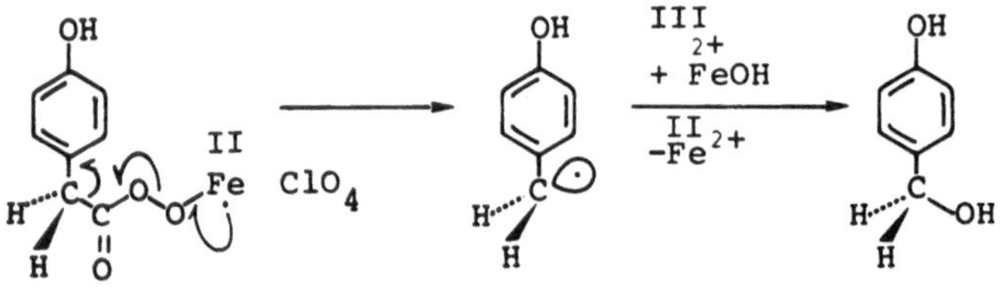

$$(78)$$

In a biomimetic-type reaction, 4-hydroxyphenylpyruvic acid is partly converted by singlet oxygen into a quinolic acid (86) which rearranges in alkali to homogentisic acid (87) (292). It was originally suggested that singlet oxygen attacks the benzene ring to give the hydroperoxy-quinol (88) (Scheme 37) which furnishes (86) by intramolecular oxidative decarboxylation. We have discovered that the peroxidation of phenols is far too slow to account for the rapidity of the reaction (287). Moreover, we have shown that the peroxy acid (84) is not an intermediate for quinol (86). We therefore conclude that a more powerful oxenoid intermediate is involved, not only for 4-hydroxyphenyl pyruvic acid, but also for α-ketocarboxylic acids in general.

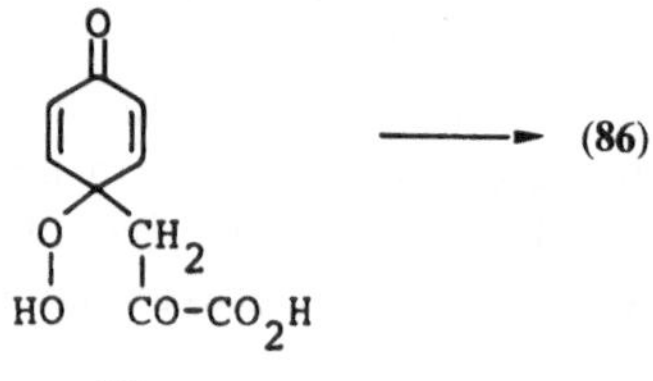

(88)

Scheme 37

An excellent candidate for this intermediate would be a trioxolanone. By analogy with the primary ozonides, this species would decompose to carbon dioxide and a peroxycarboxylic acid. However, in the presence of a suitable oxene acceptor (A), the exothermicity of the liberation of carbon dioxide and a carboxylic acid could be exploited for efficient mono-oxygenation (293, 294) (eq. 79). Similar "oxene-donating" properties have been demonstrated for analogous trioxolenes (295, 296).

$$\text{equation (79)}$$

$$R-\overset{\overset{O)}{\parallel}}{C}-\overset{\overset{O}{\parallel}}{C} \longrightarrow R-\overset{\overset{O^-}{\mid}}{C}-\overset{\overset{O}{\parallel}}{C} \longrightarrow R-\overset{\overset{O}{\parallel}}{C}-O-O^- + CO_2$$

$$\searrow^{A} \quad R-\overset{\overset{O}{\parallel}}{C}-O^- + CO_2 + AO$$

(79)

The reaction of singlet oxygen with α-ketocarboxylic acids is a non-enzymic, biomimetic model. Its utility is necessarily limited to indicating, but not defining, possible mechanisms by which an enzyme might catalyze the formation of an oxenoid intermediate. The inference, of course, is that the "oxene-donor" generated by α-ketoglutarate-decarboxylating dioxygenases is indeed the trioxolanone (**89**) or the corresponding derived carbonyl oxide or diradical. Similarly, the trioxolanone (**90**) or a related species would be the critical intermediate in the enzyme-catalyzed homogentisic acid rearrangement and the singlet oxygen-mediated formation of the quinol (**86**).

$$HO_2C-CH_2-CH_2-\overset{\overset{O^-}{\mid}}{C}-\overset{\overset{O}{\parallel}}{C} \text{(trioxolanone)}$$

(89)

(90)

The fundamental questions which now remain to be answered are how might a ferro-enzyme catalyze the combination of molecular oxygen with α-ketoglutarate to produce an oxenoid intermediate and how would the ensuing stereospecific oxidation take place.

An enzyme containing a ferrous ion with thiolate ligation would certainly have reducing properties (*297*). Moreover, since α-keto carboxylic acids readily undergo one- and two-electron reduction (*298*), chelation with ferrous ion should induce considerable radical anionic character in the α-dicarbonyl grouping. Once ferrous α-ketoglutarate is bound in the enzyme cleft (stage I, Scheme 38), these two aforementioned reduction propensities are realized in co-operative fashion. The thiolate group brings about the reversal of polarity on the α-carbonyl function by transferring negative charge through the ferrous-bound carbonyl ligand, thereby creating a carbanion or possibly a radical of tetrahedral configuration. Departure from planarity will be aided by the pinning

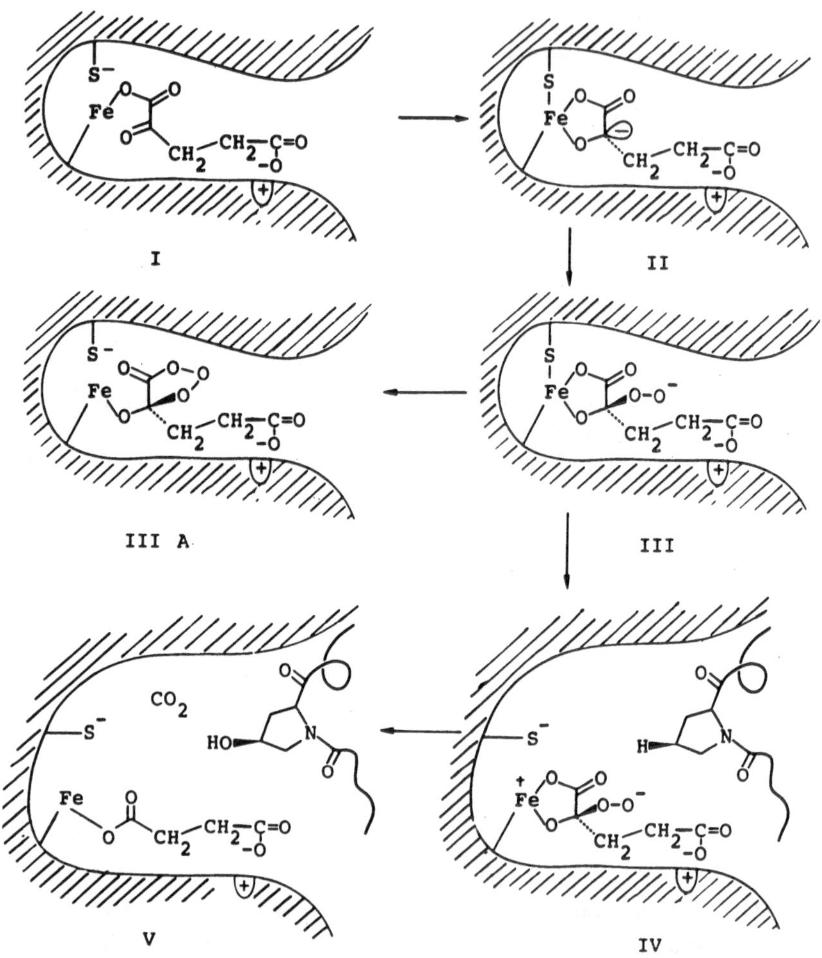

Scheme 38

down of the other carboxylate group by the enzyme (stage II). Next, oxygen enters and reacts with the waiting "soft" carbanionic center to form the hydroperoxide (stage III). The major substrate is bound last of all (stage IV). This time, on account of its size, the binding of the substrate, exemplified by peptidyl proline, will be expected to cause conformational changes in the enzyme and alter the ligand field of the iron atom (stage IV). The resulting elongation and rupture of the thiolate-iron bond will trigger decarboxylation and cause the insertion of the terminal atom of oxygen of the peroxide into the proximal carbon-hydrogen bond of the proline ring (stage V).

Variants of this mechanism can be envisaged. Cyclization to the trioxolanone (stage IIIa) could occur which then on decarboxylation delivers an oxygen atom to the substrate after it is bound. Decarboxylation could also be spontaneous, thereby producing a carbonyl oxide which brings about the hydroxylation.

In all these schemes, decarboxylation would presumably precede release of the hydroxylated proline, which is the reverse of that experienced in the enzymic process. However, the essential feature of our mechanism is co-substrate activation which fits the evidence better than "oxygen-activation" (299).

X. Summary

We have reviewed the chemical and enzymological data of mechanistic significance for fifty-seven dioxygenases and have come to the following conclusions:

(i) Most of these enzymes catalyze the dioxygenation of substrates that are already relatively reactive towards molecular oxygen. That is why the substrates readily undergo biomimetic oxygenation under the appropriate conditions, for example on free radical initiation, basic or metal-ion catalysis or reaction with electronically excited oxygen.

(ii) Nearly all the enzymes require iron or copper ion co-factors for activity. Although molecular oxygen can form complexes with the co-factors, the resulting complex does not cause significant lowering of the activation energy of dioxygenation. The metal provides at most a convenient binding site for molecular oxygen.

(iii) Direct metal activation of the substrate is better precedented. One electron oxidation of the substrate by the metal co-factor in its higher oxidation state would generate the radical cation or reduction of the substrate by the metal ion in its lower oxidation state would form the corresponding radical anion.

Radical anions could then react with molecular oxygen conveniently bound to the metal ion. Often, the free radical centers so generated, exemplified by phenoxy radicals, are too stable and do not react quickly enough with molecular oxygen. The dioxygenases appear to overcome this problem in various ways. It is probable that those dioxygenases which are responsible for intradiol cleavage of phenols develop in the substrate a localized anionic center to which the oxygen atom is eventually attached. In other words, a "soft" carbanion is generated which is sufficiently reactive to transfer charge to the π antibonding orbitals of oxygen. It is possible that the same mechanism is followed by the extradiol phenol cleaving dioxygenases, ribulose bisphosphate carboxylase-oxygenase, and also by the dioxygenases which cleave indoles, quercetin and heme.

The external flavoprotein dioxygenases function by catalyzing the transfer of electrons from reduced nicotinamides to the aromatic nuclei of the substrate. The resulting radical anions would be sufficiently "soft" to react with molecular oxygen without requiring any particular assistance from the enzymes.

In the case of luciferases, the anions formed by simple deprotonation are sufficiently reactive to undergo spontaneous peroxidation with molecular oxygen. The role of the enzymes is probably concerned with the maximization of the quantum yield of light emission.

Enzymes which catalyze peroxidation of unconjugated unsaturated fatty acids active the substrate by creating a radical cation which is easier to deprotonate than the parent olefinic substrate. The free radical so produced then reacts readily with molecular oxygen.

The α-ketocarboxylic acid-decarboxylating dioxygenases may operate by thiolate-metal complexation reversing the polarity of the α-carbonyl function, thereby rendering it more nucleophilic with respect to molecular oxygen. The initial zwitterionic hydroperoxide or its five-membered ring tautomer, since they both possess oxenoid character, are able to bring about the mono-oxygenation of aromatic rings, aldehydes and even unactivated saturated carbon atoms by oxygen atom transfer.

Lastly, we wish to re-iterate that there is neither chemical precedent nor enzymological evidence for the activation of molecular oxygen by iron and copper ion-containing enzymes for reactions in which *both* oxygen atoms are incorporated into the organic substrate. This finding contrasts with that established for some mono-oxygenases, such as the cytochrome P-450 hemoproteins, which contain an oxidizing reagent consisting of an atom of oxygen directly bound to iron in one of its higher oxidation states. This reagent is created by the exothermic cleavage of one of the initial oxygen atoms of the complexed oxygen molecule as water, thereby activating the remaining metal-bound oxygen substituent. The dioxygenases by definition cannot sacrifice one of their oxygen atoms in this way. Consequently, they are only able to catalyze the dioxygenation of substrates that are already reactive towards molecular oxygen. For this to occur, prolongation of the life-time of the collision complex between molecular oxygen and the substrate is all that is needed to bring about effective catalysis.

Acknowledgments

We are indebted to the Swiss National Science Foundation for their support (grant No. 2.882-0.77) and we thank Dr. ALAN WILLIAMS of the Department of Inorganic Chemistry, University of Geneva, for fruitful discussions.

References

1. HAYAISHI, O.: In "Molecular Mechanisms of Oxygen Activation" (O. HAYAISHI, ed.), pp. 1—28. New York-London: Academic Press, Inc. 1974.
2. HAYAISHI, O., M. NOZAKI, and M. T. ABBOTT: In "The Enzymes", Vol. 12 (P. D. BOYER, ed.), pp. 119—189. New York-London: Academic Press, Inc. 1975.
3. GEORGE, P.: In "Oxidases and Related Redox Systems", Vol. 1 (T. E. KING, H. S. MASON, and M. MORRISON, eds.), pp. 3—36. New York-London: J. Wiley & Sons, Inc. 1965.
4. HAMILTON, G. A.: In "Progress in Bioorganic Chemistry", Vol. 1 (E. T. KAISER and F. J. KÉZDY, eds.), pp. 83—157. New York-London-Toronto: Wiley-Interscience. 1971.
5. — In "Molecular Mechanisms of Oxygen Activation" (O. HAYAISHI, ed.), pp. 405—451. New York-London: Academic Press, Inc. 1974.
6. HENRICI-OLIVÉ, G., and S. OLIVÉ: Activation of Molecular Oxygen. Angew. Chem. Int. Edn (Engl.) 13, 29 (1974).
7. MARTELL, A. E., and M. M. TAQUI KHAN: In "Inorganic Biochemistry", Vol. 2 (G. L. EICHHORN, ed.), p. 645. Amsterdam: Elsevier. 1973.
8. BOYD, G. S.: In: "Biological Hydroxylation Mechanisms" (G. S. BOYD and R. M. S. SMELLIE, eds.), pp. 1—9. London-New York: Academic Press, Inc. 1972.
9. LYONS, J. E.: In "Aspects of Homogeneous Catalysis", Vol. 3 (R. UGO, ed.), pp. 1—136. Dortrecht: Reidel. 1977.
10. JONES, R. D., D. A. SUMMERVILLE, and F. BASOLO: Synthetic Oxygen Carriers Related to Biological Systems. Chem. Rev. 79, 139 (1979).
11. MATSUURA, T.: Biomimetic Oxygenation. Tetrahedron 33, 2869 (1977).
12. BAYER, E., P. KRAUSS, A. RÖDER, and P. SCHRETZMANN: In "Oxidases and Related Redox Systems", Vol. 1 (T. E. KING, H. S. MASON, and M. MORRISON, eds.), pp. 227—263. Maryland: University Park Press. 1973.
13. CRIEGEE, R.: Versuche zur Darstellung von Tetramethyl-cyclobutadien. Angew. Chem. 74, 703 (1962).
14. TURRO, N. J., V. RAMAMURTHY, K.-C. LIU, A. KREBS and R. KEMPER: Reaction of Strained Acetylenes with Molecular Oxygen. J. Amer. Chem. Soc. 98, 6758 (1976).
15. HOWARD, J. A.: In "Free Radicals", Vol. 2 (J. K. KOCHI, ed.), pp. 3—62. London: Wiley-Interscience. 1973.
16. LLOYD, W. G.: In "Methods in Free-Radical Chemistry", Vol. 4 (E. S. HUYSER, ed.), pp. 2—131. New York: Marcel Dekker, Inc. 1973.
17. DENNY, R. W., and A. NICKON: Sensitized Photooxygenation of Olefins. Org. Reactions 20, 133 (1973).
18. GOLLNICK, K.: In "Singlet Oxygen" (B. RÅNBY and J. F. RABEK, eds.), pp. 111—134. Chichester: Wiley-Interscience. 1978.
19. SCHMITT, R. J., V. M. BIERBAUM, and C. H. DePUY: Gas-Phase Reactions of Carbanions with Triplet and Singlet Molecular Oxygen. J. Amer. Chem. Soc. 101, 6443 (1979).
20. JENSEN, W. B.: The Lewis Acid-Base Definitions: A Status Report. Chem. Rev. 78, 1 (1978).

21. Nishinaga, A., T. Shimizu, and T. Matsuura: Reaction of Potassium Superoxide with Phenoxy Radicals. On the Mechanism of Base-Catalyzed Oxygenation of Phenols. Chem. Letts **1977**, 547.

22. Goto, K., H. Tamura, and M. Nagayama: The Mechanism of Oxygenation of Ferrous Ion in Neutral Solution. Inorgan. Chem. **9**, 963 (1970).

23. Ishida, H., H. Takahashi, H. Sato, and H. Tsubomura: The Interaction of Oxygen with Organic Molecules. J. Amer. Chem. Soc. **92**, 275 (1970).

24. Heidt, L. J., and A. M. Johnson: Optical Study of the Hydrates of Molecular Oxygen in Water. J. Amer. Chem. Soc. **79**, 5587 (1957).

25. Gray, H. B., and H. J. Schugar: In "Inorganic Biochemistry", Vol. 1 (G. L. Eichhorn, ed.), pp. 102—319. Amsterdam: Elsevier. 1973.

26. Reed, C. A., and S. K. Cheung: On the Bonding of FeO_2 in Hemoglobin and Related Dioxygen Complexes. Proc. Ntl. Acad. Sci. U.S.A. **74**, 1780 (1977).

27. Vaska, L.: Dioxygen-Metal Complexes: Towards a Unified View. Acc. Chem. Res. **9**, 175 (1976).

28. Basolo, F., B. M. Hoffman, and J. A. Ibers: Synthetic Oxygen Carriers of Biological Interest. Acc. Chem. Res. **8**, 384 (1975).

29. Hammond, G. S., and C.-H. S. Wu: Oxidation of Iron(II) Chloride in Nonaqueous Solvents. Adv. in Chem. **77**, 186 (1968).

30. Carter, M. J., D. P. Rillema, and F. Basolo: Oxygen Carrier and Redox Properties of Some Neutral Cobalt Chelates. Axial and In-plane Ligand Effects. J. Amer. Chem. Soc. **96**, 392 (1974).

31. Dawson, J. H., R. H. Holm, J. R. Trudell, G. Barth, R. E. Linder, E. Bunnenberg, C. Djerassi, and S. C. Tang: Oxidized Cytochrome P-450. Magnetic Circular Dichroism Evidence for Thiolate Ligation in the Substrate-Bound Form. J. Amer. Chem. Soc. **98**, 3707 (1976).

32. Valentine, J. S.: The Dioxygen Ligand in Mononuclear Group VIII Transition Metal Complexes. Chem. Rev. **73**, 235 (1973).

33. Sen, A., and J. Halpern: Role of Transition Metal-Dioxygen Complexes in Catalytic Oxidation. J. Amer. Chem. Soc. **99**, 8337 (1977).

34. Schmidt, D. D., and J. T. Yoke: Autoxidation of a Coordinated Trialkylphosphine. J. Amer. Chem. Soc. **93**, 637 (1971).

35. Hanzlik, R. P., and D. Williamson: Oxygen Activation by Transition Metal Complexes. 2. J. Amer. Chem. Soc. **98**, 6570 (1976).

36. Sutin, N., and J. K. Yandell: Autoxidation Reactions Catalyzed by Iron(III) and Iron(IV) Dithiolate Complexes. J. Amer. Chem. Soc. **95**, 4847 (1973).

37. Holland, D., and D. J. Milner: Liquid Phase Metal-Centred Autoxidation of Cyclo-octene Promoted by Rhodium Species. J. Chem. Soc. (London) Dalton Trans. **1975**, 2440.

38. Read, G., and P. J. C. Walker: Oxygenation Studies. Part 2. Rhodium(1)-catalyzed Autoxidation of Oct-1-ene at Ambient Temperature and Pressure. J. Chem. Soc. (London) Dalton Trans. **1977**, 883.

39. Mimoun, H., M. M. P. Marchirant, and I. S. de Roch: Activation of Molecular Oxygen: Rhodium-catalyzed Oxidation of Olefins. J. Amer. Chem. Soc. **100**, 5437 (1978).

40. Takao, K., H. Azuma, Y. Fujiwara, T. Imanaka, and S. Teranishi: Oxidation by Transition Metal Complexes. V. Oxidation of Vinyl Esters Catalyzed by Rhodium Complex. Bull. Chem. Soc. Japan **45**, 2003 (1972).

41. Bartlett, P. D., and J. S. McKennis: Catalyzed Decomposition of Tetramethyl-1,2-Dioxetane by Rhodium and Iridium Complexes. J. Amer. Chem. Soc. **99**, 5334 (1977).

42. Tsuji, J., and H. Takayanagi: Organic Synthesis by Means of Metal Complexes. XIII. J. Amer. Chem. Soc. **96**, 7349 (1974).

43. ROGIĆ, M. M., T. R. DEMMIN, and W. B. HAMMOND: Cleavage of Carbon-Carbon Bonds. Copper(II)-Induced Oxygenolysis of o-Quinones, Catechols and Phenols. J. Amer. Chem. Soc. **98**, 7441 (1976).

44. ROGIĆ, M. M., and T. R. DEMMIN: Cleavage of Carbon-Carbon Bonds. Copper(II)-Induced Oxygenolysis of o-Benzoquinones, Catechols and Phenols. On the Question of Nonenzymic Oxidation of Aromatics and Activation of Molecular Oxygen. J. Amer. Chem. Soc. **100**, 5472 (1978).

45. TSUJI, J., and H. TAKAYANAGI: Oxidative Cleavage Reaction of Catechol and Phenol to Monoester of cis,cis-Muconic Acid with the Oxidizing Systems of $O_2/CuCl$, $KOH/CuCl_2$ and $KO_2/CuCl_2$ in a Mixture of Pyridine and Alcohol. Tetrahedron **34**, 641 (1978).

46. — — Oxidative Reaction of 3-Methylindole Catalyzed by CuCl-Pyridine Complex under Oxygen. Chem. Letts **1980**, 65—66.

47. BROWN, D. G., L. BECKMANN, C. H. ASHBY, G. C. VOGEL, and J. T. REINPRECHT: Tetrahedron Letts **1977**, 1363. Oxygen-Dependent Ring Cleavage in a Copper Coordinated Catechol.

48. GRINSTEAD, R. R.: Metal-catalyzed Oxidation of 3,5-di-t-Butyl Pyrocatechol, and its Significance in the Mechanism of Pyrocatechase Action. Biochemistry **3**, 1308 (1964).

49. KRAMER, C. E., G. DAVIES, R. B. DAVIS, and R. W. SLAVEN: Characterization of a Novel Low Oxidation State Transition Metal Peroxide from the Reaction of Copper(I) Chloride with Oxygen in Pyridine. Chem. Commun. **1975**, 606.

50. TYSON, C. A., and A. E. MARTELL: Kinetics and Mechanism of the Metal Chelate Catalyzed Oxidation of Pyrocatechols. J. Amer. Chem. Soc. **94**, 939 (1972).

51. BUFFLE, J., and A. E. MARTELL: Metal Ion Catalyzed Oxidation of o-Dihydroxy Aromatic Compounds by Oxygen. 1. Inorgan. Chem. **16**, 2221 (1977).

52. WÜTHRICH, K., and S. FALLAB: Reaktivität von Koordinationsverbindungen. XI. Mechanismus der Kupfer(II)-katalysierten Autoxydation von o-Phenylendiamin. Helv. Chim. Acta **47**, 1440 (1964).

53. OHKATSU, Y., and O. TETSUO: The Liquid-Phase Oxidation of Aldehydes with Metal Tetra(p-tolyl)porphyrins. Bull. Chem. Soc. Japan **50**, 2945 (1977).

54. OHKATSU, Y., and T. TSURUTA: Autoxidation Reactions of Hydrocarbons Catalyzed by Co(II) Tetra(p-tolyl)porphyrin. Bull. Chem. Soc. Japan **51**, 188 (1978).

55. ABEL, W. E., J. M. PRATT, R. WHELAN, and P. J. WILKINSON: Reduction of Coordinated O_2 by Organic Substrates. J. Amer. Chem. Soc. **96**, 7119 (1974).

56. NISHINAGA, A., T. TOJO, and T. MATSUURA: A Model Catalytic Oxygenation for the Reaction of Quercetinase. Chem. Commun. **1974**, 896.

57. NISHINAGA, A., K. WATANABE, and T. MATSUURA: Oxygenation of 2,6-Di-t-Butyl-4-alkylphenols Catalyzed by Cobalt(II) Schiff's Base Complexes. Tetrahedron Letts **1974**, 1291.

58. VOGT, L. H. JR., J. G. WIRTH, and H. L. FINKBEINER: Selective Autoxidation of some Phenols Using Bis(salicylaldehyde)ethylenediiminecobalt Catalysts. J. Org. Chem. **34**, 273 (1969).

59. DANCE, I. G., R. C. CONRAD, and J. E. CLINE: Mechanism of Cobalt Dithiolene Complex Catalysis of Thiol Autoxidation in Acidic Acetonitrile Solution. Chem. Commun. **1974**, 13.

60. VOGT, L. H. JR.: Reversible Oxygen-Carrying Chelates. Chem. Rev. **63**, 269 (1963).

61. NISHINAGA, A., K. NISHIZAWA, H. TOMITA, and T. MATSUURA: Novel Peroxycobalt(III) Complexes Derived from 4-Aryl-2,6-di-tert-butylphenols. J. Amer. Chem. Soc. **99**, 1287 (1977).

62. NISHINAGA, A., H. TOMITA, and T. MATSUURA: Selective Formation of Peroxy-p-quinolato Co(III) Complexes in the Oxygenation of 4-Alkyl-2,6-di-t-butylphenols with Co(II)-Schiff's Base Complexes. Tetrahedron Letts **1979**, 2893.

63. Kamiya, Y.: The Autoxidation of α-Methylstyrene Catalyzed by Copper Phthalo-cyanine. Tetrahedron Letts **1968**, 4965.

64. — Catalysis by Metal Acetylacetonates in the Autoxidation of Hydrocarbons. J. Catalysis **24**, 69 (1972).

65. McNeal, R. J., and G. R. Cook: Photoionization of O_2 in the Metastable $^1\Delta g$ State. J. Chem. Phys. **45**, 3469 (1966).

66. Bartlett, N., and D. H. Lohmann: Dioxygenyl Hexafluoroplatinate(V). Proc. Chem. Soc. (London) **1962**, 115.

67. D'Orazio, L. A., and R. H. Wood: Thermodynamics of the Higher Oxides. 1. The Heats of Formation and Lattice Energies of the Superoxides of Potassium, Rubidium and Cesium. J. Phys. Chem. **69**, 2550 (1965).

68. Lee-Ruff, E.: The Organic Chemistry of Superoxide. J. Chem. Soc. (London) Rev. **6**, 195 (1977).

69. Sawyer, D. T., M. J. Gibian, M. M. Morrison, and E. T. Seo: On the Reactivity of Superoxide Ion. J. Amer. Chem. Soc. **100**, 627 (1978).

70. Danen, W. C., and R. J. Warner: The Remarkable Nucleophlicity of Superoxide Anion Radical. Rate Constants for Reaction of Superoxide Ion with Aliphatic Bromides. Tetrahedron Letts **1977**, 989.

71. Wilshire, J., and D. T. Sawyer: Redox Chemistry of Dioxygen Species. Acc. Chem. Res. **12**, 105 (1979).

72. Mayer, R., J. Widom, and L. Que, Jr.: Involvement of Superoxide in the Reactions of Catechol Dioxygenases. Biochem. Biophys. Res. Commun. **92**, 285 (1980).

73. Myllylä, R., L. M. Schubotz, U. Weser, and K. I. Kivirikko: Involvement of Superoxide in the Prolyl and Lysyl Hydroxylase Reactions. Biochem. Biophys. Res. Commun. **89**, 98 (1979).

74. Bhagwat, A. S., and P. V. Sane: Evidence for the Involvement of Superoxide Anions in the Oxygenase Reaction of Ribulose-1,2-diphosphate Carboxylase. Biochem. Biophys. Res. Commun. **84**, 865 (1978).

75. Kido, T., K. Soda, and K. Asada: Properties of 2-Nitropropane Dioxygenase of *Hansenula mrakii*. J. Biol. Chem. **253**, 226 (1978).

76. Hamilton, G. A., P. K. Adolf, J. deJersey, G. C. DuBois, G. R. Dyrkacz, and R. D. Libby: Trivalent Copper, Superoxide, and Galactose Oxidase. J. Amer. Chem. Soc. **100**, 1899 (1978).

77. Belluš, D.: In "Singlet Oxygen" (B. Rånby and J. F. Rabek, eds.), pp. 61—110. Chichester: Wiley-Interscience. 1978.

78. Turro, N. J., M. F. Chow, and Y. Ito: Autoxidation of Ketenes, Diradicaloid and Zwitterionic Mechanisms of Reactions of Triplet Molecular Oxygen and Ketenes. J. Amer. Chem. Soc. **100**, 5580 (1978).

79. Siegel, B., and J. Lanphear: Iron-catalyzed Oxidative Decarboxylation of Benzoyl-formic Acid. J. Amer. Chem. Soc. **101**, 2221 (1979).

80. — — Kinetics and Mechanism for the Acid-catalyzed Oxidative Decarboxylation of Benzoylformic Acid. J. Org. Chem. **44**, 942 (1979).

81. Kochi, J. K. In "Free Radicals", Vol. 1 (J. K. Kochi, ed.), pp. 529—683. London: Wiley-Interscience. 1973.

82. Jones, M. M., and J. E. Hix, Jr.: In "Inorganic Biochemistry", Vol. 1 (G. L. Eichhorn, ed.), pp. 361. Amsterdam: Elsevier. 1973.

83. Nozaki, M.: In "Molecular Mechanisms of Oxygen Activation" (O. Hayaishi, ed.), pp. 135—165. New York-London: Academic Press, Inc. 1974.

84. White, G. A., and R. M. Krupka: Ascorbic Acid Oxidase and Ascorbic Acid Oxygenase of *Myrothecium verrucaria*. Arch. Biochem. Biophys. **110**, 448 (1965).

85. Gaunt, J. K., and W. C. Evans: Metabolism of 4-Chloro-2-methylphenoxyacetate by a Soil Pseudomonad. Biochem. J. **122**, 533 (1971).

86. FUJIOKA, M., and H. WADA: The Bacterial Oxidation of Indole. Biochim. Biophys. Acta **158**, 70 (1968).
87. SHARMA, H. K., and C. S. VAIDYANATHAN: A New Mode of Ring Cleavage of 2,3-Dihydroxybenzoic Acid in *Tecoma stans* (L.). European J. Biochem. **56**, 163 (1975).
88. SEIDMAN, M. M., A. TOMS, and J. M. WOOD: Influence of Side-Chain Substituents on the Position of Cleavage of the Benezene Ring by *Pseudomonas fluorescens*. J. Bacteriol. **97**, 1192 (1969).
89. TACK, B. F., P. J. CHAPMAN, and S. DAGELY: Metabolism of Gallic Acid and Syringic Acid by *Pseudomonas putida*. J. Biological Chem. **247**, 6438 (1972).
90. QUE, L., JR., J. D. LIPSCOMB, R. ZIMMERMANN, E. MÜNCK, N. R. ORME-JOHNSON, and W. H. ORME-JOHNSON: Mössbauer and E.P.R. Spectroscopy of Protocatechuate 3,4-Dioxygenase from *Pseudomonas aeruginosa*. Biochim. Biophys. Acta **452**, 320 (1976).
91. QUE, L., JR., J. D. LIPSCOMB, E. MÜNCK, and J. M. WOOD: Protocatechuate 3,4-Dioxygenase Inhibitor Studies and Mechanistic Implications. Biochim. Biophys. Acta **485**, 60 (1977).
92. QUE, L., JR.: Non-Heme Iron Dioxygenases. Structure and Bonding **40**, 40 (1980).
93. KEYES, W. E., T. M. LOEHR, and M. L. TAYLOR: Raman Spectral Evidence for Tyrosine Coordination of Iron in Protocatechuate 3,4-Dioxygenase. Biochem. Biophys. Res. Commun. **83**, 941 (1978).
94. TATSUNO, Y., Y. SAEKI, M. IWAKI, T. YAGI, M. NOZAKI, T. KITAGAWA, and S. OTSUKA: Resonance Raman Spectra of Protocatechuate 3,4-Dioxygenase. Evidence for Coordination of Tyrosine Residue to Ferric Iron. J. Amer. Chem. Soc. **100**, 4614 (1978).
95. FELTON, R. H., L. D. CHEUNG, R. S. PHILLIPS, and S. W. MAY: A Resonance Raman Study of Substrate and Inhibitor Binding to Protocatechuate-3,4-dioxygenase. Biochem. Biophys. Res. Commun. **85**, 844 (1978).
96. MAY, S. W., and R. S. PHILLIPS: Protocatechuate 3,4-Dioxygenase: Implications of Ionization Effects on Binding and Dissociation of Halohydroxybenzoates and on Catalytic Turnover. Biochemistry **18**, 5933 (1979).
97. QUE, L., JR., and R. H. HEISTAND II: Resonance Raman Studies on Pyrocatechase. J. Amer. Chem. Soc. **101**, 2219 (1979).
98. MAY, S. W., R. S. PHILLIPS, and C. D. OLDHAM: Interaction of Protocatechuate with Substituted Hydroxybenzoic Acids and Related Compounds. Biochemistry **17**, 1853 (1978).
99. NAKATA, H., T. YAMAUCHI, and H. FUJISAWA: Studies on the Reaction Intermediate of Protocatechuate 3,4-Dioxygenase. Biochim. Biophys. Acta **527**, 171 (1978).
100. HAYAISHI, O., M. KATAGIRI, and S. ROTHBERG: Mechanism of the Pyrocatechase Reaction. J. Amer. Chem. Soc. **77**, 5450 (1955).
101. SAWAKI, Y., and Y. OGATA: Acyl Migration in the Acid-catalyzed Decomposition of α-Hydroperoxy Ketones. J. Amer. Chem. Soc. **100**, 856 (1978).
102. — — Chemiluminescence from the Base-Catalyzed Decomposition of α-Hydroperoxy Ketones. Competitive Cyclic and Acyclic Reactions. J. Amer. Chem. Soc. **99**, 5412 (1977).
103. SMITH, P. A. S.: In "Molecular Rearrangements", Vol. 1 (P. DE MAYO, ed.), pp. 457—491. New York-London: Wiley-Interscience. 1963.
104. SAWAKI, Y., and C. S. FOOTE: Acyclic Mechanism in the Cleavage of Benzils with Alkaline Hydrogen Peroxide. J. Amer. Chem. Soc. **101**, 6292 (1979).
105. JEFFORD, C. W., W. KNÖPFEL, and P. A. CADBY: Oxygenation of 3-Aryl-2-hydroxyacrylic Acids. The Question of Linear Fragmentation *vs.* Cyclization and Cleavage of Intermediates. J. Amer. Chem. Soc. **100**, 6432 (1978).
106. HASSALL, C. H.: The Baeyer-Villiger Oxidation of Aldehydes and Ketones. Org. Reactions **9**, 73 (1957).

107. Phillips, R. S., and C. D. Oldham: Fluorohydroxy Benzoic Acids as Active Site Spectral Probes for Protocatechuate 3,4-Dioxygenase. Fed. Proc., Fed. Amer. Soc. Exper. Biol. 37, 1720 (1978).
108. Nishinaga, A., T. Itahara, T. Shimizu, and T. Matsuura: Base-catalyzed Oxygenation of tert-Butylated Phenols. I. Regioselectivity in the Base-catalyzed Oxygenation of tert-Butylphenols. J. Amer. Chem. Soc. 100, 1820 (1978).
109. Nishinaga, A., T. Shimizu, and T. Matsuura: Base-catalyzed Oxygenation of tert-Butylated Phenols 3. Base-catalyzed Reaction of Peroxyquinols Derived from Oxygenation of 2,6-Di-tert-butylphenols and Mechanism of Regioselective Formation of Epoxy-o-quinol from 2,4,6-Tri-tert-butylphenol. J. Org. Chem. 44, 2983 (1979).
110. Sawaki, Y., and Y. Ogata: β Scission of Acyl Radicals in the Radical Decomposition of Various α-Hydroperoxy Ketones. J. Org. Chem. 41, 2340 (1976).
111. Fujiwara, M., L. A. Golovleva, Y. Saeki, M. Nozaki, and O. Hayaishi: Extradiol Cleavage of 3-Substituted Catechols by an Intradiol Dioxygenase, Pyrocatechase, from a Pseudomonad. J. Biol. Chem. 250, 4848 (1975).
112. Ribbons, D. W., and P. J. Senior: 2,3-Dihydroxybenzoate 3,4-Oxygenase from Pseudomonas fluorescens. Arch. Biochem. Biophys. 138, 557 (1970).
113. Gauthier, J. J., and S. C. Rittenberg: The Metabolism of Nicotinic Acid. J. Biol. Chem. 246, 3737 (1971).
114. Crandall, D. I., R. C. Krueger, F. Anan, K. Yasunobu, and H. S. Mason: Oxygen Transfer by the Homogentisate Oxidase of Rat Liver. J. Biol. Chem. 235, 3011 (1960).
115. Mehler, A. H.: In "Oxygenases" (O. Hayaishi, ed.), p. 100. New York: Academic Press, Inc. 1960.
116. Nozaki, M., K. Ono, T. Nakazawa, S. Kotani, and O. Hayaishi: Metapyrocatechase. J. Biol. Chem. 243, 2682 (1968).
117. Lipscomb, J. D., B. H. Huynh, and E. Münck: Nitric Oxide Derivatives of Fe^{2+}-EDTA and Protocatechuate Dioxygenases. Fed. Am. Soc. Exp. Biol. 63rd Ann. Meet. 1979, 2659.
118. Omo-Kamimoto, M., and S. Senoh: Studies on 3,4-Dihydroxyphenylacetate-2,3-dioxygenase. J. Biochem. (Tokyo) 75, 321 (1974).
119. Dagley, S., and P. J. Geary: The Time Sequence of Interactions of a Dioxygenase with its Substrates. Biochim. Biophys. Acta 167, 459 (1968).
120. Tai, H. H., and C. J. Sih: 3,4-Dihydroxy-9,10-secoandrost-1,3,5(10)-triene-9,17-dione-4,5-Dioxygenase from Norcardiarestrictus. J. Biol. Chem. 245, 5062 (1970).
121. Crawford, R. L., S. W. Hutton, and P. J. Chapman: Purification and Properties of Gentisate 1,2-Dioxygenase from Moraxella osloensis. J. Bacteriol. 121, 794 (1975).
122. Tokuyama, K.: Homogentisicase. I. II. III. J. Biochem. (Tokyo) 46, 1379 (1959).
123. Crandall, D. I.: Molecular Oxygenation by Fe-Activated Enzymes in Mammalian Metabolism. Oxidases Related Redox Systems, Proc. Symp., Amherst, Mass. 1, 263 (1964).
124. Koontz, W. A., and R. Shiman: Beef Kidney 3-Hydroxyanthranilic Acid Oxygenase. J. Biol. Chem. 251, 368 (1976).
125. Cain, R. B., C. Houghton, and K. A. Wright: Microbial Metabolism of the Pyridine Ring. Biochem. J. 140, 293 (1974).
126. Frydman, R. B., M. L. Tomaro, and B. Frydman: Pyrrolooxygenases. Biochim. Biophys. Acta 284, 63 (1972).
127. Fiegelson, P., and F. O. Brady: In "Molecular Mechanisms of Oxygen Activation" (O. Hayaishi, ed.), pp. 87—133. New York-London: Academic Press, Inc. 1974.
128. Makino, R., and Y. Ishimura: Negligible Amount of Copper in Hepatic L-Tryptophan 2,3-Dioxygenase. J. Biol. Chem. 251, 7722 (1976).
129. Hirata, F., T. Ohnishi, and O. Hayaishi: Indoleamine 2,3-Dioxygenase. J. Biol. Chem. 252, 4637 (1977).

130. TANIGUCHI, T., M. SONO, F. HIRATA, O. HAYAISHI, M. TAMURA, K. HAYASHI, T. IITZUKA, and Y. ISHIMURA: Indoleamine 2,3-Dioxygenase: Kinetic Studies on the Binding of Superoxide Anion and Molecular Oxygen to Enzyme. J. Biol. Chem. **254**, 3288 (1979).

131. TSUDA, H.: 5-Hydroxytryptophan Metabolism in Rat Brain. L. 5-Hydroxytryptophan Pyrrolase. Wakayama Igaku **25**, 1 (1974).

132. VANNESTE, W. H., and A. ZUBERBÜHLER: In "Molecular Mechanisms of Oxygen Activation" (O. HAYAISHI, ed.), pp. 398—399. New York-London: Academic Press, Inc. 1974.

133. BROWN, S. B., and R. F. G. J. KING: The Mechanism of Haem Catabolism. Biochem. J. **170**, 297 (1978).

134. JACKSON, A. H., M. G. LEE, R. T. JENKINS, S. B. BROWN, and B. D. CHANEY: Oxidative Ring Opening of Octaethylchlorohaemin and its *meso*-Hydroxy Derivative to Octaethylbiliverdin. Tetrahedron Letts **1978**, 5135.

135. O'CARRA, P.: In "Porphyrins and Metalloporphyrins" (K. M. SMITH, ed.), p. 123. Amsterdam: Elsevier. 1975.

136. BROWN, S. B., and R. F. G. J. KING: An ^{18}O Double-Labelling Study of Haemoglobin Catabolism in the Rat. Biochem. J. **150**, 565 (1975).

137. CHANEY, B. D., and S. B. BROWN: The Mechanism of Coupled Oxidation of Octaethylhaem to Octaethylbiliverdin. Biochem. Soc. Trans. **6**, 419 (1978).

138. BROWN, S. B., and R. F. G. J. KING: ^{18}O Studies of Haem Catabolism. Biochem. Soc. Trans. **4**, 197 (1976).

139. HO, T.-L.: The Hard Soft Acids Bases (HSAB) Principle and Organic Chemistry. Chem. Rev. **75**, 1 (1975).

140. HASTINGS, J. W., and T. WILSON: Bioluminescence and Chemiluminescence. Photochem. and Photobiol. **23**, 461 (1976).

141. DELUCA, M. A., ed.: Bioluminescence and Chemilumenescence. In "Methods in Enzymology", Vol. 57. New York: Academic Press, Inc. 1978.

142. MCELROY, W. D., and M. DELUCA: In "Chemiluminescence and Bioluminescence" (M. J. CORMIER, D. M. HERCULES, and J. LEE, eds.), pp. 285—311. New York: Plenum Press. 1973.

143. DELUCA, M.: Firefly Luciferase. Adv. in Enzymol. **44**, 37 (1976).

144. WANNLUND, J., M. DELUCA, K. STEMPEL, and P. D. BOYER: Use of ^{14}C-Carboxyl-Luciferin in Determining the Mechanism of the Firefly Luciferase Catalyzed Reactions. Biochem. Biophys. Res. Commun. **81**, 987 (1978).

145. SHIMOMURA, O., T. GOTO, and F. H. JOHNSON: Source of Oxygen in the CO_2 Produced in the Bioluminescent Oxidation of Firefly Luciferin. Proc. Natl. Acad. Sci. U.S.A. **74**, 2799 (1977).

146. KOO, J.-Y., S. P. SCHMIDT, and G. B. SCHUSTER: Bioluminescence of the Firefly: Key Steps in the Formation of the Electronically Excited State for Model Systems. Proc. Natl Acad. Sci. U.S.A. **75**, 30 (1978).

147. SCHUSTER, G. B.: Chemiluminescence of Organic Peroxides. Conversion of Ground-State Reactants to Excited-State Products by the Chemically Initiated Electron-Exchange Luminescence Mechanism. Acc. Chem. Res. **12**, 366 (1979).

148. INOUE, S., H. KAKOI, M. MURATA, T. GOTO, and O. SHIMOMURA: Complete Structure of *Renilla* Luciferin and Luciferyl Sulfate. Tetrahedron Letts **1977**, 2685.

149. INOUE, S., H. KAKOI, and T. GOTO: *Oplophorus* Luciferin, Bioluminescent Substance of the Decapod Shrimps, *Oplophorus spinosus* and *Heterocarpus laevigatus*. Chem. Commun. **1976**, 1056.

150. SHIMOMURA, O., T. MASUGI, F. H. JOHNSON, and Y. HANEDA: Properties and Reaction Mechanism of the Bioluminescent System of the Deep-Sea Shrimp *Oplophorus gracilorostris*. Biochemistry **17**, 994 (1978).

151. Inoue, S., K. Okada, H. Kakoi, and T. Goto: Fish Bioluminescence I. Isolation of a Luminescent Substance from a Myctophina Fish, *Neoscopelus microchir*, and Identification of it as *Oplophorus* Luciferin. Chem. Letts **1977**, 257.

152. Kishi, Y., T. Goto, Y. Hirata, O. Shimomura, and F. H. Johnson: Cypridina Bioluminescence I. Structure of *Cypridina* Luciferin. Tetrahedron Letts **1966**, 3427.

153. Cormier, M. J., J. Lee, and J. E. Wampler: Bioluminescence: Recent Advances. Ann. Rev. Biochem. **44**, 255 (1975).

154. Shimomura, O., and F. H. Johnson: Exchange of Oxygen Between Solvent H_2O and CO_2 Produced in *Cypridina* Bioluminescence. Biochem. Biophys. Res. Commun. **51**, 558 (1973).

155. Hart, R. C., K. E. Stempel, P. D. Boyer, and M. J. Cormier: The Mechanism of the Enzyme-Catalyzed Bioluminescent Oxidation of Coelenterate-type Luciferin. Biochem. Biophys. Res. Commun. **81**, 980 (1978).

156. Goto, T., I. Kobuta, N. Suzuki, and Y. Kishi: In "Chemiluminescence and Bioluminescence" (M. J. Cormier, D. M. Hercules, and J. Lee, eds.), pp. 325—335. New York: Plenum Press. 1973.

157. Shimomura, O., and F. H. Johnson: In "Chemiluminescence and Bioluminescence" (M. J. Cormier, D. M. Hercules, and J. Lee, eds.), pp. 337—344. New York: Plenum Press. 1973.

158. Kemal, C., T. W. Chan, and T. C. Bruice: Reaction of $^3O_2$ with Dihydroflavins. I. J. Amer. Chem. Soc. **99**, 7272 (1977).

159. Chan, T. W., and T. C. Bruice: Reactions of Nitroxides with 1,5-Dihydroflavins and $N^{3,5}$-Dimethyl-1,5-dihydrolumiflavin. J. Amer. Chem. Soc. **99**, 7287 (1977).

160. Dmitrienko, G. I., V. Snieckus, and T. Viswanatha: On the Mechanism of Oxygen by Tetrahydropterin and Dihydroflavin-dependent Mono-oxygenases. Bioorg. Chem. **6**, 421 (1977).

161. Hemmerich, P.: The Present Status of Flavin and Flavocoenzyme Chemistry. Progress in the Chemistry of Organic Natural Products **33**, 451 (1976).

162. Van Lier, J. E., G. Kan, R. Langlois, and L. L. Smith: In "Biological Hydroxylation Mechanisms" (G. S. Boyd, and R. M. S. Smellie, eds.), pp. 21—43. London-New York: Academic Press, Inc. 1972.

163. Hamberg, M., B. Samuelson, I. Bjoerkhem, and H. Danielsson: In "Molecular Mechanisms of Oxygen Activation" (O. Hayaishi, ed.), pp. 30—85. New York-London: Academic Press, Inc. 1974.

164. Matsuda, Y., T. Beppu, and K. Arima: Crystallization and Positional Specificity of Hydroperoxidation of Fusarium Lipoxygenase. Biochem. Biophys. Acta **530**, 439 (1978).

165. Gibian, M. J., and R. A. Galaway: In "Bioorganic Chemistry", Vol. 1 (E. E. Van Tamelen, ed.), pp. 117—136. New York: Academic Press, Inc. 1977.

166. Yamazaki, Y.: In "Free Radicals in Biology", Vol. 3 (W. A. Pryor, ed.), pp. 213—214. New York: Academic Press, Inc. 1977.

167. Verhagen, J., G. A. Veldink, M. R. Egmond, J. F. G. Vliegenthart, J. Boldingh, and J. Van Der Star: Steady-State Kinetics of Anaerobic Reaction of Soybean Lipoxygenase-1 with Linoleic Acid and 13-L-Hydroperoxylinoleic Acid. Biochem. Biophys. Acta **529**, 369 (1978).

168. Svingen, B. A., S. R. Tonsager, T. D. Lindstrom, and S. D. Aust: The Demonstration of the Specific Generation of Alkyl, Alkoxy and Hydroperoxy Radicals of Linoleic Acid by E.P.R. Spin Trapping Techniques. Fed. Amer. Soc. Exp. Biol. 63rd Ann. Meet. **1979**, 2211.

169. De Groot, J. J. M. C., G. J. Garssen, J. F. G. Vliegenthart, and J. Boldingh: The Detection of Linoleic Acid Radicals in the Anaerobic Reaction of Lipoxygenase. Biochem. Biophys. Acta **326**, 279 (1973).

170. ALLEN, J. C., S. NAVARATNAM, B. J. PARSONS, G. O. PHILLIPS, and A. J. SWALLOW: The Oxidation of Soybean Lipoxygenase-1. A Pulse Radiolysis Study. Biochem. Soc. Trans. **8**, 121 (1980).

171. IZUMI, Y., and A. TAI: In "Stereo-Differentiating Reactions", pp. 70—81. New York: Academic Press, Inc. 1977.

172. DE GROOT, J. J. M. C., G. A. VELDINK, J. F. G. VLIEGENTHART, J. BOLDINGH, R. WEVER, and B. F. VAN GELDER: Demonstration by EPR Spectroscopy of the Functional Role of Iron in Soybean Lipoxygenase-1. Biochim. Biophys. Acta **377**, 71 (1975).

173. EGMOND, M. R., P. M. FASELLA, G. A. VELDINK, J. F. G. VLIEGENTHART, and J. BOLDINGH: On the Mechanism of Action of Soybean Lipoxygenase-1. Eur. J. Biochem. **76**, 469 (1977).

174. EGMOND, M. R., and R. J. P. WILLIAMS: ^{1}H-NMR Study of the Conversion of 13(S)-Hydroperoxylinoleic Acid by Soya Bean Lipoxygenase-1. Biochim. Biophys. Acta **531**, 141 (1978).

175. MIYAMOTO, T., N. OGINO, S. YAMAMOTO, and O. HAYAISHI: Purification of Prostaglandin Endoperoxide Synthetase from Bovine Vesicular Gland Microsomes. J. Biol. Chem. **251**, 2629 (1976).

176. GIBSON, K. H.: Prostaglandins, Thromboxanes, PGX: Biosynthetic Products from Arachidonic Acid. Chem. Soc. (London) Rev. **6**, 489 (1977).

177. HEMLER, M., W. E. M. LANDS, and W. L. SMITH: Purification of the Cyclooxygenase that forms Prostaglandins. J. Biol. Chem. **251**, 5575 (1976).

178. SAMUELSSON, B.: Biosynthesis of Prostaglandins. Fed. Proc., Fed. Amer. Soc. Exp. Biol. **31**, 1442 (1972).

179. FIEBRICH, F., and H. KOCH: Silymarin, an Inhibitor of Lipoxygenase. Experientia **35**, 1548, 1550 (1979).

180. HEMLER, M. E., C. G. CRAWFORD, and W. E. M. LANDS: Lipoxygenation Activity of Purified Prostaglandin-forming Cyclooxygenase. Biochem. **17**, 1772 (1978).

181. HAMBERG, M., and B. SAMUELSSON: Stereochemistry in the Formation of 9-Hydroxy-10,12-octadecadienoic Acid and 13-Hydroxy-9,11-octadecadienoic Acid from Linoleic Acid by Fatty Acid Cyclooxygenase. Biochim. Biophys. Acta **617**, 545 (1980).

182. BILD, G. S., C. S. RAMADOSS, S. LIM, and B. AXELROD: Double Dioxygenation of Arachidonic Acid by Soybean Lipoxygenase. Biochem. Biophys. Res. Commun. **74**, 949 (1977).

183. BILD, G. S., C. S. RAMADOSS, and B. AXELROD: Multiple Dioxygenation by Lipoxygenase of Lipids Containing All-cis-1,4,7-octatriene Moieties. Arch. Biochem. Biophys. **184**, 36 (1977).

184. ROZA, M., and A. FRANCKE: Cyclic Peroxides from a Soya Lipoxygenase-Catalyzed Oxygenation of Methyl Linoleate. Biochim. Biophys. Acta **528**, 119 (1978).

185. BILD, G. S., S. G. BHAT, C. S. RAMADOSS, and B. AXELROD: Biosynthesis of a Prostaglandin by a Plant Enzyme. J. Biol. Chem. **253**, 21 (1978).

186. JEFFORD, C. W., and C. G. RIMBAULT: Reaction of Singlet Oxygen with a Norbornadienol Ether. Intramolecular Interception of a Zwitterionic Peroxide. J. Amer. Chem. Soc. **100**, 6515 (1978).

187. BECKWITH, A. J. L., and R. D. WAGNER: Formation of Cyclic Peroxides by Oxygenation of Thiophenol-Diene Mixtures. J. Amer. Chem. Soc. **101**, 7099 (1979).

188. HAMBERG, M., and B. SAMUELSSON: On the Mechanism of the Biosynthesis of Prostaglandins E_1 und $F_{1\alpha}$. J. Biol. Chem. **242**, 5336 (1967).

189. FLASHNER, M. S., and V. MASSEY: In "Molecular Mechanisms of Oxygen Activation" (O. HAYAISHI, ed.), pp. 245—283. New York-London: Academic Press, Inc. 1974.

190. ZIFFER, H., K. KABUTO, D. T. GIBSON, V. M. KOBAL, and D. M. JERINA: The Absolute Stereochemistry of Several cis-Dihydrodiols Microbially Produced from Substituted Benzenes. Tetrahedron **33**, 2491 (1977).

191. Gibson, D. T., J. R. Koch, and R. E. Kallio: Oxidative Degradation of Aromatic Hydrocarbons by Microorganisms. I. Enzymatic Formation of Catechol from Benzene. Biochem. **7**, 2653 (1968).

192. Gibson, D. T., G. E. Cardini, F. C. Maseles, and R. E. Kallio: Incorporation of Oxygen-18 into Benzene by *Pseudomonas putida*. Biochem. **9**, 1631 (1970).

193. Gibson, D. T., B. Gschwendt, W. K. Yeh, and V. M. Kobal: Initial Reactions in the Oxidation of Ethylbenzene by *Pseudomonas putida*. Biochem. **12**, 1520 (1973).

194. Catterall, F. A., and P. A. Williams: Some Properties of the Naphthalene Oxygenase from *Pseudomonas* sp. NCIB 9816. J. Gen. Microbiol. **67**, 117 (1971).

195. Jeffrey, A. M., H. J. C. Yeh, D. M. Jerina, T. R. Patel, J. F. Davey, and D. T. Gibson: Initial Reaction in the Oxidation of Naphthalene by *Pseudomonas putida*. Biochem. **14**, 575 (1975).

196. Yeh, W. K., D. T. Gibson, and T.-N. Liu: Toluene Dioxygenase: A Multicomponent Enzyme System. Biochem. Biophys. Res. Commun. **78**, 401 (1977).

197. Gibson, D. T., M. Hensley, H. Yoshioka, and T. J. Mabry: Formation of (+)-*cis*-2,3-Dihydroxy-1-methylcyclohexa-4,6-diene from Toluene by *Pseudomonas putida*. Biochem. **9**, 1626 (1970).

198. Sauber, K., C. Fröhner, G. Rosenberg, J. Eberspächer, and F. Lingens: Purification and Properties of Pyrazon Dioxygenase from Pyrazondegrading Bacteria. Eur. J. Biochem. **74**, 89 (1977).

199. Taniuchi, H., and O. Hayaishi: Studies on the Metabolism of Kynurenic Acid. J. Biol. Chem. **238**, 283 (1963).

200. Reiner, A. M., and G. D. Hegeman: Metabolism of Benzoic Acid by Bacteria. Biochem. **10**, 2530 (1971).

201. Kobayashi, S., S. Kuno, N. Itada, O. Hayaishi, S. Kozuka, and S. Oae: O^{18} Studies on Anthranilate Hydroxylases. A Novel Mechanism of Double Hydroxylation. Biochem. Biophys. Res. Commun. **16**, 556 (1964).

202. Taniuchi, M., M. Hatanaka, S. Kuno, O. Hayaishi, M. Nakajima, and N. Kurihara: Enzymic Formation of Catechol from Anthranilic Acid. J. Biol. Chem. **239**, 2204 (1964).

203. Kobayashi, S., and O. Hayaishi: Anthranilic Acid Conversion to Catechol *(Pseudomonas)*. Methods in Enzymol. **17A**, 505 (1970).

204. Subba Rao, P. V., N. S. Sreeleela, R. Premkumar, and C. S. Vaidyanathan: Anthranilic Acid Hydroxylase *(Aspergillus niger)*. Methods in Enzymol. **17A**, 510 (1970).

205. Kumar, R. P., N. S. Sreeleela, P. V. Subba Rao, and C. S. Vaidyanathan: Anthranilate Hydroxylase from *Aspergillus niger*: Evidence for the Participation of Iron in the Double Hydroxylation Reaction. J. Bacteriol. **113**, 1213 (1973).

206. Subramanian, V., T.-N. Liu, W. K. Yeh, and D. T. Gibson: Toluene Dioxygenase: Purification of an Iron-Sulfur Protein by Affinity Chromatography. Biochem. Biophys. Commun. **91**, 1131 (1979).

207. Axcell, B. C., and P. C. Geary: Purification and some Properties of a Soluble Benzene-oxidizing System from a Strain of *Pseudomonas*. Biochem. J. **146**, 173 (1975).

208. Crutcher, S. E., and P. J. Geary: Properties of the Iron-Sulphur Proteins of the Benzene Dioxygenase System from *Pseudomonas putida*. Biochem. J. **177**, 393 (1979).

209. Yamaguchi, M., T. Yamauchi, and H. Fujisawa: Studies on the Mechanism of Double Hydroxylation. I. Evidence for the Participation of NADH-Cytochrome *c* Reductase in the Reaction of Benzoate 1,2-Dioxygenase (Benzoate Hydroxylase). Biochem. Biophys. Res. Commun **67**, 264 (1975).

210. Ullrich, V., and W. Duppel: In "The Enzymes", Vol. 12 (P. D. Boyer, ed.), pp. 253. New York-London: Academic Press, Inc. 1975.

211. REINEKE, W., and H.-J. KNACKMUSS: Chemical Structure and Biodegradability of Halogenated Aromatic Compounds: Substituent Effects on 1,2-Dioxygenation of Benzoic Acid. Biochim. Biophys. Acta **542**, 412 (1978).

212. SPARROW, L. G., P. P. K. HO, T. K. SUNDARAM, D. ZACH, E. J. NYNS, and E. E. SNELL: The Bacterial Oxidation of Vitamin B_6. J. Biol. Chem. **244**, 2590 (1969).

213. KISHORE, G., and E. E. SNELL: Mechanism of Action of 2-Methyl-3-hydroxypyridine-5-carboxylic Acid Oxygenase. Fed. Amer. Soc. Exp. Biol. 63rd Annual Meet. **1979**, 319.

214. KISHORE, G. M., and E. E. SNELL: Reactivity of an FAD-dependent Oxygenase with Free Flavins: A New Mode of Uncoupling in Flavoprotein Oxygenases. Biochem. Biophys. Res. Commun. **87**, 518 (1979).

215. KEMAL, C., and T. C. BRUICE: Transfer of O_2 from a 4a-Hydroperoxyflavin Anion to a Phenolated Ion. A Flavin-catalyzed Dioxygenation Reaction. J. Amer. Chem. Soc. **101**, 1635 (1979).

216. SUZUKI, I.: Oxidation of Elemental Sulfur by an Enzyme System of *Thiobacillus thiooxidans*. Biochim. Biophys. Acta **104**, 359 (1965).

217. SUZUKI, I.: Incorporation of Atmospheric Oxygen-18 into Thiosulfate by the Sulfuroxidizing Enzyme of *Thiobacillus thiooxidans*. Biochim. Biophys. Acta **110**, 97 (1965).

218. CAVALLINI, D., C. DE MARCO, R. SCANDURRA, S. DUPRÉ, and M. T. GRAZIANI: The Enzymatic Oxidation of Cysteamine to Hypotaurine. J. Biol. Chem. **241**, 3189 (1966).

219. ROTILIO, G., G. FREDERICI, L. CALABRESE, M. COSTA, and D. CAVALLINI: An Electron Paramagnetic Resonance Study of the Nonheme Iron of Cysteamine Oxygenase. J. Biol. Chem. **245**, 6235 (1970).

220. EWETZ, L., and B. SÖRBO: Characteristics of the Cysteinesulfinate-forming Enzyme in Rat Liver. Biochim. Biophys. Acta **128**, 296 (1966).

221. LOMBARDINI, J. B., T. P. SINGER, and P. D. BOYER: Cysteine Oxygenase II. Studies on the Mechanism of the Reaction with [18]Oxygen. J. Biol. Chem. **244**, 1172 (1969).

222. YAMAGUCHI, K., Y. HOSOKAWA, N. KOHASHI, Y. KORI, S. SAKAKIBARA, and I. UEDA: Rat Liver Cysteine Dioxygenase (Cysteine Oxidase). J. Biochem. **83**, 479 (1978).

223. CHARALAMPOUS, F. C.: Biochemical Studies in Inositol. J. Biol. Chem. **235**, 1286 (1960).

224. — Inositol-cleaving Enzyme from Rat Kidney. Methods in Enzymol. **5**, 329 (1962).

225. REDDY, C. C., P. A. PIERZCHALA, and G. A. HAMILTON: Effects of Various Metabolites, Complexing Agents and Metal Ions on Inositol Oxygenase. Fed. Proc., Fed. Amer. Soc. Exp. Biol. **37**, 1720 (1978).

226. KIDO, T., K. SODA, T. SUZUKI, and K. ASADA: A New Oxygenase, 2-Nitropropane Dioxygenase of *Hansenula mrakii*. J. Biol. Chem. **251**, 6994 (1976).

227. SINGH, H., and H. R. CAMA: Enzymatic Cleavage of Carotenoids. Biochim. Biophys. Acta **370**, 49 (1974).

228. BOWES, G., W. L. OGREN, and R. H. HAGEMAN: Phosphoglycolate Production Catalyzed by Ribulose Diphosphate Carboxylase. Biochem. Biophys. Res. Commun. **45**, 716 (1971).

229. LORIMER, G. H., C. B. OSMOND, T. AKAZAWA, and S. ASAMI: On the Mechanism of Glycolate Synthesis by *Chromatium* and *Chlorella*. Arch. Biochem. Biophys. **185**, 49 (1978).

230. LORIMER, G. H., M. R. BADGER, and T. J. ANDREWS: The Activation of Ribulose-1,5-bisphosphate Carboxylase by Carbon Dioxide and Magnesium Ions. Equilibrium Kinetics, a Suggested Mechanism and Physiological Implications. Biochem. **15**, 529 (1976).

231. CHRISTELLER, J. T., and W. A. LAING: Effects of Manganese Ions and Magnesium Ions on the Activity of Soya-bean Ribulose Bisphosphate Carboxylase/oxygenase. Biochem. J. **183**, 747 (1979).

232. Robison, P. D., M. N. Martin, and F. R. Tabita: Differential Effects of Metal Ions on *Rhodospirillum rubrum*. Ribulosebisphosphate Carboxylase/oxygenase and Stoichiometric Incorporation of HCO₃⁻ into a Cobalt(III)-Enzyme Complex. Biochem. **18**, 4453 (1979).

233. Wildner, G. F., and J. Henkel: Differential Reactivation of Ribulose 1,5-Bisphosphate Oxygenase with Low Carboxylase Activity by Mn^{2+}. Fed. Eur. Biochem. Soc. Letts **91**, 99 (1978).

234. Bränden, R.: Ribulose-1,5-diphosphate Carboxylase and Oxygenase from Green Plants are Two Different Enzymes. Biochem. Biophys. Res. Commun. **81**, 539 (1978).

235. McCurry, S. D., N. P. Hall, J. Pierce, C. Paech, and N. E. Tolbert: Ribulose-1,5-bisphosphate Carboxylase/oxygenase from Parsely. Biochem. Biophys. Res. Commun. **84**, 895 (1978).

236. Bhagwat, A. S., J. Ramakrishna, and P. V. Sane: Specific Inhibition of Oxygenase Activity of Ribulose-1,5-diphosphate Carboxylase by Hydroxylamine. Biochem. Biophys. Res. Commun. **83**, 954 (1978).

237. Lorimer, G. H., T. J. Andrews, and N. E. Tolbert: Ribulose Oxygenase. II. Further Proof of Reaction Products. Biochem. **12**, 18 (1973).

238. Kosman, D. J.: Carbanions as Substrates in Biological Oxidation Reactions. Bioorganic Chem. **2**, 175 (1978).

239. Pierce, J., N. E. Tolbert, and R. Barker: A Mass Spectrometric Analysis of the Reaction of Ribulosebisphosphate Carboxylase/oxygenase. J. Biol. Chem. **255**, 509 (1980).

240. Sue, J. M., and J. R. Knowles: Retention of Oxygens at C-2 and C-3 of D-Ribulose 1,5-Bisphosphate in the Reaction Catalyzed by Ribulose-1,5-bisphosphate Carboxylase. Biochem. **17**, 4041 (1978).

241. Cardinale, G. J., and S. Udenfriend: Prolyl Hydroxylase. Adv. in Enzymol. **41**, 245 (1974).

242. Risteli, J., K. Tryggvason, and K. I. Kivirikko: Prolyl 3-Hydroxylase: Partial Characterization of the Enzyme from Rat Kidney Cortex. Eur. J. Biochem. **73**, 485 (1977).

243. Tryggvason, K., K. Majamaa, J. Risteli, and K. I. Kivirikko: Partial Purification and Characterization of Chick-Embryo Prolyl 3-Hydroxylase. Biochem. J. **183**, 303 (1979).

244. Miller, R. L., and H. H. Varner: Purification and Enzymic Properties of Lysyl Hydroxylase from Fetal Porcine Skin. Biochem. **18**, 5928 (1979).

245. Turpeenniemi, T. M., U. Puistola, H. Anttinen, and K. I. Kivirikko: Affinity Chromatography of Lysyl Hydroxylase on Concanavalin A-Agarose. Biochim. Biophys. Acta **483**, 215 (1977).

246. Henderson, L. L., and L. M. Henderson: Purification and Properties Trimethyllysine Hydroxylase. Fed. Amer. Soc. Exp. Biol. 63rd Annual Meet. **1979**, 2032.

247. Hulse, J. D., S. R. Ellis, and L. M. Henderson: Carnitine Biosynthesis. J. Biol. Chem. **253**, 1654 (1978).

248. Bankel, L., G. Lindstedt, and S. Lindstedt: Thymine 7-Hydroxylase from *Neurospora crassa*. Substrate Specificity Studies. Biochim. Biophys. Acta **481**, 431 (1977).

249. Turner, M. K., J. E. Farthing, and S. J. Brewer: Oxygenation of (3-methyl-³H) Desacetoxycephalosporin C to (3-hydroxymethyl-³H) Desacetylcephalosporin C by 2-Oxoglutarate-linked Dioxygenases from *Acremonium chrysogenum* and *Steptomyces clavuligerus*. Biochem. J. **173**, 839 (1978).

250. Hook, D. J., L. T. Chang, R. P. Elander, and R. B. Morin: Stimulation of the Conversion of Penicillin N to Cephalosporin by Ascorbic Acid, α-Ketoglutarate, and Ferrous Ions in Cell-Free Extracts of Strains of *Cephalosporium acremonium*. Biochem. Biophys. Res. Commun. **87**, 258 (1979).

251. STEVENS, C. M., E. P. ABRAHAM, F.-C. HUANG, and C. J. SIH: Incorporation of Molecular Oxygen at C-17 of Cephalosporin C during its Biosynthesis. Fed. Proc. Fed. Amer. Soc. Exp. Biol. **34**, 625 (1975).

252. TANIGUCHI, K., T. KAPPE, and M. D. ARMSTRONG: Further Studies on Phenylpyruvate Oxidase. J. Biol. Chem. **239**, 3389 (1964).

253. ABBOTT, M. T., and S. UDENFRIEND: In "Molecular Mechanisms of Oxygen Activation" (O. HAYISHI, ed.), pp. 168—214. New York-London: Academic Press, Inc. 1974.

254. TUDERMAN, L., R. MYLLYLÄ, and K. I. KIVIRIKKO: Mechanism of Prolylhydroxylase Reaction I. Eur. J. Biochem. **80**, 341 (1977).

255. COUNTS, D. F., G. J. CARDINALE, and S. UDENFRIEND: Prolyl Hydroxylase Half Reaction: Peptidyl Prolyl-independent Decarboxylation of α-Ketoglutarate. Proc. Natl Acad. Sci. U.S.A. **75**, 2145 (1978).

256. RAO, N. V., and E. ADAMS: Partial Reaction of Prolyl Hydroxylase. J. Biol. Chem. **253**, 6327 (1978).

257. PUISTOLA, U., T. M. TURPEENNIEMI-HUJANEN, R. MYLLYLÄ, and K. I. KIVIRIKKO: Studies on the Lysyl Hydroxylase Reaction. I. Initial Velocity Kinetics and Related Aspects. Biochim. Biophys. Acta **611**, 40 (1980).

258. HOLME, E., G. LINDSTEDT, and S. LINDSTEDT: Partial Reaction of Thymine 7-hydroxylase. Acta Chem. Scand. **B33**, 621 (1979).

259. LINDSTEDT, G., S. LINDSTEDT, and I. NORDIN: Purification and Some Properties of γ-Butyrobetaine Hydroxylase from *Pseudomonas* sp. AK 1. Biochem. **16**, 2181 (1977).

260. SHAFFER, P. M., R. P. MCCROSKEY, R. D. PALMATIER, R. J. MIDGETT, and M. T. ABBOTT: The Cell-free Conversion of a Deoxyribonucleoside to a Ribonucleoside without Detachment of the Deoxyribose. Biochem. Biophys. Res. Commun. **33**, 806 (1968).

261. HAUSMANN, E.: Cofactor Requirements for the Enzymatic Hydroxylation of Lysine in a Polypetide Precursor of Collagen. Biochim. Biophys. Acta **133**, 591 (1967).

262. ABBOTT, M. T., E. K. SCHANDL, R. F. LEE, T. S. PARKER, and R. J. MIDGETT: Cofactor requirements for Thymine 7-Hydroxylase. Biochim. Biophys. Acta **132**, 525 (1967).

263. ABBOTT, M. T., T. A. DRAGILA, and R. P. MCCROSKEY: The Formation of 5-Formyluracil by Cell-Free Preparation from *Neurospora crassa*. Biochim. Biophys. Acta **169**, 1 (1968).

264. WATANABE, M. S., R. P. MCCROSKEY, and M. T. ABBOTT: The Enzymatic Conversion of 5-Formyluracil to Uracil 5-Carboxylic Acid. J. Biol. Chem. **245**, 2023 (1970).

265. WADA, G. H., J. H. FELLMAN, T. S. FUJITA, and E. S. ROTH: Purification and Properties of Avian Liver p-Hydroxyphenylpyruvate Hydroxylase. J. Biol. Chem. **250**, 6720 (1975).

266. LINDBLAD, B., G. LINDSTEDT, S. LINDSTEDT, and M. RUNDGREN: Purification and Some Properties of Human 4-Hydroxylphenylpyruvate Dioxygenase (I). J. Biol. Chem. **252**, 5073 (1977).

267. LINDSTEDT, S., B. ODELHÖG, and M. RUNDGREN: Purification and Some Properties of 4-Hydroxyphenylpyruvate Dioxygenase from *Pseudomonas* sp. P. J. 874. Biochem. **16**, 3369 (1977).

268. LINDBLAD, B., G. LINDSTEDT, M. TOFFT, and S. LINDSTEDT: The Mechanism of α-Ketoglutarate Oxidation in Coupled Enzymatic Oxygenations. J. Amer. Chem. Soc. **91**, 4604 (1968).

269. CARDINALE, G. J., R. E. RHOADS, and S. UDENFRIEND: Simultaneous Incorporation of ^{18}O into Succinate and Hydroxyproline Catalyzed by Collagen Prolyl Hydroxylase. Biochem. Biophys. Res. Commun. **43**, 537 (1971).

270. HOLME, E., G. LINDSTEDT, S. LINDSTEDT, and M. TOFFT: ^{18}O Studies of the 2-Keto-glutarate-dependent Sequential Oxygenation of Thymine to 5-Carboxyuracil. J. Biol. Chem. **246**, 3314 (1971).

271. LINDBLAD, B., G. LINDSTEDT, and S. LINDSTEDT: The Mechanism of Enzymic Formation of Homogentisate from p-Hydroxyphenylpyruvate. J. Amer. Chem. Soc. **92**, 7446 (1970).

272. MYLLYLÄ, R., L. TUDERMAN, and K. I. KIVIRIKKO: Mechanism of the Prolyl Hydroxylase Reaction 2. Eur. J. Biochem. **80**, 349 (1977).

273. HOLME, E.: A Kinetic Study of Thymine 7-Hydroxylase from *Neurospora crassa*. Biochem. **14**, 4999 (1975).

274. PUISTOLA, U., T. M. TURPEENNIEMI, R. MYLLYLÄ, and K. I. KIVIRIKKO: Studies on the Lysyl Hydroxylase Reaction. II. Inhibition Kinetics and the Reaction Mechanism. Biochim. Biophys. Acta **611**, 51 (1980).

275. RUNDGREN, M.: Steady State Kinetics of 4-Hydroxyphenylpyruvate Dioxygenase from Human Liver II. J. Biol. Chem. **252**, 5094 (1977).

276. HURYCH, J., P. HOBZA, J. RENCOVA, and R. ZAHRADNIK: In "The Biology of Fibroplasts" (E. KULONEN, ed.), pp. 365—372. New York: Academic Press, Inc. 1973.

277. LIU, T. Z., and R. S. BHATNAGAR: Mechanism of Hydroxylation of Proline. Fed. Proc., Fed. Amer. Soc. Exp. Biol. **32**, 613 (1973).

278. HALME, J., K. I. KIVIRIKKO, and K. SIMONS: Isolation and Partial Characterization of Highly Purified Protocollagen Proline Hydroxylase. Biochim. Biophys. Acta **198**, 460 (1970).

279. POPENOE, E. A., R. B. ARONSON, and D. D. VAN SLYKE: The Sulfhydryl Nature of Collagen Proline Hydroxylase. Arch. Biochem. Biophys. **133**, 286 (1969).

280. ENGLARD, S., and C. F. MIDELFORT: Stereochemical Course of γ-Butyrobetaine Hydroxylation to Carnitine. Fed. Proc., Fed. Amer. Soc. Exp. Biol. **37**, 1806 (1978).

281. TRYGGVASON, K., J. RISTELI, and K. I. KIVIRIKKO: Separation of Prolyl 3-Hydroxylase and 4-Hydroxylase Activities and the 4-Hydroxyproline Requirement for Synthesis of 3-Hydroxyproline. Biochem. Biophys. Res. Commun. **76**, 275 (1977).

282. ISBELL, H. S., H. L. FRUSH, and Z. ORHANOVIC: Reaction of Carbohydrates with Hydroperoxides III. Oxidation of Sodium Salts of Alduronic and Glyulosonic Acids by Sodium Peroxide. Carbohydrate Res. **36**, 283 (1974).

283. SANFILIPPO, J., JR., C.-I. CHERN, and J. S. VALENTINE: Oxidative Cleavage of α-Keto, α-Hydroxy- and α-Halo Ketones, Esters and Carboxylic Acids by Superoxide. J. Org. Chem. **41**, 1077 (1976).

284. IL'INA, L. M., S. A. BORISENKOVA, A. P. RUDENKO, and E. V. LAVROVA: Transition Metal Phthalocyanines as Pyruvic Acid Decarboxylation Catalysts. Vestn. Mosk. Univ. Khim. **13**, 249 (1972).

285. JEFFORD, C. W., A. F. BOSCHUNG, T. A. B. M. BOLSMAN, R. M. MORIARTY, and B. MELNICK: Reaction of Singlet Oxygen with α-Ketocarboxylic Acids. J. Amer. Chem. Soc. **98**, 1017 (1976).

286. JEFFORD, C. W., A. EXARCHOU, and P. A. CADBY: The Role of Singlet Oxygen as Reagent in the Dye-Sensitized Photo-oxygenation of α-Ketocarboxylic Acids. Tetrahedron Letts **1978**, 2053.

287. JEFFORD, C. W., and P. A. CADBY, in press.

288. BRUICE, T. C., and P. Y. BRUICE: Solution Chemistry of Arene Oxides: Acc. Chem. Res. **9**, 379 (1976).

289. GROVES, J. T., and G. A. McCLUSKY: Aliphatic Hydroxylation via Oxygen Rebound. Oxygen Transfer Catalyzed by Iron. J. Amer. Chem. Soc. **98**, 859 (1976).

290. GROVES, J. T., and M. VAN DER PUY: Stereospecific Aliphatic Hydroxylation by Iron-Hydrogen Peroxide. Evidence for a Stepwise Process. J. Amer. Chem. Soc. **98**, 5290 (1976).

291. JEFFORD, C. W., and P. A. CADBY, unpublished results.
292. SAITO, I., Y. CHUJO, H. SHIMAZU, M. YAMANE, T. MATSUURA, and H. J. CAHNMANN: Non Enzymic Oxidation of p-Hydroxyphenylpyruvic Acid with Singlet Oxygen to Homogentisic Acid. A Model for the Action of p-Hydroxyphenylpyruvate Hydroxylase. J. Amer. Chem. Soc. 97, 5272 (1975).
293. MORIARTY, R. M., A. CHIN, and M. P. TUCKER: Dioxygen Fixation. Oxene Transfer in the Reaction of Singlet Dioxygen with α-Keto Acids. J. Amer. Chem. Soc. 100, 5578 (1978).
294. MORIARTY, R. M., K. B. WHITE, and A. CHIN: Ozonation of Ketenes. Nature of Intermediates. J. Amer. Chem. Soc. 100, 5582 (1978).
295. YANG, N. C., and J. LIBMAN: Ozonation of Acetylenes and Related Compounds in the Presence of Tetracyanoethylene and Pinacolone. J. Org. Chem. 39, 1782 (1974).
296. KEAY, R. E., and G. A. HAMILTON: Alkene Epoxidation by Intermediates Formed During the Ozonation of Alkynes. J. Amer. Chem. Soc. 98, 6578 (1976).
297. ANTHOLINE, W. E., and D. H. PETERING: On the Reaction of Iron Bleomycin with Thiols and Oxygen. Biochem. Biophys. Res. Commun. 90, 384 (1979).
298. OHMORO, M., and M. TAKAGI: Polarography of α-Keto Acids in Aqueous and Nonaqueous Solutions. Bull. Chem. Soc. (Japan). 50, 773 (1977).
299. HOBZA, P., J. HURYCH, and R. ZAHRADNIK: Quantum Chemical Study of the Mechanism of Collagen Proline Hydroxylation. Biochim. Biophys. Acta 304, 466 (1973).

(Received July 30, 1980)

Author Index

Subject Index

By

A. SIEGEL, Wien

Fortschritte der Chemie organischer Naturstoffe

Progress in the Chemistry of Organic Natural Products

All Volumes and Cumulative Index 1—20 available / Alle Bände und Generalregister 1—20 lieferbar.

Price reduction for subscribers / Preisermäßigung für Subskribenten: 10%.

Special reduced price (20% reduction) for the complete Series Vols. 1—40 incl. the Cumulative Index to Vols. 1—20 / Vorzugspreis (20% Nachlaß) bei Bezug der Bände 1—40 inklusive Generalregister (Band 1—20).

Volume 36: 11 figures. VII, 425 pages. 1979. ISBN 3-211-81472-8.

Contents: F. W. WEHRLI and T. NISHIDA, The Use of Carbon-13 Nuclear Magnetic Resonance Spectroscopy in Natural Products Chemistry — G. OHLOFF and I. FLAMENT, The Role of Heteroatomic Substances in the Aroma Compounds of Foodstuffs — A. J. WEIN-HEIMER, C. W. J. CHANG, and J. A. MATSON, Naturally Occurring Cembranes — Author Index — Subject Index.

Volume 37: 8 figures. IX, 367 pages. 1979. ISBN 3-211-81528-7.

Contents: J. M. BRAND, J. CHR. YOUNG, and R. M. SILVERSTEIN, Insect Pheromones: A Critical Review of Recent Advances in Their Chemistry, Biology, and Application — M. McNEIL, A. G. DARVILL, and P. ALBERSHEIM, The Structural Polymers of the Primary Cell Walls of Dicots — U. SCHMIDT, J. HÄUSLER, ELISABETH ÖHLER, and H. POISEL, Dehydroamino Acids, α-Hydroxy-α-amino Acids and α-Mercapto-α-amino Acids — Author Index — Subject Index.

Volume 38: 5 figures. VII, 430 pages. 1979. ISBN 3-211-81529-5.

Contents: R. W. FRANCK, The Mitomycin Antibiotics — N. H. FISCHER, E. J. OLIVIER, and H. D. FISCHER, The Biogenesis and Chemistry of Sesquiterpene Lactones — Author Index — Subject Index.

Volume 39: 5 figures. XI, 316 pages. 1980. ISBN 3-211-81530-9.

Contents: B. FRASER-REID and R. C. ANDERSON, Carbohydrate Derivatives in the Asymmetric Synthesis of Natural Products — H. JONES and G. H. RASMUSSON, Recent Advances in the Biology and Chemistry of Vitamin D — S. LIAAEN-JENSEN, Stereochemistry of Naturally Occurring Carotenoids — T. KASAI and P. O. LARSEN, Chemistry and Biochemistry of γ-Glutamyl Derivatives from Plants Including Mushrooms (Basidiomycetes) — Author Index — Subject Index.

Springer-Verlag Wien · New York